MIF

MOST INTERESTING FACTOR

EDITOR

RICHARD BUCALA

Yale University, USA

MIF

MOST INTERESTING FACTOR

World Scientific

NEW JERSEY • LONDON • SINGAPORE • BEIJING • SHANGHAI • HONG KONG • TAIPEI • CHENNAI

Published by

World Scientific Publishing Co. Pte. Ltd.
5 Toh Tuck Link, Singapore 596224
USA office: 27 Warren Street, Suite 401-402, Hackensack, NJ 07601
UK office: 57 Shelton Street, Covent Garden, London WC2H 9HE

Library of Congress Cataloging-in-Publication Data
MIF : most interesting factor / edited by Richard Bucala.
p. ; cm.
Includes bibliographical references and index.
ISBN-13: 978-981-270-648-5 (hardcover : alk. paper)
ISBN-10: 981-270-648-8 (hardcover : alk. paper)
1. Macrophage migration inhibitory factor. I. Bucala, Richard.
[DNLM: 1. Macrophage Migration-Inhibitory Factors--physiology. 2. Macrophage Migration-Inhibitory Factors--genetics. 3. Macrophage Migration-Inhibitory Factors--immunology. QW 568 M633 2007]

QR185.8.M28M54 2007
616.07'995--dc22

2007013739

British Library Cataloguing-in-Publication Data
A catalogue record for this book is available from the British Library.

Typeset by Stallion Press
Email: enquiries@stallionpress.com

Printed by Mainland Press Pte Ltd

Contents

List of Contributors

Gillian S. Ashcroft
Faculty of Life Sciences
University of Manchester, Michael Smith Building
Oxford Road, Manchester, M13 9PT
E-mail: Gillian.s.ashcroft@manchester.ac.uk

Robert C. Atkins
Department of Nephrology, Monash Medical Centre
Clayton, Victoria, Australia

Monash University Department of Medicine
Monash Medical Centre, Clayton, Victoria, Australia
E-mail: Bob.Atkins@med.monash.edu.au

John Baugh
The Conway Institute of Biomolecular and Biomedical Research
Department of Medicine and Medical Science
University College Dublin
Belfield, Dublin 4, Ireland
E-mail: John.baugh@ucd.edu

Jürgen Bernhagen
Department of Biochemistry and Molecular Cell Biology
Institute of Biochemistry, University Hospital RWTH Aachen
Pauwelsstrasse 30, 52074 Aachen, Germany
E-mail: jbernhagen@ukaachen.de

Richard Bucala
Pathology, and Epidemiology and Public Health
Yale University School of Medicine
The Anlyan Center, S525 300 Cedar Street
New Haven, CT 06520
E-mail: Richard.Bucala@Yale.edu

Thierry Calandra
Infectious Diseases Service, Department of Medicine
Centre Hospitalier Universitaire Vaudois
CH-1011 Lausanne, Switzerland
E-mail: Thierry.Calandra@chuv.ch

Gregg Crichlow
Yale University School of Medicine
Sterling Hall of Medicine 345B
333 Cedar Street, New Haven, CT 06520

John R. David
Department of Immunology and Infectious Diseases
Harvard School of Public Health
Harvard Medical School
665 Huntington Avenue, Boston MA 02115
E-mail: jdavid@hsph.harvard.edu

Claudia M. Denkinger
Department of Medicine
Beth Israel Deaconess Medical Center
330 Brookline Avenue, Boston, MA 02215

Rachelle Donn
ARC/EU & Centre for Molecular Medicine
Stopford Building, Oxford Road
Manchester M13 9PT, UK
E-mail: Rachelle.Donn@manchester.ac.uk

Thomas G. Forsthuber
Department of Biology
University of Texas at San Antonio
6900 North Loop 1604 West
San Antonio, TX 78249
E-mail: Thomas.Forsthuber@utsa.edu

Matthew J. Hardman
Faculty of Life Sciences
University of Manchester, Michael Smith Building
Oxford Road, Manchester, M13 9PT
E-mail: Matthew.j.hardman@manchester.ac.uk

Chloé Kaech
Infectious Diseases Service, Department of Medicine
Centre Hospitalier Universitaire Vaudois
CH-1011 Lausanne, Switzerland
E-mail: Chloe.Kaech@chuv.ch

Hui Y. Lan
Professor of Medicine, Department of Medicine
The University of Hong Kong
L839 Laboratory Block
21 Sassoon Road, Hong Kong
E-mail: hylan@hkucc.hku.hk

Michelle Leech
Centre for Inflammatory Diseases
Monash Institute for Medical Research/Monash University
Department of Medicine, 246 Clayton Road
Clayton 3168 VIC Australia

Lin Leng
Department of Medicine, Yale University School of Medicine
The Anlyan Center, S525, 300 Cedar Street
New Haven, CT 06520
E-mail: Lin.Leng@Yale.edu

Claudio Lezama-Davila
Department of Microbiology, Ohio State University
484 West 12th Avenue, Columbus, OH 43210, USA

Elias Lolis
Associate Professor of Pharmacology
Yale University School of Medicine
Sterling Hall of Medicine 345B
333 Cedar Street, New Haven, CT 06520
E-mail: Elias.Lolis@Yale.edu

Robert Mitchell
J. G. Brown Cancer Center, University of Louisville
Delia Baxter Research Bldg.
580 S. Preston St. Louisville, KY 40202
E-mail: robert.mitchell@louisville.edu

Eric F. Morand
Centre for Inflammatory Diseases
Monash Institute for Medical Research/Monash University
Department of Medicine, 246 Clayton Road
Clayton 3168 VIC Australia
E-mail: eric.morand@med.monash.edu.au

David J. Nikolic-Paterson
Department of Nephrology, Monash Medical Centre
Clayton, Victoria, Australia

Monash University Department of Medicine
Monash Medical Centre, Clayton, Victoria, Australia

David Ray
ARC/EU & Centre for Molecular Medicine
Stopford Building, Oxford Road
Manchester M13 9PT, UK
E-mail: David.W.Ray@manchester.ac.uk

Miriam Rodríguez-Sosa
Unidad de Biomedicina, FEZ-Iztacala, U.N.A.M.
Tlalnepantla, Edo. De Mexico 54090, Mexico

Thierry Roger
Infectious Diseases Service, Department of Medicine
Centre Hospitalier Universitaire Vaudois
CH-1011 Lausanne, Switzerland

Abhay R. Satoskar
Department of Microbiology, Ohio State University
484 West 12th Avenue, Columbus, OH 43210, USA
E-mail: satoskar.2@osu.edu

Andreas Schober
Institute of Molecular Cardiovascular Research
RWTH Aachen University, 52074 Aachen, Germany

Present address: Medizinische Poliklinik
Ludwig Maximilans-Universität München, Germany

Heidi M. Snider
Department of Microbiology, Ohio State University
484 West 12th Avenue, Columbus, OH 43210, USA

Luis I. Terrazas
Unidad de Biomedicina, FEZ-Iztacala, U.N.A.M.
Tlalnepantla, Edo. De Mexico 54090, Mexico

Gregory H. Tesch
Department of Nephrology, Monash Medical Centre
Clayton, Victoria, Australia

Monash University Department of Medicine
Monash Medical Centre, Clayton, Victoria, Australia
E-mail: gtesch@hotmail.com

Christian Weber
Institute of Molecular Cardiovascular Research
RWTH Aachen University, 52074 Aachen, Germany

Institut für Kardiovaskuläre Molekularbiologie
Universitätsklinikum der RWTH Aachen
Pauwelsstrasse 30, 52074 Aachen, Germany
E-mail: cweber@ukaachen.de

1

Historical Overview

*John R. David, MD**

The story of MIF began in 1961 when I came to NYU to work with Lewis Thomas and H. Sherwood Lawrence. Thomas gave me an unpublished manuscript he had received from George and Vaughn describing an *in vitro* method of assessing cellular immunity using capillary tubes. The tubes were filled with guinea pig peritoneal exudates cells and antigens, and then placed in Mackaness-like chambers made of plastic with a shallow chamber sealed at the top and bottom by a round coverslip, and the cells were allowed to migrate onto the bottom coverslip. Cells from animals sensitized to a specific antigen were prevented from migrating by that antigen. This was easy to quantify and thus a great advantage over the inhibition of migration from pieces of lymph nodes that had been previously described many years before by Rich and Lewis.[1] John Vaughn had mentioned to Thomas that he hoped this capillary tube method could be repeated and urged him to try. I found out later that several scientists at NYU had failed in their attempt to repeat it.

Salah Al Askari, a talented urologist from Iraq, and I worked together to try to get this capillary tube method to work. We made a modification simplifying the method by putting only the cells in the tubes and the various antigens in the media filling the chambers, and after a few months had

*Richard Pearson Strong Professor Emeritus, Department of Immunology and Infectious Diseases, Harvard School of Public Health, Professor of Medicine, Harvard Medical School, Email: jdavid@hsph.harvard.edu

it working well. Our early studies were directed to finding out if this migration inhibition assay was related to delayed hypersensitivity. We quickly found that cells were only inhibited when they came from an animal that displayed delayed-type hypersensitivity (DHS) and not if they only produced an antibody.[2] At this point we urged George and Vaughn to publish their paper, assured it could be repeated.[3] We found that, as with the DHS skin test shown by Benacerraf, the migration of macrophages was only inhibited if the DNP protein conjugate was the same as that used to sensitize the animal.[4] This differed from the reaction of an antibody, which bound to the DNP no matter to what protein it was attached. Later, as Schlossman had shown with the skin test, a DNP-lysine conjugate had to have seven or more lysines to induce and elicit a DHS test or inhibit cell migration while needing only one lysine to react with an antibody.[5,6]

In trying to understand the mechanism of the migration inhibition, we mixed cells from guinea pigs sensitized with complete Freund's adjuvant (thus sensitive to *M. tuberculosis* antigen or PPD) with non-sensitized cells and found that if as few as 2.5% of the cells came from the sensitized guinea pigs, the migration of the whole population was inhibited by the specific antigen, PPD[7] (see Fig. 1). As the peritoneal exudates used contained various types of cells, including macrophages, lymphocytes and neutrophils, we wondered what type of cell was the antigen-sensitized cell. It should be recalled that this was just at the time when Gowans was showing the importance of lymphocytes as immunological reactive cells. In subsequent studies, we found that sensitized lymph node lymphocyte preparations, when stimulated by specific antigen, produced a soluble factor that would inhibit the migration of normal peritoneal exudate macrophages.[8] Independently, and at that time, Bloom and Bennett observed the same phenomenon.[9] The soluble factor was called MIF for migration inhibitory factor and was the first lymphokine. These studies demonstrated that antigen sensitized lymphocytes produced a soluble factor that mediates a delayed hypersensitivity reaction involving macrophages *in vitro*. Later, the same factor could be produced by stimulating lymphocytes with a mitogen such as concavalin A.[10] It should be stated that MIF did not kill the macrophages; indeed, when observed under high power, one could see that the macrophages were markedly spread out on the glass coverslip, as activated macrophages have been described.

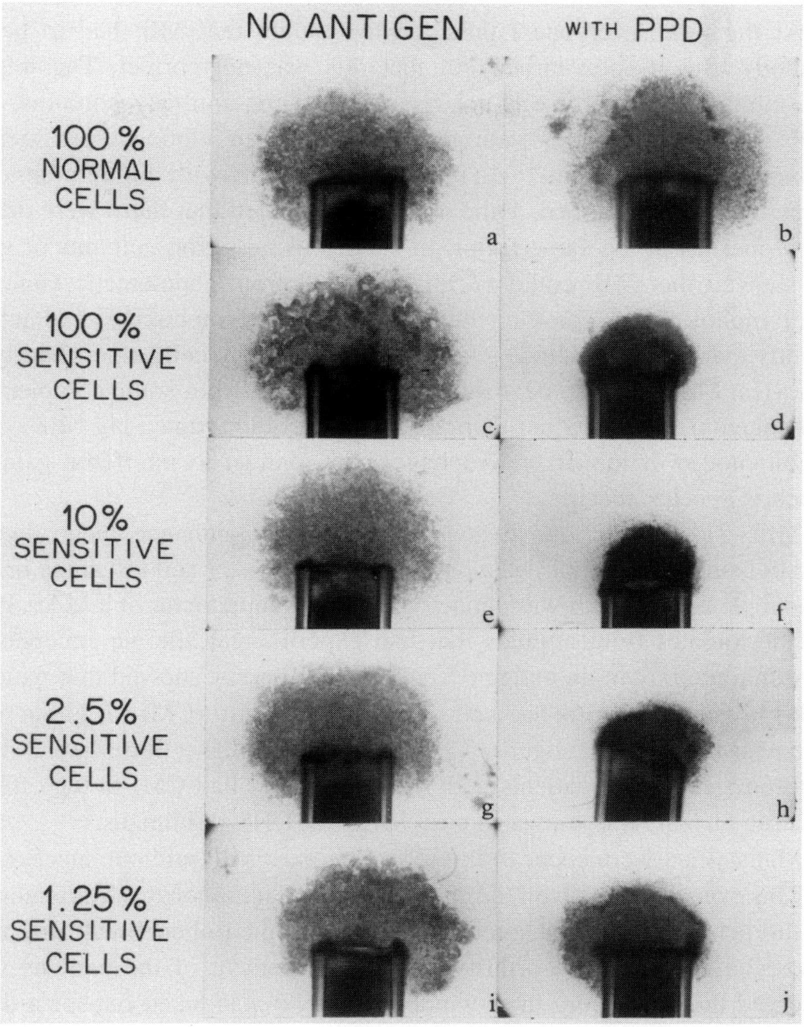

Fig. 1 Capillary tubes filled with guinea pig peritoneal exudates cells. Cells migrate unto coverslip in Mackaness-type chamber. On left, no antigen in the media; on the right PPD (tuberculin antigen). Note that when as little 2.5% of the cells in the mixture come from PPD sensitized cells, the whole population is inhibited in migration by the antigen (Ref. 7).

At the time, everyone I talked with thought that MIF had to be an antibody. But it soon turned out that this was not correct. The migration inhibitory property was not removed by anti-antibody columns, and MIF was much smaller than an antibody when eluting from sizing-chromatography columns.[11] As more soluble factors with different biologic properties were described, some skeptics postulated that these were different properties of the same factor. In studies using sizing columns or gels, we showed that MIF could be differentiated from chemotactic factors[12] and lymphotoxin.[13] These mediators were called lymphokines and subsequently cytokines to include similar factors made by cells other than lymphocytes. MIF was also heat stable, thus differing from some complement components, and it was not animal species specific (guinea pig MIF could inhibit mouse or human macrophages), in contrast to interferon-gamma, which is species specific.

MIF was shown to activate macrophages to enhance their glucose metabolism,[14] kill tumor cells, but not normal cells,[15] and kill some bacteria and parasites.[16,17] It was a macrophage activating factor or a MAF. Peritoneal exudates from animals that had experimental allergic encephalitis were inhibited by brain antigen[18]; and the MIF assay showed that patients with glomerulonephritis had cell-mediated immunity (CMI) to glomerular basement membrane antigens,[19] patients with Guillame Barre had CMI to brain antigens[20] and patients with drug sensitivity had CMI to the offending drug.[21] MIF was shown to correlate with DHS in humans.[22]

Many years were spent trying to isolate pure MIF without success.

This even once involved trying to get enough lymphocytes by canulating the thoracic duct of a cow housed at the back of the Robert Breck Brigham, a hospital committed to arthritis, to the amazement of the patients who observed the beast from their windows when it was taken out for a daily walk. But we never got enough material. For a review of the studies on MIF prior to its being cloned, see Ref. 23.

A breakthrough came in 1989 when a cDNA encoding a human macrophage migration inhibitory factor (MIF) was isolated, through functional expression cloning in COS-1 cells from a cDNA library prepared from a lectin-stimulated T-cell hybridoma, T-CEMB.[24] The 115-amino acid polypeptide encoded by the MIF cDNA (p7-1) was effectively released from the transfected COS-1 cells and yielded readily detectable MIF activity in the culture supernatant despite the apparent lack of a classical protein secretory sequence. It was different from all proteins in the gene banks at the time. Insertional mutational analysis and elution of MIF activity from polyacrylamide gel slices demonstrated that the 12.5 Kd protein with MIF activity released by the COS-1 cells was encoded by p7-1 (see Fig. 2). The p7-1 cDNA hybridized with a 700-base mRNA expressed by Con-A-stimulated lymphocytes but not unstimulated lymphocytes. It was thought that the availability of the MIF cDNA clone and recombinant MIF would facilitate the analysis of the role of this cytokine in cell-mediated immunity, immunoregulation and inflammation.

The first indication that MIF was not limited to lymphocytes came from a report by Lanahan *et al.* in 1992; they were studying growth factor-induced response genes in mouse 3T3 fibroblasts.[25] Some early genes are expressed within minutes after the addition of the growth factor and some,

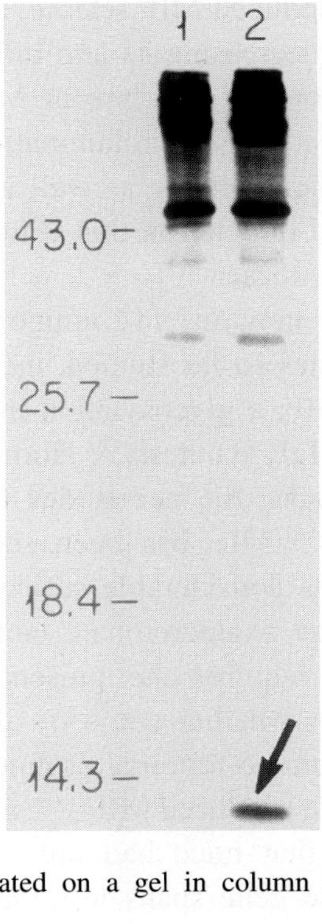

Fig. 2 MIF cloned and separated on a gel in column 2 (Ref. 24). MIF found to be 12.5 K.

called delayed early response genes, are expressed a few hours later. The cDNAs of 13 delayed early response genes were cloned and one, representing 23% of the cDNA, was homologous to human MIF. In fact, it was 100% identical to a mouse MIF we had cloned. Lanahan *et al.* found MIF cDNA in all tissue assayed, including brain, heart, intestine, kidney, liver, lung, ovary seminal vesicles, spleen, testes, thymus and uterus. MIF was also found in differentiating cells of the eye lens.[26]

Studies from Bucala's group resulted in an avalanche of knowledge about MIF and its role in inflammation and autoimmunity. Bernhagen *et al.* reported in 1993 that pituitary-derived MIF, released by endotoxin, potentiates lethal endotoxemia.[27] Further, administration of anti-MIF antibodies neutralized the lethal effect of endotoxin. Subsequently, macrophages were shown to be an important source of MIF.[28] LPS and TNF-α increased macrophage release of MIF and MIF in turn increased the release of TNF-α. A fascinating finding was that glucocortioids, generally known to dampen inflammation, induced MIF release, and that MIF regulated the action of glucocorticoids, decreasing its anti-inflammatory activity.[29]

The extensive studies on the structure of MIF genes and protein, the role of MIF in infectious diseases, inflammation, autoimmunity, cancer, cell differentiation and atherogenesis, as well as the various mechanisms of MIF action, are covered in detail in this volume. This introduction will briefly summarize some of these.

Structure of MIF. MIF contains 114 amino acids and is 90% homologous in mammalian species so far studied, including human, mouse and cow, and is identical to GIF, a glycosylation inhibiting factor, which has been reported to suppress IgE synthesis.[30] Homologues of MIF have been found in fish, chickens, ticks, the nematodes *C. elegans* and *B. malayi*, cyanobacteria and plants.[31] MIF has been shown to be a homotrimer by crystallography and has considerable structural homology with some microbial enzymes such as oxalocrotonate tautomerase.[32] MIF has tautomerase activity, which requires the presence of the amino-terminal proline.[33] It is not known whether some or all of the biologic activity of MIF also requires this amino-terminal proline.

MIF Knockout mice. We produced MIF$^{-/-}$ mice in 1999[34] (see Fig. 3). Previous studies showed that mice had only one MIF gene but many pseudogenes.[33] The mouse gene spans less than 0.7 kb of chromosomal

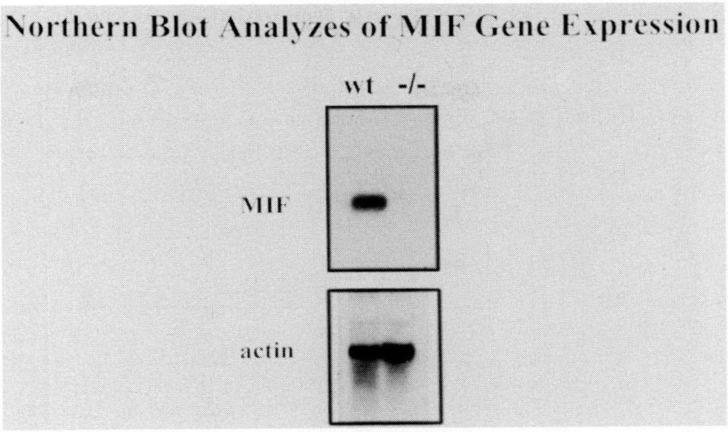

Fig. 3 Northern blot lymphocytes MIF$^{-/-}$ and wild type mice stimulated by concanavalin A. MIF$^{-/-}$ have no MIF mRNA (Ref. 34).

DNA in the middle of chromosome 10 and is composed of three exons. The MIF$^{-/-}$ mouse lacked part of exon 2 and all of exon 3. Analysis of the role of MIF during sepsis showed that the MIF$^{-/-}$ mice were resistant to the lethal effects of high doses of bacteria lipopolysaccharide (LPS; see Fig. 4), or *Staphylococcus aureus* enterotoxin B (SEB) with D-galactosanine and such mice had lower plasma levels of TNF-α than did wild-type mice but normal levels of IL-6 and IL-12.

MIF and infection. To our surprise, MIF$^{-/-}$ mice cleared gram-negative bacteria *Pseudomonas aeruginosa* instilled into the trachea better than wild type mice, and had diminished neutrophil accumulation in their bronchoalveolar fluid compared to wild-type mice.[34] Thioglycollate elicited peritoneal exudates in uninfected MIF$^{-/-}$ mice, however, showed normal neutrophil accumulation. MIF knockout mice were shown to clear *E. coli* better than the wild type, as Bozza found that when *E. coli* were injected into the peritoneum of MIF$^{-/-}$ mice, the amount of *E. coli* recovered six hours later was less than one fourth of those found in wild type mice (personal communication; see Fig. 5). Of special interest, Calandra *et al.* found that mice could recover from lethal peritonitis induced by caecal ligation and puncture (CPL) even when the anti-MIF antibody was administered 8 hours after CPL.[36] The finding that anti-MIF antibody will prevent the lethal effects of LPS and peritonitis and that MIF$^{-/-}$ mice

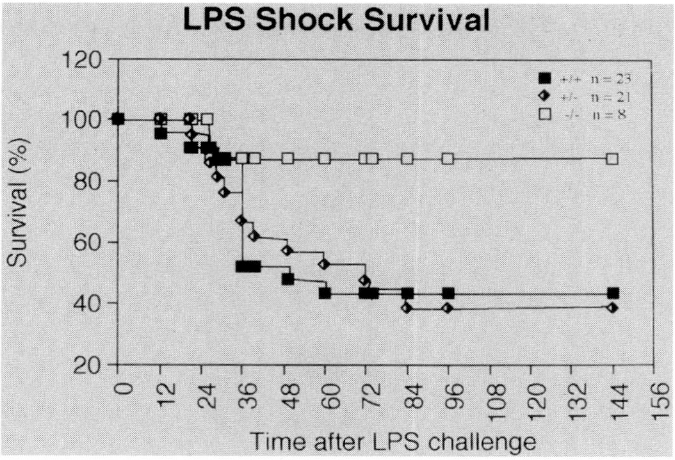

Fig. 4 Survival of $MIF^{-/-}$, $MIF^{-/+}$ and MIF+/+ mice given LPS (Ref. 34). $MIF^{-/-}$, open squares. This is similar to the survival of mice given anti-MIF in Ref. 27.

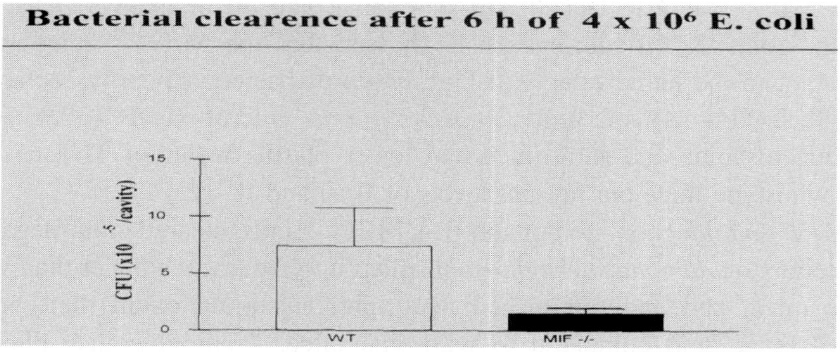

Fig. 5 Number of bacteria in peritoneum 6hrs after injection 4×10^6 *E. coli*. $MIF^{-/-}$ demonstrate much better clearance (Bozza M., personal communication).

resist the lethal effects of LPS and clear the Gram-negative bacteria *E. coli* and *Pseudomonas. aeruginosa* better than the wild type suggests that neutralizing MIF in such cases of sepsis may be useful. It should be noted that this is not the case for all Gram-negative organisms, as $MIF^{-/-}$ mice were shown to be more susceptible to *Salmonella typhimurium* than wild type mice.[37] One explanation for this is that immunity to *Salmonella* requires activated macrophages. As seen below, $MIF^{-/-}$ mice are also more susceptible to other intracellular infectious organisms such as *Leishmania* and *M. tuberculosis*.

MIF may also play a role in the anemia associated with malaria. *Plasmodium chabaudi*-infected erythrocytes or malarial pigment (hemozoin) was shown to induce the release of MIF from macrophages. MIF was detected in the sera of *P. chabaudi*-infected BALB/c mice, and circulating levels of this cytokine correlated with disease severity. MIF, in concentrations found in patients with malaria, suppressed erythropoietin-dependent erythroid colony formation.[38] Further, $MIF^{-/-}$ mice infected with *P. chabaudi* have less severe anemia, improved erythroid development and increased survival compared to wild-type controls.[38] Of interest, human mononuclear cells, which had the *MIF* alleles that highly express MIF, produced more MIF when stimulated with hemozoin than cells with *MIF* alleles of low expression; this polymorphism may be partly responsible for determining who gets severe anemia when infected with malaria parasites.[38]

As described in detail in this volume, $MIF^{-/-}$ mice are more susceptible to *Leishmania major*, *Trypanosoma cruzi* and *Taenia crassiceps*.[39–41] For example, compared to wild-type mice, $MIF^{-/-}$ mice infected with *L. major* have larger lesions that persist longer and contain over 100-fold more parasites; their macrophages produce much more IL-6, a cytokine that inhibits the activation of macrophages by interferon-gamma,[42] and they produce less superoxide and NO (substances important in killing microorganisms), all of which may partly explain the increase susceptibility of these MIF knockout mice.[39]

In collaboration with Alexandra Sousa and Barry Bloom, we showed that BALB/c $MIF^{-/-}$ (5 generation crossovers) were more susceptible to *Mycobaterium tuberculosis* than wild type mice, the mean survival being 85 days in $MIF^{-/-}$ mice compared to 128 days in wild type controls. ($p < 0.001$) (see Fig. 6).

MIF in autoimmunity and inflammations. Studies, using $MIF^{-/-}$ mice and/or neutralizing anti-MIF antibodies, have shown that MIF plays a proinflammatory role in many inflammatory and autoimmune states. MIF levels have been shown to correlate with fatal outcomes in sepsis.[43] Animal models of immune glomerulonephritis,[44] adjuvant arthritis,[45] immune type-1-like diabetes,[46] allergic encephalitis,[47] acute lung injury or asthma,[48]

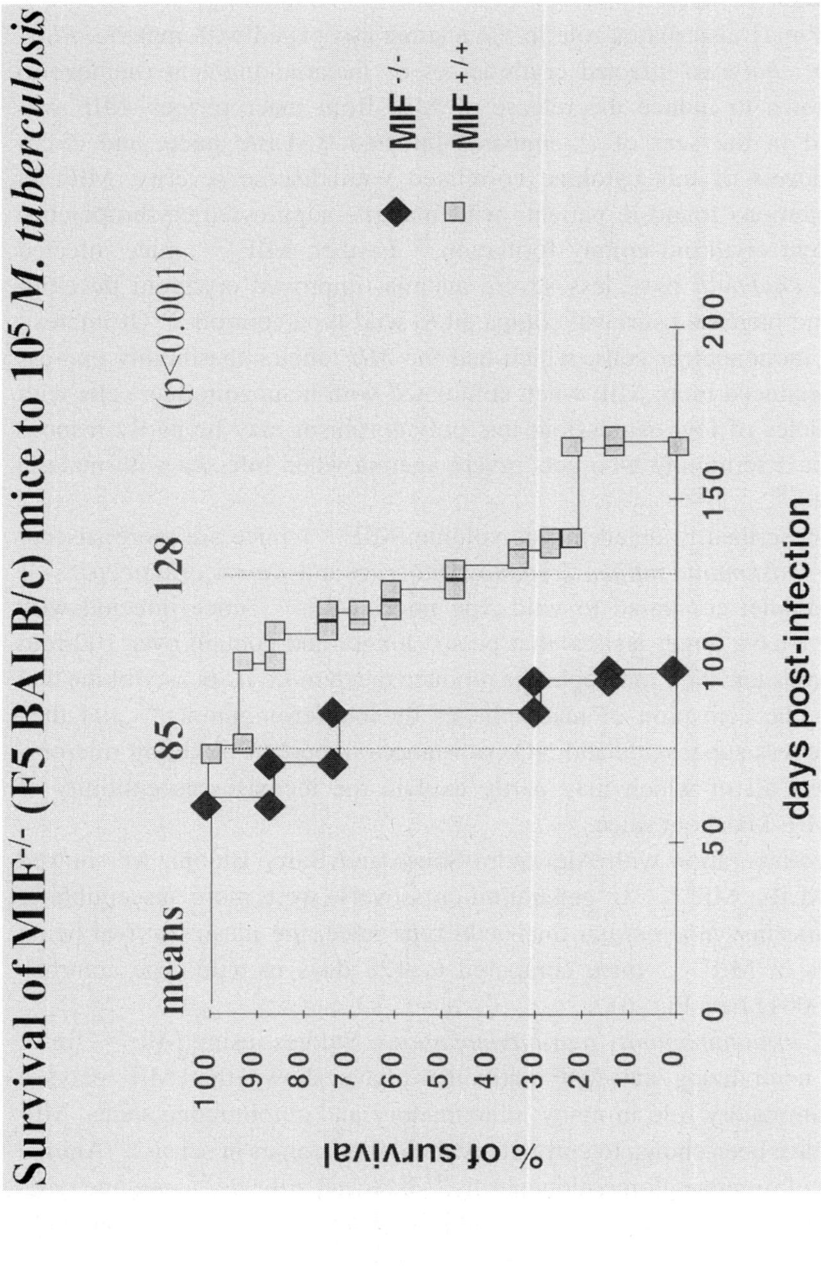

Fig. 6 Survival of MIF$^{-/-}$ and wild type mice after given 10^5 *M. tuberculosis* (A. Sousa, B. Bloom, JR David, unpublished observations).

allografts,[49] wound healing[50] and inflammatory bowel disease[51] have all much milder disease when carried out in $MIF^{-/-}$ animals or when given anti-MIF antibodies (see review, Ref. 31). For example, Rag-$2^{-/-}$ mice, when given $CD45^{hi}$ T cells, develop inflammatory bowel disease, a model for Crohn's disease. However, Rag-2/MIF double knockouts get only minimal disease.[51] Reconstitution of MIF deficient mice with wild-type innate immune cells restored colitis. Further, anti-MIF antibodies could prevent IBD when given at the time of T cell administration and markedly diminish IBD already present when given three weeks after challenge.[51] In contrast, transgenic over-expression of MIF renders mice markedly more susceptible to IBD.[52] These studies, along with the finding of elevated MIF in patients with Crohn's diseases, suggest that MIF is a new target for intervention in this disease.

MIF and tumor genesis. MIF also plays a role in cancer. It has been shown to increase angiogenesis in the growth of murine lymphomas[53] and in sarcoma.[54] MIF-transfected mice initially had larger tumors than control, but subsequently were shown to cure their tumors. Presumably this was because CTL and NK activity was found to be increased. MIF has been associated with prostatic cancer and been shown to act as a marker for disease progression.[55,56]

MIF and atherosclerosis. Marked upregulation of MIF was found in vascular endothelial cells of rabbits undergoing atherogenesis in response to a hypercholesterol diet.[57] Subsequent studies showed that $MIF^{-/-}$ low-density lipoprotein receptor-deficient ($LDLr^{-/-}$) double knockout mice that consumed an atherogenic diet had much smaller atherosclerotic lesions than similarly fed $LDLr^{-/-}$ mice[58] (see Fig. 7).

Mechanisms of action of MIF. A number of studies have shed some light on the mechanism of action of MIF. MIF signal transduction is initiated by binding to CD74, a type II transmembrane protein,[59] and this binding is required for activation of the extracellular signal regulated kinase-mitogen-activated protein kinase cascade (ERK1/ERK2-1/2 MAPK),[60] for cell proliferation and prostaglandin E2 production. Soluble CD74 will inhibit these MIF mediated reactions.

MIF sustains macrophage proinflammatory function by inhibiting p53.[61] p53 mediates growth arrest and apoptosis, and thus, by suppressing it, MIF may aid in wound healing, but also plays a part in tumor genesis.

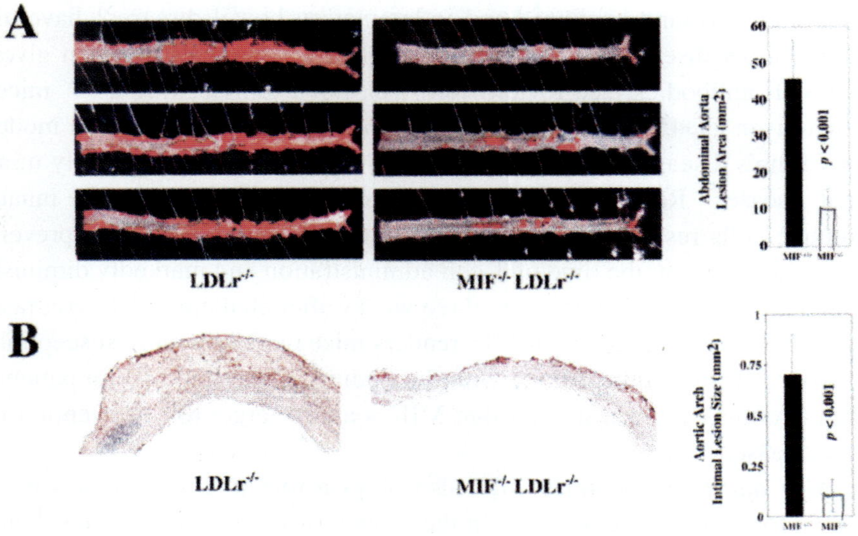

Fig. 7 Much less atheromatous lesions in MIF$^{-/-}$ LDR$^{-/-}$, than in LDR$^{-/-}$ mice given an atherogenic diet (Ref. 58).

MIF is an essential regulator of macrophage response to LPS and Gram-negative bacteria; it upregulates Toll-like receptor 4 (TLR4) expression, the part of the receptor complex for LPS necessary for signal transduction.[62] Thus, MIF acts on the transcription factor PU.1, which activates the *Tlr4* gene. LPS acts on the expressed TLR4 on the surface, inducing signal transduction involving NFκB that subsequently activates the *Tnf-α* gene, causing the macrophage to produce more TNF-α as a response to LPS (see Fig. 8). MIF$^{-/-}$ mice are hyporesponsive to LPS, have less TLR-4 on their cell surface, diminished PU.1 and produce less TNF-α in response to LPS than do wild type controls.

Now to the end of this introduction. Little did we know in the 1960s that MIF, the first lymphocyte mediator/lymphokine/cytokine, or IL-0 if you wish, an *in vitro* measure of delayed hypersensitivity, would have such an impact on so many areas of interest, including inflammation, sepsis and shock, immunity to infectious agents, autoimmunity, wound healing, artherogenesis and cancer. It is now considered an important target for the production of antibodies and anti-MIF drugs and within the next decade,

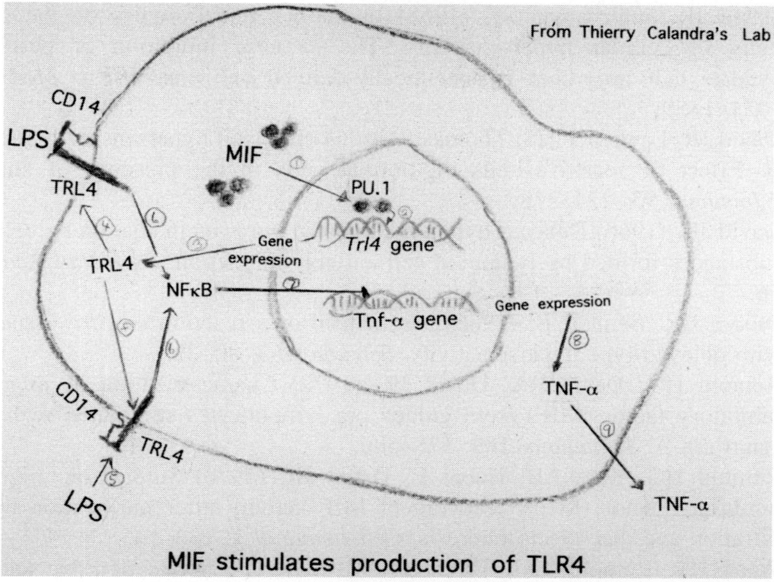

Fig. 8 Diagram of mechanism of MIF stimulating TLR4 (Ref. 62).

we may see if the promise of this Most Interesting Factor, as Richard Bucala likes to call MIF, comes to fruition.

REFERENCES

1. Rich AR, Lewis MR. (1932) The nature of allergy in tuberculosis as revealed by tissue culture studies. *Bull Johns Hopkins Hosp* **50**: 115.
2. David JR, Al-Askari S, Lawrence HS, Thomas L. (1964) Delayed hypersensitivity *in vitro*. I. The specificity of inhibition of cell migration by antigens. *J Immunol* **93**: 264–273.
3. George M, Vaughn JH. (1962) *In vitro* cell migration as a model for delayed sensisivity. *Proc Soc Exp Biol Med* **111**: 514–521.
4. David JR, Lawrence HS, Thomas L. (1964) Delayed hypersensitivity *in vitro*. III. The specificity of hapten-protein conjugates in the inhibition of cell migration. *J Immunol* **93**: 279–282.
5. Schlossman SF, Yaron A, Ben-Efraim S, Sober HA. (1965) Immunogenicity of a series of alpha,N-DNP-L-lysines. *Biochemistry* **4**(8): 1638–1645.

6. David JR, Schlossman SF. (1968) Immunochemical studies on the specificity of cellular hypersensitivity. The *in vitro* inhibition of peritoneal exudate cell migration by chemically-defined antigens. *J Exp Med* **128**: 1451–1459.
7. David JR, Lawrence HS, Thomas L. (1964) Delayed hypersensitivity *in vitro*. II. Effect of sensitive cells on normal cells in the presence of antigen. *J Immunol* **93**: 274–278.
8. David JR. (1966) Delayed hypersensitivity *in vitro*: its mediation by cell-free substances formed by lymphoid cell-antigen interaction. *Proc Natl Acad Sci* **56**: 72–77.
9. Bloom BR, Bennett B. (1966) Mechanism of a reaction *in vitro* associated with delayed-type hypersensitivity. *Science*. **153**: 80–82.
10. Remold HG, David RA, David JR. (1972) Characterization of migration inhibitory factor (MIF) from guinea pig lymphocytes stimulated with concanavalin A. *J Immunol* **109**: 578–586.
11. Remold HG, Katz AB, Haber E, David JR. (1970) Studies on migration inhibitory factor (MIF): recovery of MIF activity after purification by gel filtration and disc electrophoresis. *Cell Immunol* **1**: 133–145.
12. Ward PA, Remold HG, David JR. (1970) The production by antigen-stimulated lymphocytes of a leukotactic factor distinct from migration inhibitory factor. *Cell Immunol* **1**: 162–174.
13. Coyne JA, Remold HG, Rosenberg SA, David JR. (1973) Guinea pig lymphotoxin (LT). II. Physicochemical properties of LT produced by lymphocytes stimulated with antigen or concanavalin A its differentiation from migration inhibitory factor (MIF). *J Immunol* **110**: 1630–1637.
14. Nathan CF, Karnovsky ML, David JR. (1971) Alterations of macrophage functions by mediators from lymphocytes. *J Exp Med* **133**: 1356–1376.
15. Piessens WF, Churchill WH, David JR. (1975) Macrophage activated *in vitro* with lymphocyte mediators kill neoplastic but not normal cells. *J. Immunol.* **114**: 293–299
16. Fowles RE, Fajardo IM, Leibowitch JL, David JR. (1973) The enhancement of macrophage bacteriostasis by products of activated lymphocytes. *J Exp Med* **138**: 952–964.
17. Bout DT, Joseph M, David JR, Capron AR. (1981) *In vitro* killing of *S. mansoni* schistosomula by lymphokine-activated mouse macrophages. *J Immunol* **127**: 1–5.
18. David JR, Patterson PY. (1965) *In vitro* demonstration of cellular sensitivity in allergic encephalomyelitis. *J Exp Med* **122**: 1161–1171.
19. Rocklin RE, Lewis EJ, David JR. (1970) *In vitro* evidence for cellular hypersensitivity to glomerular-basement-membrane antigens in human glomerulonephritis. *N Engl J Med* **283**: 497–501.

20. Rocklin RE, Sheremata WA, Feldman RG, *et al.* (1971) The Guillain-Barre syndrome and multiple sclerosis *in vitro* cellular responses to nervous-tissue antigens. *New Engl J Med* **284**: 803–808.
21. Rocklin RE, David JR. (1971) Detection *in vitro* of cellular hypersensitivity to drugs. *J Allergy Clin Immunol* **48**: 276–282.
22. Rocklin RE, Rosen FS, David JR. (1970) *In vitro* lymphocyte response of patients with immunologic deficiency diseases. Correlation of production of macrophage inhibitory factor with delayed hypersensitivity. *N Engl J Med* **282**: 1340–1343.
23. David JR, David RA. (1972) Cellular hypersensitivity and immunity. Inhibition of macrophage migration and the lymphocyte mediators. In: Kallos P, Waksman BH, deWeck A (eds.) *Progress in Allergy, Vol. 16*, S. Karger, Basel, pp. 300–449.
24. Weiser WY, Temple PA, Witek-Giannotti JS, *et al.* (1989) Molecular cloning of a cDNA encoding a human macrophage migration inhibitory factor (MIF). *Proc Natl Acad Sci USA* **86**: 7522–7526.
25. Lanahan A, Williams JB, Sanders LK, Nathans D. (1992) Growth factor-induced delayed early response genes. *Mol Cell Biol* **12**: 3919–3929.
26. Wistow GJ, Shaughnessy MP, Lee DC, *et al.* (1993) A macrophage migration inhibitory factor is expressed in the differentiating cells of the eye lens. *Proc Natl Acad Sci USA* **90**: 1272–1275.
27. Bernhagen J, Calandra T, Mitchell RA, *et al.* (1993) MIF is a pituitary-derived cytokine that potentiates lethal endotoxaemia. *Nature* **365**(6448): 756–759.
28. Bernhagen J, Calandra T, Mitchell RA (1994) The macrophage is an important and previously unrecognized source of macrophage migration inhibitory factor. *J Exp Med* **179**: 1895–1902.
29. Calandra T, Bernhagen J, Metz CN, *et al.* (1995) MIF as a glucocorticoid-induced modulator of cytokine production. *Nature* **377**: 68–71.
30. Ishizaka K, Ishii Y, Nakano T, Sugie K. (2000) Biochemical basis of antigen-specific suppressor T cell factors: controversies and possible answers. *Adv Immunol* **74**: 1–60.
31. Calandra T, Roger T. (2003) Macrophage migration inhibitory factor: a regulator of innate immunity. *Nat Rev Immunol* **3**(10): 791–800.
32. Sun HW, Bernhagen J, Bucala R, Lolis E. (1996) Crystal structure at 2.6-A resolution of human macrophage migration inhibitory factor. *Proc Natl Acad Sci USA* **28**(93): 5191–5196.
33. Bendrat K, Al-Abed Y, Callaway DJ, *et al.* (1997) Biochemical and mutational investigations of the enzymatic activity of macrophage migration inhibitory factor. *Biochemistry* **36**: 15356–15362.
34. Bozza M, Satoskar AR, Lin G, *et al.* (1999) Targeted disruption of MIF gene revealed its critical role in sepsis. *J Exp Med* **189**: 341–346.

35. Bozza M, Kolakowski LF, Jenkins NA, *et al.* (1995) Structural characterization and chromosomal location of mouse macrophage migration inhibitory factor gene and pseudogenes. *Genomics* **27**: 412–419.
36. Calandra T, Echtenacher B, Roy DL, *et al.* (2000) Protection from septic shock by neutralization of macrophage migration inhibitory factor. *Nat Med* **6**: 164–170.
37. Koebernick H, Grode L, David JR, *et al.* (2002) Macrophage migration inhibitory factor (MIF) plays a pivotal role in immunity against almonella typhimurium. *Proc Natl Acad Sci USA* **99**: 13681–13686.
38. McDevitt MA, Xie J, Shanmugasundaram G, *et al.* (2006) A critical role for the host mediator macrophage migration inhibitory factor in the pathogenesis of malarial anemia. *J Exp Med.*
39. Satoskar AR, Bozza M, Sosa MR, *et al.* (2001) Migration-inhibitory factor gene-deficient mice ar susceptible to cutaneous *Leishmania major* infection. *Infect Immun* **69**: 906–911.
40. Reyes JL, Terrazas LI, Espinoza B, *et al.* (2006) Macrophage migration-inhibitory factor contributes to host defense against acute *Trypanosoma cruzi. Infect Immun* **74**:
41. Rodríguez-Sosa M, Rosas LE, David JR, *et al.* (2003) Macrophage migration inhibitory factor plays a critical role in mediating protection against the helminth parasite *Taenia crassiceps. Infect. Immun* **71**: 1247–1254.
42. Hatzigeorgiou DE, He S, Sobel J, *et al.* (1993) IL-6 down-modulates the cytokine-enhanced antileishmanial activity in human macrophages. *J Immunol* **15**: 3682–3692.
43. Bozza FA, Gomes RN, Japiassu AM, *et al.* (2004) Macrophage migration inhibitory factor levels correlate with fatal outcome in sepsis. *Shock* **22**: 309–313.
44. Yang N, Nikolic-Paterson DJ, Ng YY, *et al.* (1998) Reversal of established rat crescentic glomerulonephritis by blockade of macrophage migration inhibitory factor (MIF): potential role of MIF in regulating glucocorticoid production. *Mol Med* **4**: 413–424.
45. Leech M, Lacey D, Xue JR, *et al.* (2000) Regulation of p53 by macrophage migration inhibitory factor in inflammatory arthritis. *Arthritis Rheum* **48**: 1881–1889.
46. Cvetkovic I,. Al-Abed Y, Miljkovic D, *et al.* (2005) Critical role of macrophage migration inhibitory factor activity in experimental autoimmune diabetes. *Endocrinology* **146**: 2942–2951.
47. Powell ND, Papenfuss TL, McClain MA, *et al.* (2005) Cutting edge: macrophage migration inhibitory factor is necessary for progression of experimental autoimmune encephalomyelitis. *J Immunol* **1**(175): 5611–5614.
48. Mizue Y, Ghani S, Leng L, *et al.* (2005) Role for macrophage migration inhibitory factor in asthma. *Proc Natl Acad Sci USA* **102**: 14410–14415.

49. Fukuyama S, Yoshino I, Yamaguchi M, *et al.* (2005) Blockage of the macrophage migration inhibitory factor expression by short interference RNA inhibited the rejection of an allogeneic tracheal graft. *Transpl Int* **18**: 1203–1209.
50. Zhao Y, Shimizu T, Nishihira J, *et al.* (2005) Tissue regeneration using macrophage migration inhibitory factor-impregnated gelatin microbeads in cutaneous wounds. *Am J Pathol* **167**: 1519–1529.
51. Jong YP, Abadia-Molina AC, Satoskar R, *et al.* (2001) Development of chronic colitis is dependent on the cytokine MIF. *Nat Immunol* **10**: 1038.
52. Ohkawara T, Miyashita K, Nishihira J, *et al.* (2005) Transgenic over-expression of macrophage migration inhibitory factor renders mice markedly more susceptible to experimental colitis. *Clin Exp Immunol* **140**: 241–248.
53. Chesney J, Metz C, Bacher M, *et al.* (1999) An essential role for macrophage migration inhibitory factor (MIF) in angiogenesis and the growth of a murine lymphoma. *Mol Med* **5**: 181–191.
54. Fukushima T, Nishihira J, Yoshiki T, *et al.* (2005) Evidence of dual function of macrophage migration inhibitory factor relevant to tumor progression and regression. *Int J Mol Med* **16**: 119–126.
55. Meyer-Siegler KL, Bellino MA, Tannenbaum M. (2002) Macrophage migration inhibitory factor evaluation compared with prostate specific antigen as a biomarker in patients with prostate carcinoma. *Cancer* **94**: 1449–1456.
56. Muramaki M, Miyake H, Yamada Y, Hara I. (2006) Clinical utility of serum macrophage migration inhibitory factor in men with prostate cancer as a novel biomarker of detection and disease progression. *Oncol Rep* **15**: 253–257.
57. Lin SG, Yu XY, Chen YX, *et al.* (2000) De novo expression of macrophage migration inhibitory factor in atherogenesis in rabbits. *Circ Res* **87**: 1202–1208.
58. Pan JH, Sukhova GK, Yang JT, *et al.* (2004) Macrophage Migration inhibitory factor deficiency impairs atherosclerosis in low-density lipoprotein receptor-knockout mice. *Circulation* **109**: 3149–3153
59. Leng L, Metz CN, Fang Y, *et al.* (2003) MIF signal transduction initiated by binding to CD74. *J Exp Med* **197**: 1467–1476.
60. Mitchell RA, Metz CN, Peng T, Bucala R. (1999) Sustained mitogen-activated protein kinase (MAPK) and cytoplasmic phospholipase A2 activation by macrophage migration inhibitory factor (MIF). Regulatory role in cell proliferation and glucocorticoid action. *J Biol Chem* **274**: 18100–18106.
61. Mitchell RA, Liao H, Chesney J, *et al.* (2002) Macrophage migration inhibitory factor (MIF) sustains macrophage proinflammatory function by inhibiting p53: regulatory role in the innate immune response. *Proc Natl Acad Sci USA* **99**: 345–350.
62. Roger T, David J, Glauser MP, Calandra T. (2001) MIF regulates innate immune responses through modulation of Toll-like receptor 4. *Nature* **414**: 920–924.

2

MIF "Rediscovered"

*Richard Bucala**

INTRODUCTION

The historical position of MIF as the first cytokine is well-established based on reports as far back as 1932 by Rich and Lewis describing the inhibition of immune cell migration *in vitro*,[1] and the historical information provided by John David in the preface to this volume. Cytokine biology is now a mature field, and the term "cytokine" has come to describe a functional class of protein mediators, originally considered to be immunologic hormones, which are produced in a regulated fashion to effect the activation and differentiation of the immune response. Host immunity is both highly coordinated and tightly regulated, and it necessitates close interaction between diverse humoral and cellular components. Cytokines regulate epithelial barrier and endothelial activation responses, the release of anti-microbial mediators, and the recruitment of the diverse cell types that comprise the cellular elements of host immunity. The ensuing activation of the innate (dendritic cell, monocytes/macrophages) or the adaptive (T and B cell) response is characterized by the production of an array of immunoregulatory cytokines that regulates the innate effector response and

*Professor of Medicine, Pathology, and Epidemiology and Public Health, Yale University School of Medicine, The Anlyan Center, S525 300 Cedar Street, New Haven, CT 06520. E-mail: Richard.Bucala@Yale.edu

T and B lymphocyte differentiation. Cytokines display the cardinal properties of pleiotropism, synergy, antagonism, and redundancy. These properties, when overlaid on the cell and tissue specific patterns of cytokine expression, have often times hindered investigations into the precise function of these mediators in different physiologic and pathologic contexts.

Despite over 40 years of descriptive studies, the protein responsible for MIF activity resisted cloning and expression until relatively recently. [2,3] MIF's eponymic "migration" activity has been supplanted by a panoply of cellular functions that are described in the subsequent chapters. Certain of these actions, such as glucocorticoid antagonism or protection from apoptosis, may be considered central to immunologic function. Nevertheless, the biologic importance of migration "arrest" should not be gainsaid. The persistence of monocytes/macrophages within inflammatory lesions such as the granuloma, or within inappropriate tissue site, such as the vascular wall, contribute importantly to the initiation and progression of clinical disease. The molecular mechanisms that underlie MIF's ability to affect migration, which now is appreciated to include both positive (chemotactic) and negative (inhibitory) actions, are slowly becoming revealed.[3,4]

A discrete gene sequence responsible for MIF "activity" was first reported by Weiser *et al.* in 1989,[2] but scientific progress was slow initially because of difficulties in producing a pure, recombinant protein product. In the early 1990s, my recently formed laboratory at Rockefeller University began experiments aimed at identifying mediators that might control, or counter-regulate, the immunosuppressive action of glucocorticoids. This interest developed during my clinical training at the Brigham & Womens' Hospital in Boston, and originated in the powerful impression conveyed by the rapidity and effectiveness by which the pharmacological application of steroids shut down immunologic responses. Glucococorticoids in high doses are truly life-saving therapy for overwhelming inflammatory and autoimmune reactions, and it is only their untoward, dose-dependent side effects that limits their clinical efficacy.[5]

Our studies were further prompted by the observation that in distinction to other regulators of carbohydrate (insulin-glucagon), mineral (calcitonin-PTH), or blood pressure (acetylcholine/epinephrine) homeostasis, no circulating mediators regulating the systemic, immunologic action of glucocorticoids had been described. Glucocorticoid levels increase markedly as a result of the systemic response to tissue invasion, pain, or other stressors, and their constitutive presence is considered to act to control the setpoint for the release of pro-inflammatory mediators. The central role of glucocorticoids in immune homeostasis also can be appreciated by the compromise exhibited by experimental animals that have had their hypothalamic-pituitary-adrenal axis interrupted. Rodents that have been hypophysectomized to remove their source of ACTH secretion, for instance, readily succumb to nanogram quantities of bacterial endotoxin, whereas up to 1000-fold higher doses are needed to achieve comparable lethality in intact mice. Similarly, adrenalectomy or pharmacologic antagonism of glucocorticoids can convert models of a relapsing and remitting autoimmune disease to ones that are rapidly fulminant and lethal.[6]

In the course of searching for glucocorticoid regulatory factors, which took the form of examining neuroendocrine cell types for novel, secreted mediators, three scientists in my group played an especially important role: Robert Mitchell identified by microsequencing and then by cloning a murine MIF secreted by the corticotrophic, AtT-20 pituitary cell line; Jurgen Bernhagen expressed MIF in pure form and first characterized its molecular properties; and Thierry Calandra defined MIF's glucocorticoid regulating properties. In 1993, we reported that the systemic response to endotoxin (LPS) is accompanied by the pituitary release of MIF — notably from the same cells that release ACTH, which drives the adrenal production of corticosteroids.[7] Immunoneutralization of MIF was found to fully protect mice from endotoxic shock, and by a pathway later shown to be consistent with MIF counter-regulation of the immunosuppressive and protective action of glucocorticoids.[8] These observations had significant implications for our understanding of the neuroendocrine control of host immunity, and led to a model whereby the magnitude of the macrophage activation response — which underlies the pathogenesis of the severe systemic, inflammatory reaction responsible for endotoxic shock — was the result of an interplay between glucocorticoid and MIF levels (Fig. 1).

Subsequent studies in hypophysectomized mice led to the conclusion that the monocyte/macrophage, which had long been considered the target of MIF action, is in fact a major source of MIF production.[9] Stimulation of macrophages with gram-negative endotoxin (LPS), gram-positive exotoxins (toxic shock syndrome protein-1, streptococcal pyrogenic exotoxin A),[10]

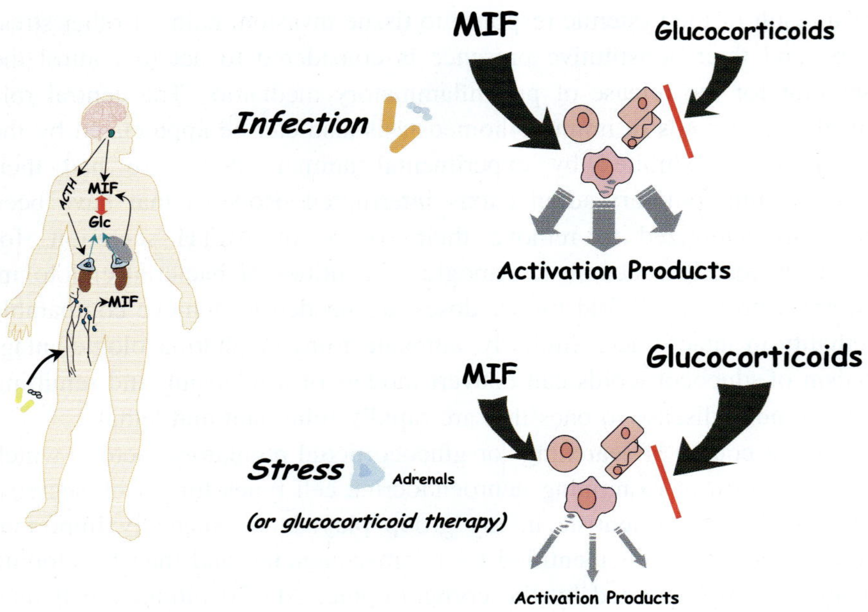

Fig. 1 Relationship between MIF production and its functional interaction with glucocortiocids in regulating immune responsiveness to invasive stimuli.

mycobacterial products,[11] or malarial pigment (hemozoin)[12] induces the rapid release of MIF from macrophages. Autocrine activation by MIF is necessary for an optimal inflammatory and cytokine response by these cells. MIF promotion of TNFα expression by macrophages leads to further MIF release, producing a "re-entrant" activation pathway that is necessary for the optimal expression of TNFα by macrophages.[9] That MIF has a central role in the autocrine/paracrine activation of macrophages has been best validated by the fact that genetic deletion of MIF in mice results in a global decrease in the production of monocyte/macrophage mediators, such as TNFα, IL-1β and PGE_2.[13,14] These data, which are in accord with immunoneutralization studies,[7] place MIF in a "high" upstream position within the cytokine cascade.

Relevant to MIF's upstream, pro-inflammatory function is that the protein is present in pre-formed, cytoplasmic pools within the macrophage (and in other cell types),[15,16] and that it is rapidly released in response to microbial products.[9] This rapid, cellular release of MIF by activating stimuli has been confirmed by tissue staining studies in experimental animals,[17,18] and it is distinct from other monocyte/macrophage cytokines that require transcriptional activation and mRNA translation before release can occur.

MIF'S UNIQUE, UPSTREAM POSITION IN IMMUNITY

The powerful immunologic actions of such monocyte/macrophage-derived cytokines as TNFα and IL-1β are well-appreciated, and their role in pathologic responses verified by their successful therapeutic targeting.[19] Studies in a variety of experimental systems over the last decade have affirmed MIF's action in promoting cytokine (TNFα, IL-1β, IL-2, IL-6, IL-8, IL-12, IFNγ),[9,13,15,20,21] nitric oxide,[3] matrix metalloproteinase,[22,23] and PGE_2 production.[14,24] Many of these upstream actions have been linked to a broad, counter-regulating action of MIF on glucocorticoid suppression of cellular responses.[8]

MIF has distinctive structural and functional properties that are further elaborated in succeeding chapters. It is encoded by a single gene (in mammals); its 3-dimensional structure and its protein fold are unique[25–27]; its receptor (CD74) is distinct from other cytokine receptor superfamilies[28]; and its downstream signaling properties, which involve sustained, ERK-1/2 activation[29,30] and the regulation of Jab1[31] and p53 transcriptional activity[14,32]; reflect features more central to growth regulation, apoptosis, and cell cycle control than that of inflammatory signal transduction. Indeed, emerging data suggest that many, if not most, of MIF's pro-inflammatory actions may be due to its unique ability to regulate activation-induced, p53-dependent apoptosis.[32] An important mechanism for resolving inflammation and the innate immune response is the timely removal of activated monocytes/macrophages by apoptosis. In the presence of high concentrations of MIF, this response is suppressed by a mechanism that prevents p53 accumulation in the cytoplasm[14] (Fig. 2). This effect allows for enhanced monocyte/macrophage survival, increased TNFα, IL-1β, and PGE_2 production, and a sustained pro-inflammatory response. If generalized, this effect can precipitate shock, as verified by studies of endotoxemia in mice genetically-deficient in MIF.[13,14]

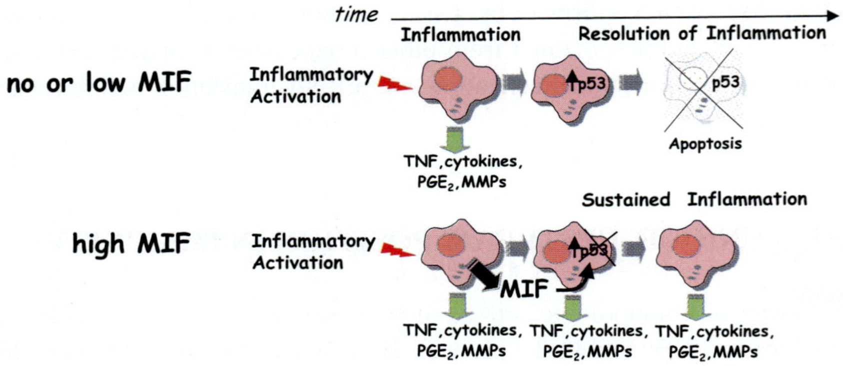

Fig. 2 Scheme for MIF's action in sustaining inflammatory responses by inhibiting activation-induced, p53-dependent apoptosis.

MIF shows a striking degree of conservation across species — greater than that for any "cytokine," and closely related genes have been described in nematodes, protozoa, and even plant species.[33] MIF's three-dimensional structure was solved in 1996 by three research groups,[25–27] and these studies revealed MIF to define a new structural superfamily. MIF was also found to show structural — but not primary sequence — homology with two prokaryotic isomerases (tautomerases). Contemporaneously with these findings came the report that MIF can tautomerize a non-physiological stereoisomer of the tyrosine metabolite, *D*-dopachrome.[34] The *L*-dopachromes are substrates for "melanotic encapsulation," which is a primitive innate response by invertebrates to microbial invasion.[35] The catalytic activity of mammalian MIF is most likely vestigial, but may point to the cytokine's ancestral origin in innate immunity.

The catalytic values for the tautomerization of *D*-dopachrome, or for a second substrate, phosphophenylpyruvate, indicate that enzymatic action is unlikely to be a natural function for mammalian MIF.[36,37] Site-directed mutagenesis and X-ray crystallographic studies, nevertheless, have defined that MIF's N-terminal proline and a hydrophobic pocket mediate substrate binding and catalytic function.[4,36,38–40] Recent data indicate that the MIF receptor makes critical contact with the MIF N-terminal region, which encompasses the tautomerization site, thus providing an explanation for the sensitivity of this site to pharmacologic or structural modification.[41,42]

MIF — HUMAN GENETICS

The human gene for MIF (*MIF*) lies on chromosome 22q11.2. The *MIF* gene appears to be constitutively expressed at a basal level in most cell types; however, its expression can be upregulated by diverse, activating stimuli. The level of constitutive or induced *MIF* expression also varies with the precise cellular and tissue context. Several years ago, our group searched for genetic polymorphisms in human *MIF* by sequencing the entire MIF gene in a group of our laboratory workers. This effort led to the discovery of a tetranucleotide repeat unit, CATT, at position −794 of the promoter region that is represented between 5–8 times (the 8 repeat is extremely rare) (Fig. 3).[43] This repeat unit falls within a potential Pituitary-1 (Pit-1) transcription factor binding site, which may be relevant for MIF expression in the anterior pituitary gland and, perhaps, for its role in the neuroendocrine control of inflammatory phenomena.[44] In model gene reporter assays performed in monocytic, epithelial and fibroblast cell lines, there is an almost proportional relationship between CATT repeat number and basal and stimulated promoter activity. Initial studies investigating the potential clinical relevance of this repeat unit revealed that the presence of the 5-CATT repeat confers low susceptibilty to rheumatoid arthritis and correlates with low disease severity in those patients with established disease.[43]

Independent investigations by the group of Donn *et al.* identified a significant disease association between a single nucleotide polymorphisms in MIF (−173 G/C) and the incidence of juvenile inflammatory arthritis.[45] Given the close proximity of the two polymorphic sites (−794 CATT and −173 G/C), it is not surprising that there is strong linkage disequilibrium between these regions, and indeed there is a strong association between

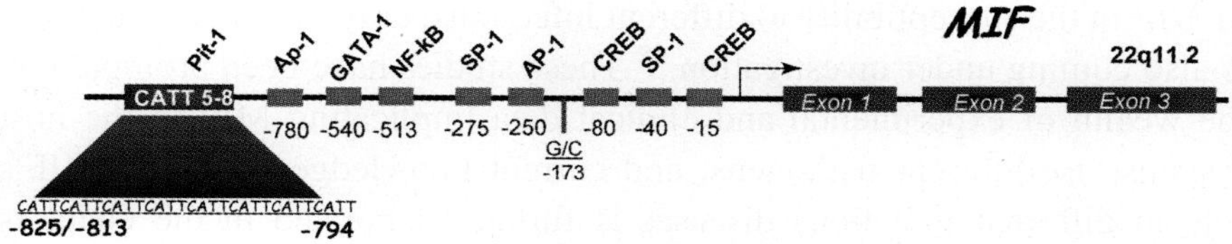

Fig. 3 Polymorphic structure of the human MIF gene (*MIF*), showing the −794 CATT repeat, and the −173 G/C single nucleotide polymorphism.

−173C and the high expression, 7-CATT repeat. Subsequent studies have shown that the −173C SNP correlates with increased serum and synovial fluid MIF protein levels, and to predict response to intra-articular glucocorticoid therapy.[46,47] While the vast majority of non-coding SNPs lack functionality, there is some evidence from *in vitro* transcription assays that the −173C allele may be associated with enhanced promoter activity in selected cell lines.[47] In other studies, however, disease associations were observed only for the CATT repeat variants.[48,49] Genetic epidemiology studies are beginning to define more closely the association between the allelic variants of *MIF* and different inflammatory and infectious diseases. Both disease susceptibility and severity are being examined, which is necessary for identifying disease modifier effects that account for the contribution of different gene variants to particular diseases.[47,50] At present, the conclusions from genetic association studies appear to support a more prominent role for *MIF* alleles in affecting the clinical expression rather than susceptibility to disease.[43,48,49,51–53] While conclusions to date may be limited by study size or by locus heterogeneity, particularly at the CATT repeat site, it is nevertheless consonant with the upstream role of MIF in regulating the effector responses important for vascular and connective tissue damage.[54] The primary effect of *MIF* alleles therefore may be to influence the phenotypic expression of inflammation and immunity once disease has developed in genetically susceptible individuals exposed to the appropriate environmental triggers.

To date, the *MIF* SNP or CATT allelic variants have been positively associated with juvenile and rheumatoid arthritis.[43,45,46,52] sarcoidosis,[51] asthma,[49,51] scleroderma,[53] psoriasis,[55] and ulcerative colitis.[56] In a study following up on the linkage of chromosome 22q11 with atopy-related phenotypes, Hizawa *et al.* identified a significant association between high expression MIF variants and atopic dermatitis.[57] The role of allelic variants of *MIF* in the susceptibility to different infections, or to their complications, is also coming under investigation.[48] These studies have been prompted by the wealth of experimental and clinical data implicating MIF in the host response to different pathogens, and current knowledge regarding MIF's role in different infectious diseases is further elaborated in the chapters authored by Thierry Calandra and Abhay Satoskar. Immunoneutralization or genetic deletion of MIF, for instance, has long been known to protect mice from endotoxin or exotoxin shock,[7,10,13] although the generalizability of these finding to live, bacterial infection remains to be clarified. Anti-MIF also protects mice from polymicrobial sepsis,[58] but MIF appears necessary for host control of infection by the gram-negative (and facultative, intracellular) bacterium, *Salmonella typhimurium*.[59] Given the complexity of the immunologic response necessary for clearing particular pathogens, it is likely that the precise role of MIF in host immunity will vary with the nature of the particular infectious agent. Indeed, the response of MIF-KO mice to infection by *Taenia crassiceps*, which causes neurocystercecosis, varies importantly with the genetic background of the host.[60] MIF-KO mice are relatively resistant to the inflammatory sequelae of experimental malaria,[61] but they may be more susceptible to cutaneous infection by the intracellular parasite, *Leishmania major*.[62]

Polymorphic genes constitute a critical basis for host immunity by mechanisms that act at both the individual and the population level. The HLA system, which is the most polymorphic gene system known, is a well-studied example of the phenotypic power of a polygenic response system. Variation in the HLA genes is widely considered to have evolved in response to infections that have afflicted different human populations over time.[63] The prevalence and distribution of different *MIF* promoter polymorphisms in our initial studied groups suggested to us that the *MIF* locus does not constitute a rare, susceptibility allele, but instead represents a balanced polymorphism that developed in response to different selective pressures. Emerging evidence in support of MIF in the innate response to malaria[12,61,64] and other parasitic infections that have long afflicted human populations support this view. When we genotyped a control, disease-free African-American population, we found that there was a significant over-representation of the low-expression, 5-CATT promoter repeat when compared to Caucasians. This observation prompted us to better investigate the distribution of variant *MIF* alleles in different populations, with the aim of better understanding the role of selective pressures on the evolution of the *MIF* locus. Figure 4 summarizes the data that we have obtained for the frequency of the *MIF* alleles in different populations around the world. It is evident that with respect to the low-expression, 5-CATT variant repeat, there is a significant over-representation of this allele in Africa, with the proportion of the frequency of this allele then decreasing in a pattern that

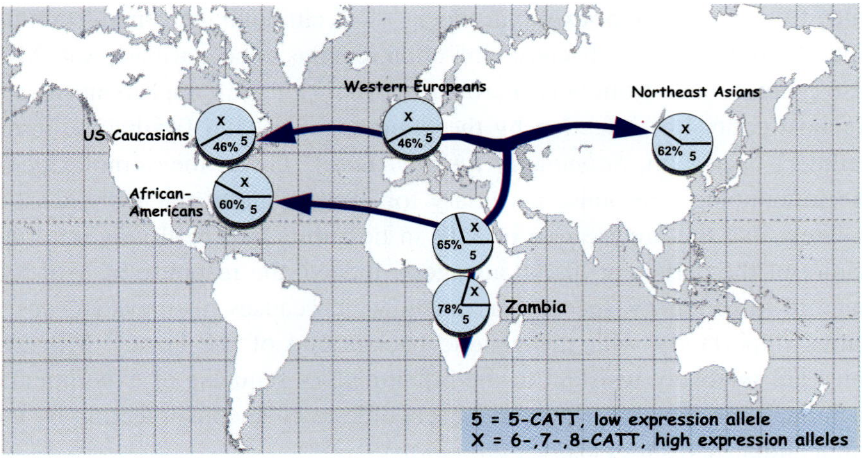

Fig. 4 Diagram of regional variation in the *MIF* CATT promoter variants (5 = 5-CATT, and X = 6-, 7-, or 8-CATT) superimposed on human population migration patterns. From Ref. 65.

may reflect human migration patterns.[65] Ethnic and regional variation in the distribution of different *MIF* alleles is an important finding, given the emerging association between these alleles and different inflammatory and infectious diseases. Racial or geographic variations in the population distribution of *MIF* alleles also suggest the possibility of different selective pressures acting on the *MIF* locus. Given the high prevalence of low expression *MIF* alleles in Africa, these pressures could include historical exposure to tropical infections such as malaria, in which the innate response is considered to play an important role with respect to severe complications and mortality.[66,67]

MIF AND THE GENETIC BASIS OF MONOCYTE/MACROPHAGE RESPONSES

Functionally relevant polymorphisms in a number of genes important for the development of innate immunity have been described in the last several years.[68] MIF's high, upstream role in the regulation of innate responses, its predominantly regulatory rather than effector role, and its allelic structure that is suggestive of a transcriptional "rheostat," place this gene in a unique position to control the responsiveness of monocytes/macrophages

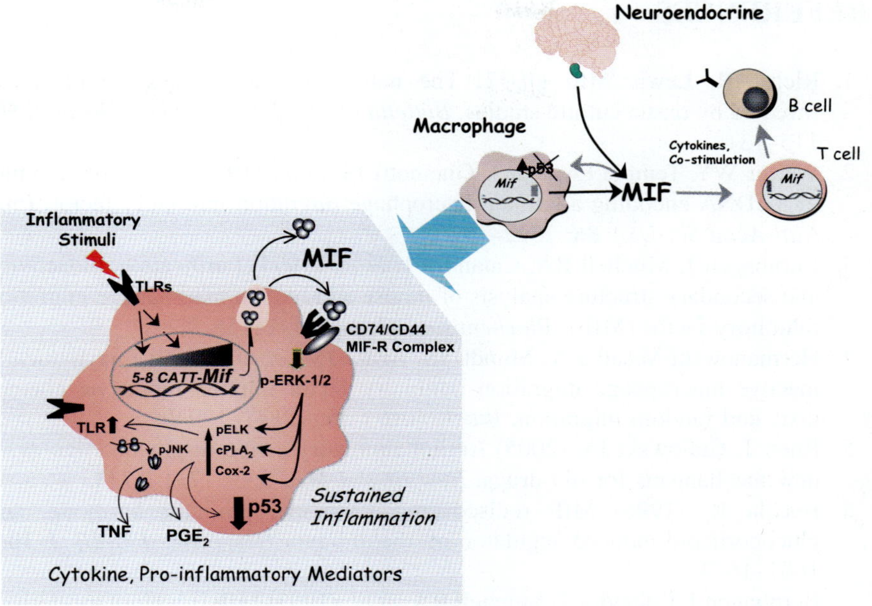

Fig. 5 Integrated scheme for the local and systemic production of MIF, selected intracellular signaling pathways emanating from the MIF-Receptor Complex (CD74/CD44), and the genetic "rheostat" defined by the MIF promoter polymorphisms.

to diverse, invasive stimuli. Using the tools of genomic medicine — genetic epidemiology coupled with mechanistic studies in genetic and conditionally-defined mouse strains — elucidation of the role of these alleles in host immunity to infectious pathogens and the complementary responses producing inflammatory and autoimmune sequelae will be a fertile area for future investigations. These studies, together with the more precise definition of the protein's unique structure-function relationships, will ensure that MIF retains its reputation as a "most interesting factor" for some time to come.

ACKNOWLEDGMENTS

We acknowledge the support of the Arthritis Foundation, the Alliance for Lupus Research, and the US National Institutes of Health, which has supported MIF-studies in the author's laboratory since 1990.

REFERENCES

1. Rich AR, Lewis MR. (1932) The nature of allergy in tuberculosis as revealed by tissue culture studies. *Bulletin of The Johns Hopkins Hospital* **50**: 115–131.
2. Weiser WY, Temple PA, Witek-Giannotti JS, *et al.* (1989) Molecular cloning of a cDNA encoding a human macrophage migration inhibitory factor. *Proc Natl Acad Sci USA* **86**: 7522–7526.
3. Bernhagen J, Mitchell RA, Calandra T, *et al.* (1994) Purification, bioactivity, and secondary structure analysis of mouse and human macrophage migration inhibitory factor (MIF). *Biochemistry* **33**: 14144–14155.
4. Hermanowski-Vosatka A, Mundt SS, Ayala JM, *et al.* (1999) Enzymatically inactive macrophage migration inhibitory factor inhibits monocyte chemotaxis and random migration. *Biochemistry* **38**: 12841–12949.
5. Rhen T, Cidlowski JA, (2005) Antiinflammatory action of glucocorticoids — new mechanisms for old drugs. *New Engl J Med* **353**: 1711–1723.
6. Bucala R. (1996) MIF rediscovered: cytokine, pituitary hormone, and glucocorticoid-induced regulator of the immune response. *FASEB J* **10**: 1607–1613.
7. Bernhagen J, Calandra T, Mitchell RA, *et al.* (1993) MIF is a pituitary-derived cytokine that potentiates lethal endotoxaemia. *Nature* **365**: 756–759.
8. Calandra T, Bernhagen J, Metz CN, *et al.* (1995) MIF as a glucocorticoid-induced modulator of cytokine production. *Nature* **377**: 68–71.
9. Calandra T, Bernhagen J, Mitchell RA, Bucala R. (1994) The macrophage is an important and previously unrecognized source of macrophage migration inhibitory factor. *J Exp Med* **179**: 1895–1902.
10. Calandra T, Spiegel LA, Metz CN, Bucala R. (1998) Macrophage migration inhibitory factor is a critical mediator of the activation of immune cells by exotoxins of Gram-positive bacteria. *Proc Natl Acad Sci USA* **95**: 11383–11388.
11. Bernhagen J, Bacher M, Calandra T, *et al.* (1996) An essential role for macrophage migration inhibitory factor in the tuberculin delayed-type hypersensitivity reaction. *J Exp Med* **183**: 277–282.
12. Martiney JA, Sherry B, Metz CN, *et al.* (2000) Macrophage migration inhibitory factor release by macrophages after ingestion of *Plasmodium chabaudi*-infected erythrocytes: possible role in the pathogenesis of malarial anemia. *Infect Immun* **68**: 2259–2267.
13. Bozza M, Satoskar AR, Lin G, *et al.* (1999) Targeted disruption of migration inhibitory factor gene reveals its critical role in sepsis. *J Exp Med* **189**: 341–346.
14. Mitchell RA, Liao H, Chesney J, *et al.* (2002) Macrophage migration inhibitory factor (MIF) sustains macrophage proinflammatory function by

inhibiting p53: regulatory role in the innate immune response. *Proc Natl Acad Sci USA* **99**: 345–350.
15. Bacher M, Metz CN, Calandra T, *et al.* (1996) An essential regulatory role for macrophage migration inhibitory factor in T-cell activation. *Proc Natl Acad Sci USA* **93**: 7849–7854.
16. Benigni F, Atsumi T, Calandra T, *et al.* (2000) The proinflammatory mediator macrophage migration inhibitory factor (MIF) induces glucose catabolism in muscle. *J Clin Invest* **106**: 1291–1300.
17. Bacher M, Meinhardt A, Lan HY, *et al.* (1997) Migration inhibitory factor expression in experimentally induced endotoxemia. *Am J Pathol* **150**: 235–246.
18. Fingerle-Rowson G, Koch P, Bikoff R, *et al.* (2003) Regulation of macrophage migration inhibitory factor expression by glucocorticoids *in vivo*. *Am J Pathol* **162**: 47–56.
19. Feldmann M, Brennan FM, Paleolog E, *et al.* (2004) Anti-TNFα therapy of rheumatoid arthritis: what can we learn about chronic disease? *Novartis Found Symp* **256**: 53–69.
20. Donnelly SC, Haslett C, Reid PT, *et al.* (1997) Regulatory role for macrophage migration inhibitory factor in acute respiratory distress syndrome. *Nature Med* **3**: 320–323.
21. de Jong YP, Abadia-Molina AC, Satoskar AR, *et al.* (2001) Development of chronic colitis is dependent on the cytokine MIF. *Nat. Immunol* **2**: 1061–1066.
22. Onodera S, Kaneda K, Mizue Y, *et al.* (2000) Macrophage migration inhibitory factor up-regulates expression of matrix metalloproteinases in synovial fibroblasts of rheumatoid arthritis. *J Biol Chem* **275**: 444–450.
23. Onodera S, Nishihira J, Iwabuchi K, *et al.* (2002) Macrophage migration inhibitory factor up-regulates matrix metalloproteinase-9 and -13 in rat osteoblasts Relevance to intracellular signaling pathways. *J Biol Chem* **277**: 7865–7874.
24. Sampey AV, Hall PH, Mitchell RA, *et al.* (2001) Regulation of synoviocyte phospholipase A2 and cyclooxygenase 2 by macrophage migration inhibitory factor. *Arthritis Rheum* **44**: 1273–1280.
25. Suzuki M, Sugimoto H, Nakagawa A, *et al.* (1996) Crystal structure of the macrophage migration inhibitory factor from rat liver. *Nature Struct Biol* **3**: 259–66.
26. Sun HW, Bernhagen J, Bucala R, Lolis E. (1996) Crystal structure at 26–A resolution of human macrophage migration inhibitory factor. *Proc Natl Acad Sci USA* **93**: 5191–5196.
27. Kato Y, Muto T, Tomura T, *et al.* (1996) The crystal structure of human glycosylation-inhibiting factor is a trimeric barrel with three 6-stranded beta-sheets. *Proc Natl Acad Sci USA* **93**: 3007–10.

28. Leng L, Metz C, Fang Y, *et al.* (2003) MIF signal transduction initiated by binding to CD74. *J Exp Med* **197**: 1467–1476.
29. Mitchell RA, Metz CN, Peng T, Bucala R. (1999) Sustained mitogen-activated protein kinase (MAPK) and cytoplasmic phospholipase A2 activation by macrophage migration inhibitory factor (MIF). Regulatory role in cell proliferation and glucocorticoid action. *J Biol Chem* **274**: 18100–18106.
30. Roger T, Chanson AL, Knaup-Reymond M, Calandra T. (2005) Macrophage migration inhibitory factor promotes innate immune responses by suppressing glucocorticoid-induced expression of mitogen-activated protein kinase phosphatase-1. *Eur J Immunol* **35**: 3405–3413.
31. Kleemann R, Hausser A, Geiger G, Mischke R, *et al.* (2000) Intracellular action of the cytokine MIF to modulate AP-1 activity and the cell cycle through Jab1. *Nature* **408**: 211–216.
32. Hudson JD, Shoaibi MA, Maestro R, *et al.* (1999) A proinflammatory cytokine inhibits p53 tumor suppressor activity. *J Exp Med* **190**: 1375–1382.
33. Fingerle-Rowson G, Satoskar A, Bucala R. (2003) Macrophage migration inhibitory factor (MIF)-deficient mice. In: G Fantuzzi (ed.) *Cytokine Knockouts*, Humana Press Totowa, NJ, pp. 361–378.
34. Rosengren E, Bucala R, Aman P, *et al.* (1996) The immunoregulatory mediator macrophage migration inhibitory factor (MIF) catalyzes a tautomerization reaction. *Molecul Med* **2**: 143-149.
35. Infanger LC, Rocheleau TA, Bartholomay LC, *et al.* (2004) The role of phenylalanine hydroxylase in melanotic encapsulation of filarial worms in two species of mosquitoes. *Insect Biochem Molecul Biol* **34**: 1329–1338.
36. Stamps SL, Fitzgerald MC, Whitman CP. (1998) Characterization of the role of the amino-terminal proline in the enzymatic activity catalyzed by macrophage migration inhibitory factor. *Biochemistry* **37**: 10195–10202.
37. Swope MD, Lolis E. (1999) Macrophage migration inhibitory factor: cytokine hormone or enzyme. *Rev Physiol Biochem Pharmacol* **139**: 1–32.
38. Bendrat K, Al-Abed Y, Callaway DJ, *et al.* (1997) Biochemical and mutational investigations of the enzymatic activity of macrophage migration inhibitory factor. *Biochemistry* **36**: 15356–15362.
39. Taylor AB, Johnson WH, Czerwinski RM, *et al.* (1998) Crystal structure of macrophage migration inhibitory factor complexed with (E)-2-fluoro-p-hydroxycinnamate at 1.8 A resolution: implications for enzymatic catalysis and inhibition. *Biochemistry* **38**: 7444–7454.
40. Swope M, Sun HW, Blake PR, Lolis E (1998) Direct link between cytokine activity and a catalytic site for macrophage migration inhibitory factor. *EMBO J* **17**: 3534–3541.
41. Senter PD, Al-Abed Y, Metz CN, *et al.* (2002) Inhibition of macrophage migration inhibitory factor (MIF) tautomerase and biological activities by acetaminophen metabolites. *Proc Natl Acad Sci USA* **99**: 144–149.

42. Lolis E, Bucala R. (2003) Macrophage migration inhibitory factor. *Expert Opin Therap Targets* **7**: 153–164.
43. Baugh JA, Chitnis S, Donnelly SC, *et al.* (2002) A functional promoter polymorphism in the macrophage migration inhibitory factor (MIF) gene associated with disease severity in rheumatoid arthritis. *Genes Immun* **3**: 170–176.
44. Fingerle-Rowson GR, Bucala R. (2001) Neuroendocrine properties of macrophage migration inhibitory factor (MIF). *Immunol Cell Biol* **79**: 368–375.
45. Donn RP, Shelley E, Ollier WE, Thomson W. (2001) A novel 5′-flanking region polymorphism of macrophage migration inhibitory factor is associated with systemic-onset juvenile idiopathic arthritis. *Arthritis Rheum* **44**: 1782–1785.
46. De Benedetti F, Meazza C, Vivarelli M, *et al.* (2003) Functional and prognostic relevance of the -173 polymorphism of the MIF gene in systemic juvenile idiopathic arthritis. *Arthritis Rheum* **48**: 1398–1407.
47. Donn R, Alourfi Z, Zeggini E, *et al.* (2004) A functional promoter haplotype of macrophage migration inhibitory factor is linked and associated with juvenile idiopathic arthritis. *Arthritis Rheum.* **50**: 1604–1610.
48. Plant BJ, Gallagher CG, Bucala R, *et al.* (2005) Cystic fibrosis disease severity and a macrophage migration inhibitory factor polymorphism. *Am J Respir Crit Care Med* **172**: 1412–1415.
49. Mizue Y, Ghani S, Leng L, *et al.* (2005) Role for macrophage migration inhibitory factor (MIF) in asthma. *Proc Natl Acad Sci USA* **102**: 14410–14415.
50. Gregersen P, Bucala R. (2003) MIF *Mif* alleles, and the genetics of inflammatory disorders: incorporating disease outcome into the definition of phenotype. *Arthritis Rheum* **48**: 1171–1176.
51. Amoli M, Donn R, Thomson W, *et al.* (2002) Macrophage migration inhibitory factor gene polymorphism is associated with sarcoidosis in biopsy proven erythema nodosum, *J Rheumatol* **29**: 1671–1673.
52. Radstake TRDJ, Sweep FCGJ, Welsing P, *et al.* (2005) Correlation of rheumatoid arthritis severity with the genetic functional variants and circulating levels of macrophage migration inhibitory factor. *Arthritis Rheum* **52**: 3020–3029.
53. Wu S-P, Leng L, Feng Z, *et al.* (2006) *MIF* promoter polymorphisms influence the clinical expression of scleroderma. *Arthritis Rheum* **54**: 3661–3669.
54. Calandra T, Roger T. (2003) Macrophage migration inhibitory factor: a regulator of innate immunity. *Nat Rev Immunol* **3**: 791–800.
55. Donn RP, Ray DW. (2004) Macrophage migration inhibitory factor: molecular, cellular and genetic aspects of a key neuroendocrine molecule. *J Endocrinol* **182**: 1–9.

56. Nohara H, Okayam N, Inoue N, *et al.* (2004) Association of the −173G/C polymorphism of the macrophage migration inhibitory factor gene with ulcerative colitis. *J Gastroenterol* **39**: 242–246.
57. Hizawa N, Yamaguchi E, Takahashi D, *et al.* (2004) Functional polymorphisms in the promoter region of macrophage migration inhibitory factor and atopy. *Am J Resp Crit Care Med* **169**: 1014–1018.
58. Calandra T, Echtenacher B, Le Roy D, *et al.* (2000) Protection from septic shock by neutralization of macrophage migration inhibitory factor. *Nature Med.* **6**: 164–169.
59. Koebernick H, Grode L, David JR, *et al.* (2002), Macrophage migration inhibitory factor (MIF) plays a pivotal role in immunity against Salmonella typhimuriumi. *Proc Nat Acad Sci USA* **99**: 13681–13686.
60. Rodriguez-Sosa M, Rosas LE, David JR, *et al.* (2003) Macrophage migration inhibitory factor plays a critical role in mediating protection against the helminth parasite *Taenia crassiceps. Infect Immun* **71**: 1247–1254.
61. McDevitt MA, Xie J, Shanmugasundaram G, *et al.* (2006) A critical role for the host mediator macrophage migration inhibitory factor in the pathogenesis of malarial anemia. *J Exp Med* **203**: 1185–1196.
62. Satoskar A, Bozza M, Rodriguez Sosa M, *et al.* (2001) Migration inhibitory factor gene-deficient mice are susceptible to cutaneous *Leishmania major* infection. *Infect Immun* **69**: 906–911.
63. Janeway CA, Travers P, Walport M, Shlomchik M. (2005) Antigen presentation to T lymphocytes. In: *Immunobiology*, pp.169–201, Garland, New York.
64. Chaisanvaneeyakorn S, Moore J, Othoro C, *et al.* (2002) Immunity to placental malaria. IV. Placental malaria is associated with up-regulation of macrophage migration inhibitory factor in intervillous blood. *J Infect Dis* **186**: 1371–1375.
65. Zhong XB, Leng L, Beitin A, *et al.* (2005) Simultaneous detection of microsatellite repeats and SNPs in the macrophage migration inhibitory factor (MIF) gene by thin-film biosensor chips and application to rural field studies. *Nucl Acids Res* **33**.
66. Stevenson MM, Riley EM. (2004) Innate immunity to malaria. *Nature Rev. Immuno.* **4**: 169–180.
67. McDevitt M, Xie J, Gordeuk V, Bucala R. (2004) The anemia of malaria infection: role of inflammatory cytokines. *Curr Hematol Rep* **3**: 97–106.
68. Cooke GS, Hill AV. (2001) Genetics of susceptibility to human infectious disease. *Nat Rev Genet* **2**: 967–977.

3

Immunopathologic Mechanisms Relevant to Diseases

*Hui Y Lan**

INTRODUCTION

Macrophage migration inhibitory factor (MIF) is a pleiotropic cytokine and mediates a number of immune and inflammatory diseases, including septic shock,[1–3] rheumatoid arthritis (RA),[4–6] delayed-type hypersensitivity (DTH),[7] acute respiratory distress syndrome (ARDS),[8,9] glomerulonephritis[10–13] and cancer.[14] MIF is produced during immune responses by activated T-cells, macrophages, and a variety of nonimmune cells. Constitutive expression of MIF mRNA and protein is found in various nonimmune cells within normal tissue, such as anterior pituitary cells; cardiac myocytes; vascular cells; parenchymal cells within the liver, brain, kidneys; and pancreatic islet cells. It is noted that MIF is significantly upregulated and released under various pathological conditions, such as septic shock, infectious disease, ARDS, ischemic cardiac injury, atherosclerosis, glomerulonephritis, multiple sclerosis, colitis, type 2 diabetes and

*Professor of Medicine, Department of Medicine, The University of Hong Kong, L839 Laboratory Block, 21 Sassoon Road, Hong Kong. E-mail: hylan@hkucc.hku.hk

pancreatitis. These implicate a role for MIF in the disease process. Studies using neutralizing antibodies or MIF-deficient animals demonstrated that MIF is a crucial mediator of several immune and inflammatory disorders in rodents, including gram-negative and gram-positive sepsis,[15] DTH,[7] glomerulonephritis,[10–13] arthritis,[4–6] experimental autoimmune encephalomyelitis,[16] experimental autoimmune myocarditis,[17] and gastritis and colitis.[18–21]

In this chapter, the main aim is to discuss the current understanding of the general immunopathologic mechanisms of MIF in immune and inflammatory diseases, with emphasis on the activation of T-cells and macrophages, inflammatory cytokine expression, and matrix metalloproteinase (MMP)-mediated tissue injury under various pathologic conditions. The specific role and the pathogenic mechanisms of MIF in individual diseases will be described in relevant chapters of this book.

MECHANISMS OF MIF IN REGULATION OF T-CELL-MEDIATED IMMUNE DISEASE

MIF is produced by activated T-cells and plays a role in T-cell-mediated immune response in a number of immunologically-induced diseases in human and experimental animal models. These include renal allograft rejection,[22,23] graft-versus-host-disease (GVHD),[24] glomerulonephritis,[10–13] arthritis,[4–6] DTH,[7] experimental autoimmune encephalomyelitis[16] and experimental autoimmune myocarditis.[17] In these diseases, T-cells produce MIF, and up-regulation of MIF in the diseased tissues is significantly associated with the severity and progression of the disease.

There are several possible mechanisms for the regulation of T-cell-mediated immune injury involving MIF. First, MIF may have a role in recruiting T-cells to the sites of immune and inflammatory injury directly or indirectly. The latter may be associated with up-regulation of chemokines and adhesion molecules induced by MIF. Indeed, immunohistochemical studies demonstrate that up-regulation of MIF is associated with the local accumulation of T-cells in renal allograft rejection,[22,23] GVHD,[24] and the development of T-cell-mediated glomerulonephritis in human and experimental animal models.[10–13] Furthermore, the accumulation of T-cells at the sites of immune injury is prevented by blocking of MIF activities with a neutralizing antibody,[11,12] demonstrating the critical role for MIF in T-cell accumulation in the sites of the disease.

Second, MIF may also act by stimulating T-cell proliferation and activation to mediate the disease. It is known that MIF acts as a critical cofactor in T-cell activation in response to both mitogenic and antigenic stimuli.[25] Antigen-driven-T-cell proliferation, IL-2 production, and antibody production are inhibited by a neutralizing anti-MIF antibody *in vitro* and *in vivo*.[25] More importantly, the findings that T-cells also release MIF in response to glucocorticoid stimulation and MIF acts to override glucocorticoid inhibition of T-cell proliferation and activation to produce Th-1 cytokines including IL-2 and interferon gamma (INF-γ) indicate that MIF acts in concert with glucocorticoids to control T-cell activation and proliferation, and plays a critical role in antigen-specific immune responses.[2,25] In MIF knockout mice, increased susceptibility to infection by *Salmonella typhimurium* is accompanied by a reduced Th1 response, as demonstrated by decreased levels of IL-12, IFN-γ and TNF-α.[26] The inhibition appears to be related to blockade of the signaling pathway via the T-cell receptor (TCR), but not the IL-2 receptor (IL-2R). However, the anti-MIF monoclonal antibody shows no inhibitory effect on NK-T-cell responses stimulated through TCR.

A recent study has shown that T-cell proliferation and activation is also implicated in the development of experimental autoimmune diabetes, which is MIF-dependent.[27] MIF protein is significantly elevated in islet cells during the development of diabetes. Blockage of MIF activity with either a neutralizing antibody or the pharmacological inhibitor (S, R)-3-(4-hydroxyphenyl)-4,5-dihydro-5-isoxazole acetic acid methyl ester, produces marked inhibitory effects on hyperglycemia and insulitis, which are associated with inhibition of islet antigen-specific T-cell proliferative responses. These findings suggest a critical role for MIF in the immune-mediated β-cell destruction in an animal model of type 1 diabetes mellitus and demonstrate a new therapeutic strategy for the prevention and treatment of diabetic disease in humans that is based on the selective inhibition of MIF activity.

MIF is important in the T-cell-mediated DTH response, a key mechanism in the pathogenesis of many immune diseases. Indeed, MIF is produced by activated T-cells and is the first T-cell cytokine identified in association with DTH response. Direct evidence for a role of MIF in DTH response comes from the findings that recombinant MIF exacerbates, and a neutralization of MIF antibody inhibits, a specific antigen-challenged skin DTH reaction in primed anti-GBM glomerulonephritis in rats and in a classic DTH response in tuberculosis in mice.[7,11,12] *In vivo* blockade of MIF inhibits CD4+ cell-mediated skin graft destruction, demonstrating a critical role of MIF in inhibition of alloreactive DTH response.[28] However, although MIF antibody treatment is able to reduce skin DTH, the absence of the MIF gene ($MIF^{-/-}$) had no effect on the survival of skin graft, or kidney graft, cardiac graft,[29,30] suggesting that acute graft rejection is not dependent on local MIF production. Thus, although MIF is a key mediator in DTH response, it appears that MIF may not be required for the allograft rejection. These findings also suggest that DTH may not be a key mechanism in allograft rejection. The mechanisms by which MIF inhibits DTH response without prevention of allograft rejection remain largely unclear.

MIF IS AN IMPORTANT MEDIATOR OF MACROPHAGE-MEDIATED INNATE IMMUNE RESPONSE

Macrophages are the major cells that produce MIF in both physiological and pathological conditions. Macrophages are found to contain significant amounts of preformed MIF protein within intracellular pools.[31] A number of proinflammatory stimuli that are able to initiate the release of macrophage MIF include gram-negative endotoxin (LPS) and gram-positive exotoxins, including the superantigen toxic shock syndrome protein-1 (TSST-1), the malarial product hemazoin, and the proinflammatory mediators TNF-α and IFN-γ.[15] Once released, MIF could act in both a paracrine and an autocrine fashion to activate critical, proinflammatory responses in macrophages, T-cells, and many intrinsic cells within the diseased tissues/organs. Thus, MIF is a key mediator in macrophage-mediated innate immune responses.

MIF in Macrophage-mediated Innate Immunity

Emerging evidence has shown that MIF plays an important role in the regulation of innate immunity.[15] Using MIF-deficient cells, MIF-specific antibodies or recombinant MIF, investigators have provided evidence that MIF directly or indirectly promotes the production or expression of a large panel of proinflammatory molecules, including cytokines (such as TNF-α, IFN-γ, IL-1β, IL-2, IL-6, IL-8, and macrophage inflammatory protein 2), nitric oxide, COX2, and products of the arachidonic acid pathway (such as PGE_2) by macrophages.[15,32,33] In the mouse macrophage line, RAW 264.7, or isolated mouse peritoneal macrophages, marked MIF secretion was induced by the staphylococcal toxic shock syndrome toxin 1 (TSST-1) and the streptococcal pyrogenic exotoxin A.[15] In addition, a number of proinflammatory cytokines such as TNF-α and IFN-γ are also strong inducers of MIF production by macrophages.[31] After MIF is released in the tissue or in the circulation, it acts as a classic proinflammatory cytokine to promote innate responses through the activation of macrophages. In a lethal mouse model of TSST-1-induced shock, anti-MIF antibody significantly enhances survival from 8% to 54%.[15] These studies indicate that gram-positive exotoxins are extremely potent inducers of MIF secretion and establish a critical role for MIF and the macrophage in the pathogenesis of the TSST and in the innate immune response. This is further confirmed by the finding that mice null for MIF are resistant to endotoxic shock.[15]

In addition to bacterial sepsis and endotoxic shock, MIF has been implicated in the pathogenesis of parasitic (malaria, cysticercosis and leishmaniasis) and viral (cytomegalovirus and influenza virus) infections. *In vivo* administration of MIF reduces the severity of infection, while mice lacking MIF are susceptible to leishmaniasis and cysticercosis.[34,35] The mechanisms of MIF in antiparasitic infection remain largely unclear. It has been reported that recombinant MIF activates murine macrophages to kill *Leishmania major*, via the iNOS and TNFα-dependent mechanisms.[34] It is also possible that MIF-induced macrophage phagocytosis is involved in antiparasitic properties of MIF. Indeed, MIF is released from murine macrophage-like cells (RAW 264.7) in response to phagocytosis of fluorescent-latex beads in a particle dose-dependent manner.[36] Pretreatment of RAW 264.7 cells with rat recombinant MIF increases the extent of phagocytosis by 1.6-fold compared with that in the control.[36] These results suggest that MIF plays an important role by activating macrophages in an autocrine and paracrine fashion to phagocytose foreign particles as well as parasitic pathogens.

MIF also contributes to the development of atherosclerosis. In both human disease and in an animal model of atherosclerosis, macrophages are the rich source of MIF.[37,38] Upregulation of MIF by both inflammatory macrophages and intrinsic vascular cells, including endothelial cells and vascular smooth muscle cells, stimulates macrophages, leading to their further accumulation and activation to produce inflammatory cytokines and adhesion molecules, phagocytosis of lipids, and transformation to foam cells.[37,38] These inflammatory processes result in atheromatous plaque formation and progression. A critical role of MIF in atherogenesis is further demonstrated in recent studies. In atherosclerosis-susceptible mice, vascular inflammation and neointimal thickening after angioplasty are reduced by anti-MIF treatment.[39,40] Furthermore, LDL-$R^{-/-}$ mice null for MIF have impaired atherosclerosis.[41] All these findings suggest that MIF may play a causal role in atherogenesis.

Apoptosis has been suggested to limit the half-life of activated macrophages and to restrict the macrophage-mediated inflammatory activities. MIF has been shown to sustain macrophage survival and function by suppressing activation-induced, p53-dependent apoptosis.[42] The administration of endotoxin to $MIF^{-/-}$ mice results in inhibition of macrophage viability, proinflammatory function, and an increase in apoptosis.[42] In contrast, inhibition of p53 in endotoxin-treated, MIF-deficient macrophages suppresses apoptosis and restores their proinflammatory properties.[42] MIF inhibits p53 activity in macrophages via an autocrine regulatory pathway, resulting in a decrease in cellular p53 accumulation and subsequent function. It is now clear that inhibition of p53 by MIF is associated with upregulation of arachidonic acid metabolism and cyclooxygenase-2 (COX-2). Thus, deficiency in the regulation of p53-dependent apoptotic pathways in macrophages could explain the sensitivity of $MIF^{-/-}$ mice to endotoxemia.

MIF may also participate in gastrointestinal diseases. In *Helicobacter pylori*-induced gastritis, we found that MIF is markedly upregulated in macrophages as well as intrinsic gastric cells and is associated with the severity of gastritis associated with *Helicobacter* infection.[18] Further, we found that upregulation of MIF by macrophages and gastric epithelial cells is associated with the development of gastric ulcers induced by acid and the administration of a neutralizing anti-MIF antibody prevents both neutrophil and macrophage accumulation, activation, and gastric inflammation via TNF-α, iNOS, and ICAM-1-dependent mechanisms.[43] In necrotizing enterocolitis, MIF is able to stimulate macrophages to produce IL-6 and IL-8 during the development of necrotizing inflammation.[19] This is further confirmed by the finding that dextran sulphate sodium (DSS)-induced colitis is exacerbated in mice with transgenic expression of MIF, but inhibited by a neutralizing anti-MIF antibody.[20,21]

MIF in Macrophage-mediated Immune Injury

Similar to the role of MIF in T-cell-mediated adaptive immunity, many studies have demonstrated that MIF plays a critical role in macrophage-mediated immune diseases. We found that macrophages are the major cell types generating MIF in both renal allograft and bone marrow transplant rejection in humans,[22,24] as well as in an animal model of kidney transplantation.[23] Indeed, upregulation of MIF is tightly associated with macrophage accumulation and the severity of tissue rejection, suggesting that local production of MIF may act by stimulating macrophage localization and activation to produce cytokines (IL-1, TNF-α, IL-2, INF-γ), chemokines (MCP-1), and adhesion molecules (ICAM-1), which cause tissue injury. However, the findings that deletion of MIF could not protect against renal and cardiac allograft rejection in MIF$^{-/-}$ mice indicate that MIF is not required for acute or chronic allograft rejection in mice.[29,30] Although MIF blockade reduces the skin DTH response, the findings that neither local nor systemic MIF are required for the rejection of fully mismatched skin, renal, and cardiac allografts raise questions about the essential role of DTH as an important mediator of allograft injury.[28–30]

Evidence to support a critical role of MIF in macrophage-mediated immune disease also comes from the findings that MIF participates in both human and experimental glomerulonephritis and arthritis.[4–13] We found that upregulation of MIF is associated with local accumulation of macrophages and T-cells and the development of severe kidney injury, including glomerular crescent formation and tubulointerstitial injury.[10–13] The functional role of MIF in glomerulonephritis and arthritis is further confirmed by the ability of administration of a neutralizing anti-MIF antibody to prevent or reverse progressive renal injury and arthritis.[6] Inhibition of macrophage and T-cell accumulation and activation, expression of proinflammatory cytokines (IL-1, TNF-α), adhesion molecules (I-CAM-1 and VCAM-1), and iNOS are key mechanisms by which anti-MIF treatment blocks immunologically-induced glomerulonephritis and arthritis.

MIF AS A CRITICAL PROINFLAMMATORY CYTOKINE IN IMMUNE AND INFLAMMATORY DISEASES

It is now well established that MIF has a direct proinflammatory role in inflammatory diseases, such as sepsis, rheumatoid arthritis, and glomerulonephritis. Unlike proinflammatory cytokines IL-1 and TNF-α that are inducible, MIF is pre-formed and stored within the cytoplasm in macrophages and many other intrinsic cells within the body, including in the hypothalamic-pituitary-adrenal (HPA) axis. MIF is released upon stimulation by stress, endotoxin, exotoxin, infection, and inflammatory and immune stimuli. Once released, MIF acts as a proinflammatory cytokine to induce expression of other inflammatory cytokines, including IL-1, TNF-α, IL-2, IL-6, IL-8, INF-γ, and iNOS,[33] which promote inflammatory and immune responses. Unlike other proinflammatory cytokines that are generally suppressed by glucocorticoids, MIF has been proposed to be the physiologic counter-regulator of glucocorticoid action within the immune system. MIF is induced by physiological glucocorticoids *in vitro* as evident by findings that MIF is secreted in response to glucocorticoids in monocytes and T-cells, pancreatic islet cells, neuronal cells and synovial fibroblasts.[15,31–33] In experimental arthritis, adrenalectomy results in the reduction of synovial MIF expression, indicating that MIF is under the control of endogenous glucocorticoids.[6] In contrast, the administration of exogenous glucocorticoids in the rat leads to the post-translational upregulation of MIF.[6] In patients with SLE, higher serum MIF levels, compared with those in controls, were strongly associated with the administration of glucocorticoids.[44] MIF also plays an important role in regulating the inflammatory response by counter-regulating the anti-inflammatory and immunosuppressive effects of glucocorticoids. Specifically, MIF counteracts glucocorticoid-induced inhibition of inflammatory cytokine secretion in macrophages (IL-1, TNF-α, IL-6, and IL-8), in T-cells (IL-2 and IFN-γ), and in synovial fibroblasts (TNF-α).[2,6,15,32,33]

The important role of MIF within the proinflammatory cytokine cascade is demonstrated by a number of studies using a neutralizing anti-MIF antibody.[2,6,15,32,33] MIF is one of several proinflammatory cytokines over-expressed in the alveolar air space during ARDS.[8,9] The *ex vivo* treatment of alveolar macrophages from patients with ARDS with exogenous MIF results in an elevated expression of TNF-α and IL-8. Conversely, treatment with anti-MIF antibodies inhibits the expression of these cytokines.[9] Recent clinical studies confirm these findings and show that increased MIF expression correlates with the development of ARDS in septic patients.[9] In a rat model of lung injury, anti-MIF treatment is shown to reduce the accumulation of activated neutrophils in the lungs.[45] Bronchoalveolar lavage samples from these animals are found to contain significantly reduced levels of the neutrophil chemoattractant macrophage inflammatory protein-2.[45] Because MIF is not itself a chemoattractant for neutrophils, these data suggest that MIF may play a role in ARDS via upregulation of the neutrophil chemoattractant macrophage inflammatory protein-2. In immunologically-induced anti-GBM glomerulonephritis, treatment with an anti-MIF antibody prevents and reverses anti-GBM glomerulonephritis in association with marked reduction of the proinflammatory cytokine IL-1 and TNF-α, demonstrating a critical and upstream role of MIF in inflammatory cytokine expression in the pathogenesis of glomerulonephritis.[11,12] Furthermore, blockade of TNF-α action in the same disease model by preventing TNF-α binding to its receptor also inhibits upregulation of MIF and progressive renal injury.[46] This indicates that MIF is also within the inflammatory cytokine cascade in the disease. Thus, MIF acts not only upstream of proinflammatory cytokines to induce IL-1 and TNF-α production, but also acts as a proinflammatory cytokine in response to other inflammatory stimulations. More importantly, MIF is also an immunomodulator of glucocorticoids and plays a critical role in counter-regulating the antiinflammatory effects of glucocorticoids in immune and inflammatory diseases.

The mechanisms by which MIF counter-regulates glucocorticoid actions remains largely unclear. The relationships between hypothalamic-pituitary-adrenal (HPA) hormones and immunity have received special attention due to the importance of these hormones in the stress response and in autoimmune diseases. Stress, inflammation and immune responses can stimulate release of MIF, which activates the inflammatory cytokine cascade to override the physiological function of glucocorticoids as a consequence of acute activation of the HPA axis. The mechanisms by which peripherally released cytokines (MIF, IL-1, TNF-α, IL-6) act on the brain to activate the HPA system have been investigated. However, very little information is available about the cytokine effects on glucocorticoid action and glucocorticoids-receptor interactions.

Activation of NF-κB may be another possible mechanism by which MIF may modulate inflammatory responses and counter-regulate the effects of glucocorticoids in inhibition of proinflammatory cytokine expression. NF-κB is an important regulator of inflammatory cytokine gene expression and several studies suggest that glucocorticoids may inhibit the production of inflammatory mediators, such as TNF-α and IL-1β, via modulation of NF-κB activity. Glucocorticoids have been proposed to inhibit the binding of the p65 subunit of NF-κB to the transcriptional machinery of target genes and to induce IκBα synthesis, thereby inhibiting NF-κB activation and preventing the expression of NF-κB-driven proinflammatory cytokine genes. Recent studies have shown that MIF inhibits the ability of glucocorticoids to induce IκB synthesis in LPS-stimulated human peripheral blood mononuclear cells.[47] Thus, by blocking glucocorticoid-induced IκB synthesis, MIF promotes the translocation of NF-κB into the nucleus, thereby activating the proinflammatory cytokine cascade and adhesion molecule expression.

MIF INDUCES MATRIX METALLOPROTEINASES TO MEDIATE TISSUE INJURY

Increasing evidence has shown that upregulation of MIF is associated with the pathogenesis of rheumatoid arthritis (RA).[4–6] Degradation of extracellular matrix components by neutral matrix metalloproteinases (MMPs)

(MMPs) is often seen as a typical pathological feature of RA. It has been shown that MIF is able to stimulate MMP-1 and MMP-3 mRNA expression by synovial fibroblasts via tyrosine kinase-, protein kinase C-, and AP-1-dependent pathways,[48] and MMP-13 mRNA expression by rat osteoblasts via the Src-related tyrosine kinase-, Ras-, ERK1/2-, and AP-1-dependent pathway.[49] The importance of MMPs released from synoviocytes in RA is well established. In a passive-transfer model induced by anti-CII Ab/LPS-induced arthritis, histopathological arthritis scores for synovial inflammation are significantly decreased in anti-MIF Ab-treated mice and in MIF-deficient mice when compared with that in wild-type mice.[50] In addition, mRNA levels of MMP-13 and MIP-2 in anti-CII Ab/LPS-induced arthritis joint tissues are significantly reduced in MIF-deficient mice compared with those in wild-type control mice. These findings suggest that MIF may act as an inducer of MMPs to mediate joint tissue destruction.

Apart from synoviocytes and synoviocyte-derived MMPs, neutrophils are also important for cartilage destruction. Neutrophils are attracted by chemokines such as MIP-2 and IL-8 and cause cartilaginous matrix and chondrocyte damage by releasing lysosomal enzymes and toxic radicals. Findings that anti-MIF antibody prevents LPS-induced pulmonary neutrophil accumulation[45] as well as acetic acid-induced macrophage and neutrophil accumulation in gastric ulcer,[43] indicate a profound relationship between MIF and neutrophil infiltration in the inflammatory state. Inhibition of neutrophils within the synovial tissues in RA induced in MIF-deficient mice support the notion that neutrophil-mediated joint destruction is largely mediated by MIF.

Our recent studies also find that MIF is able to induce MMP-1 and MMP-9 expression by both vascular and/or macrophages *in vitro* and that upregulation of MIF in atherosclerotic plaque lesions is associated with macrophage accumulation and marked MMP-1 and MMP-9 expression in human vulnerable atheromatous plaques.[51,52] These findings suggest that MIF-induced MMP expression may be a key mechanism in the destabilization of atheromatous plaques. It is now well accepted that a atherosclerosis complicated by plaque rupture is the major cause of potentially lethal acute coronary syndromes and stroke. The integrity of the fibrous cap is important for the stability of atherosclerotic plaques, which depends largely on the collagenous extracellular matrix produced by VSMCs or myofibroblasts. An increased inflammatory response, decreased ECM synthesis, and increased ECM degradation contribute significantly to plaque instability. Increased MMP activity appears to be the most important risk factor for the instability of vulnerable atheromatous plaques or plaque rupture. Several MMPs, including MMP-1, -2, -3, -9, -13 and -14 may be involved. Among them, MMP-1 and MMP-9 have been considered as highly specific proteases capable of degrading matrix, including Type I and Type III collagens, the most important components of the ECM in fibrous caps, resulting in the development of vulnerable atheromatous plaques. Our recent studies found that upregulation of MIF is associated with activation of MMP-1 and MMP-9 in human vulnerable atheromatous plaques. These are associated with marked collagenolysis and plaque rapture, suggesting a causal role of MIF in the destablization of the plaque.[51,52] Indeed, macrophages are the key inflammatory cells in vulnerable atheromatous plaques and are a major source of MIF and MMPs in the plaques,[51,52] suggesting that MIF-derived from macrophages acts in an autocrine and/or paracrine manner to induce MMP expression by macrophages, resulting in matrix degradation and fibrous cap weakening or rupture. The findings of marked upregulation of MIF and MMP-1 and MMP-9 in vulnerable atheromatous plaques, together with the ability of MIF to directly induce MMP-1 and MMP-9 expression by macrophages and VSMCs, indicate that MIF is a potentially important upstream activator of the MMP system. Direct evidence for the role of MIF in destabilization of atheromatous plaques comes from a recent study in apolipoprotein E-deficient mice.[40] In this study, inhibition of MIF with a neutralizing MIF antibody resulted in a shift in the cellular composition of neointimal plaques toward a stabilized phenotype with reduced macrophage/foam cell content and increased SMC content.[40] These observations underscore a potentially important role for MIF in atherosclerosis.

In summary, MIF is a stress factor, an inflammatory cytokine, a growth factor, and a hormone in both physiological and pathological conditions. MIF plays an important role in many diseases, including sepsis, infectious and inflammatory diseases, autoimmune diseases, graft rejection, atherosclerosis and cancer. MIF acts through many mechanisms to mediate diseases, including innate and adaptive immunity, induction of cytokines, chemokines and adhesion molecules, interactions with glucocorticoids and HAP axis, influence in cell proliferation, apoptosis and phagocytosis, and activation of the MMP system. However, most of these studies are based on *in vitro* and animal models. Further studies in human disease conditions are needed.

REFERENCES

1. Bernhagen J, Calandra T, Mitchell RA, *et al.* (1993) MIF is a pituitary-derived cytokine that potentiates lethal endotoxaemia. *Nature* **365**: 756–759.
2. Calandra T, Bernhagen J, Metz CN, *et al.* (1995) MIF as a glucocorticoid-induced modulator of cytokine production. *Nature* **377**: 68–71.
3. Calandra T, Echtenacher B, Roy DL, *et al.* (2000) Protection from septic shock by neutralization of macrophage migration inhibitory factor. *Nat Med* **6**: 164–170.
4. Leech M, Metz C, Hall P, *et al.* (1999) Macrophage migration inhibitory factor (MIF) in rheumatoid arthritis: evidence for pro-inflammatory function and regulation by glucocorticoids. *Arthritis Rheum* **42**: 1601–1608.
5. Santos LL, Hall P, Metz CN, *et al.* (2001) Role of macrophage migration inhibitory factor (MIF) in murine antigen-induced arthritis: interaction with glucocorticoids. *Clin Exp Immunol* **123**: 309–314.
6. Morand EF, Leech M. (2005) Macrophage migration inhibitory factor in rheumatoid arthritis. *Front Biosci* **10**: 12–22.
7. Bernhagen J, Bacher M, Calandra T, *et al.* (1996) An essential role for macrophage migration inhibitory factor in the tuberculin delayed-type hypersensitivity reaction. *J Exp Med* **183**: 277–282.
8. Lai KN, Leung JC, Metz CN, *et al.* (2003) Role for macrophage migration inhibitory factor in acute respiratory distress syndrome. *J Pathol* **199**: 496–508.
9. Donnelly SC, Haslett C, Reid PT, *et al.* (1997) Regulatory role for macrophage migration inhibitory factor in acute respiratory distress syndrome. *Nat Med* **3**: 320–323.
10. Lan HY, Mu W, Yang N, *et al.* (1996) De novo renal expression of macrophage migration inhibitory factor during the development of rat crescentic glomerulonephritis. *Am J Pathol* **149**: 1119–1127.
11. Lan HY, Bacher M, Yang N, *et al.* (1997) The pathogenic role of macrophage migration inhibitory factor in immunologically induced kidney disease in the rat. *J Exp Med* **185**: 1455–1465.
12. Yang N, Nikolic-Paterson DJ, Ng YY, *et al.* (1998) Reversal of established rat crescentic glomerulonephritis by blockade of macrophage migration

inhibitory factor (MIF) — potential role of MIF in regulating glucocorticoid production. *Mol Med* **4**: 413–424.
13. Lan HY, Yang N, Nikolic-Paterson DJ, *et al.* (2000) Expression of macrophage migration inhibitory factor in human glomerulonephritis. *Kidney Int* **57**: 499–509.
14. Mitchell RA. (2004) Mechanisms and effectors of MIF-dependent promotion of tumourigenesis. *Cell Signal* **16**: 13–19.
15. Calandra T, Roger T. (2003) Macrophage migration inhibitory factor: a regulator of innate immunity. *Nat Rev Immunol* **3**: 791–800.
16. Denkinger CM, Denkinger M, Kort JJ, *et al.* (2003) *In vivo* blockade of macrophage migration inhibitory factor ameliorates acute experimental autoimmune encephalomyelitis by impairing the homing of encephalitogenic T cells to the central nervous system. *J Immunol* **170**: 1274–1282.
17. Matsui Y, Okamoto H, Jia N, *et al.* (2004) Blockade of macrophage migration inhibitory factor ameliorates experimental autoimmune myocarditis. *J Mol Cell Cardiol* **37**(2): 557–566.
18. Xia HH, Lam SK, Huang XR, *et al.* (2004) *Helicobacter pylori* infection is associated with increased expression of macrophage migratory inhibitory factor — by epithelial cells, T cells, and macrophages — in gastric mucosa. *J Infect Dis* **190**: 293–302.
19. Ren Y, Lin CL, Li Z, *et al.* (2005) Up-regulation of macrophage migration inhibitory factor in infants with acute neonatal necrotizing enterocolitis. *Histopathology* **46**: 659–667.
20. Ohkawara T, Miyashita K, Nishihira J, *et al.* (2005) Transgenic overexpression of macrophage migration inhibitory factor renders mice markedly more susceptible to experimental colitis. *Clin Exp Immunol* **140**: 241–248.
21. Ohkawara T, Nishihira J, Takeda H, *et al.* (2002) Amelioration of dextran sulfate sodium-induced colitis by anti-macrophage migration inhibitory factor antibody in mice. *Gastroenterology* **123**: 256–270.
22. Lan HY, Yang N, Brown FG, *et al.* (1998) Macrophage migration inhibitory factor expression in human renal allograft rejection. *Transplantation* **66**: 1465–1471.
23. Brown FG, Nikolic-Paterson DJ, Metz C, *et al.* (1999) Up-regulation of macrophage migration inhibitory factor in acute renal allograft rejection in the rat. *Clin Exp Immunol* **118**: 329–336.
24. Lo JW, Leung AY, Huang XR, *et al.* (2002) Macrophage migratory inhibitory factor (MIF) expression in acute graft-versus-host disease (GVHD) in allogeneic hemopoietic stem cell transplant recipients. *Bone Marrow Transplant* **30**: 375–380.
25. Bacher M, Metz CN, Calandra T, *et al.* (1996) An essential regulatory role for macrophage migration inhibitory factor in T-cell activation. *Proc Natl Acad Sci USA* **93**: 7849–7854.

26. Koebernick H, Grode L, David JR, *et al.* (2002) Macrophage migration inhibitory factor (MIF) plays a pivotal role in immunity against *Salmonella typhimurium. Proc Natl Acad Sci USA* **99**: 13681–13686.
27. Cvetkovic I, Al-Abed Y, Miljkovic D, *et al.* (2005) Critical role of macrophage migration inhibitory factor activity in experimental autoimmune diabetes. *Endocrinology* **146**: 2942–2951.
28. Hou G, Valujskikh A, Bayer J, *et al.* (2001) *In vivo* blockade of macrophage migration inhibitory factor prevents skin graft destruction after indirect allorecognition. *Transplantation* **72**: 1890–1897.
29. Demir Y, Chen Y, Metz C, *et al.* (2003) Cardiac allograft rejection in the absence of macrophage migration inhibitory factor. *Transplantation* **76**: 244–247.
30. Jose MD, David JR, Atkins RC, Chadban SJ. (2003) Blockade of macrophage migration inhibitory factor does not prevent acute renal allograft rejection. *Am J Transplant* **3**: 1099–1106.
31. Calandra T, Bernhagen J, Mitchell RA, Bucala R. (1994) The macrophage is an important and previously unrecognized source of macrophage migration inhibitory factor. *J Exp Med* **179**: 1895–1902.
32. Gregersen PK, Bucala R. (2003) Macrophage migration inhibitory factor, MIF alleles, and the genetics of inflammatory disorders: incorporating disease outcome into the definition of phenotype. *Arthritis Rheum* **48**: 1171–1176.
33. Baugh JA, Bucala R. (2002) Macrophage migration inhibitory factor. *Crit Care Med* **30**(Suppl): S27–S35.
34. Juttner S, Bernhagen J, Metz CN, *et al.* (1998) Migration inhibitory factor induces killing of *Leishmania major* by macrophages: dependence on reactive nitrogen intermediates and endogenous TNF-alpha. *J Immunol* **161**: 2383–2390.
35. Rodriguez-Sosa M, Rosas LE, David JR, *et al.* (2003) Macrophage migration inhibitory factor plays a critical role in mediating protection against the helminth parasite *Taenia crassiceps. Infect Immun* **71**: 1247–1254.
36. Onodera S, Suzuki K, Matsuno T, *et al.* (1997) Macrophage migration inhibitory factor induces phagocytosis of foreign particles by macrophages in autocrine and paracrine fashion. *Immunology* **92**: 131–137.
37. Lin SG, Yu XY, Chen YX, *et al.* (2000) De novo expression of macrophage migration inhibitory factor in atherogenesis in rabbits. *Circ Res* **87**: 1202–1208.
38. Burger-Kentischer A, Goebel H, Seiler R, *et al.* (2002) Expression of macrophage migration inhibitory factor in different stages of human atherosclerosis. *Circulation* **105**: 1561–1566.
39. Chen Z, Sakuma M, Zago AC, *et al.* (2004) Evidence for a role of macrophage migration inhibitory factor in vascular disease. *Arterioscler Thromb Vasc Biol* **24**: 709–714.

40. Schober A, Bernhagen J, Thiele M, *et al.* (2004) Stabilization of atherosclerotic plaques by blockade of macrophage migration inhibitory factor after vascular injury in apolipoprotein E-deficient mice. *Circulation* **109**: 380–385.
41. Pan JH, Sukhova GK, Yang JT, *et al.* (2004) Macrophage migration inhibitory factor deficiency impairs atherosclerosis in low-density lipoprotein receptor-deficient mice. *Circulation* **109**: 3149–3153.
42. Mitchell RA, Liao H, Chesney J, *et al.* (2002) Macrophage migration inhibitory factor (MIF) sustains macrophage proinflammatory function by inhibiting p53: regulatory role in the innate immune response. *Proc Natl Acad Sci USA* **99**: 345–350.
43. Huang XR, Chun Hui CW, Chen YX, *et al.* (2001) Macrophage migration inhibitory factor is an important mediator in the pathogenesis of gastric inflammation in rats.*Gastroenterology* **121**: 619–630.
44. Hoi A, Morand EF, Leech M. (2003) Is macrophage migration inhibitory factor a therapeutic target in systemic lupus erythematosus? *Immunol Cell Biol* **81**: 367–373.
45. Makita H, Nishimura M, Miyamoto K, *et al.* (1998) Effect of anti-macrophage migration inhibitory factor antibody on lipopolysaccharide-induced pulmonary neutrophil accumulation. *Am J Respir Crit Care Med* **158**: 573–579.
46. Lan HY, Yang N, Metz C, *et al.* (1997) TNF-alpha up-regulates renal MIF expression in rat crescentic glomerulonephritis. *Mol Med* **3**: 136–144.
47. Daun JM, Cannon JG. (2000) Macrophage migration inhibitory factor antagonizes hydrocortisone-induced increases in cytosolic IkappaBalpha. *Am J Physiol Regul Integr Comp Physiol* **279**: R1043–R1049.
48. Onodera S, Kaneda K, Mizue Y, *et al.* (2000) Macrophage migration inhibitory factor up-regulates expression of matrix metalloproteinases in synovial fibroblasts of rheumatoid arthritis. *J Biol Chem* **275**: 444–450.
49. Onodera S, Nishihira J, Iwabuchi K, *et al.* (2002) Macrophage migration inhibitory factor up-regulates matrix metalloproteinase-9 and -13 in rat osteoblasts. Relevance to intracellular signaling pathways. *J Biol Chem* **277**: 7865–7874.
50. Ichiyama H, Onodera S, Nishihira J, *et al.* (2004) Inhibition of joint inflammation and destruction induced by anti-type II collagen antibody/lipopolysaccharide (LPS)-induced arthritis in mice due to deletion of macrophage migration inhibitory factor (MIF). *Cytokine* **26**: 187–194.
51. Kong YZ, Yu X, Tang JJ, *et al.* (2005) Macrophage migration inhibitory factor induces MMP-9 expression: implications for destabilization of human atherosclerotic plaques. *Atherosclerosis* **178**: 207–215.
52. Kong YZ, Huang XR, Ouyang X, *et al.* (2005) Evidence for vascular macrophage migration inhibitory factor in destabilization of human atherosclerotic plaques. *Cardiovasc Res* **65**: 272–282.

4

Structural Studies of MIF

Elias Lolis and Gregg Crichlow*

INTRODUCTION

Macrophage migration inhibition was one of the first established cytokine activities and was associated with delayed-type hypersensitivity reactions[1] due to a non-dialyzable product from activated T cells.[2,3] This activity was the topic of hundreds of manuscripts from the mid-1960s through the late 1980s. The cDNA for macrophage migration inhibitory factor (MIF) was finally cloned in 1989.[4] At that time, the only known homology was to a handful of other mammalian MIF proteins that had been identified. It is now known that MIF is distributed more widely than originally anticipated for a cytokine. MIF is present in a variety of infectious protozoa (e.g. *Leishmania major* and *Plasmodium falciparum*) where it is proposed to interfere with the immune response of the mammalian host. However, it is also present in organisms (e.g. *Oryza sativa*, *Caenorhabditis elegans*, and *Arabidopsis thaliana)* that do not infect mammals and have few, if any, cytokines. The presence of MIF in these other organisms should be useful in exploring other functions not associated with inhibiting the immune response. Interestingly, no homologues have been identified in *Drosophila melanogaster* or *Saccharomyces cerevisiae*, indicating that whatever role

*Associate Professor of Pharmacology, Yale University School of Medicine, Sterling Hall of Medicine 345B, 333 Cedar Street, New Haven, CT 06520. E-mail: Elias.Lolis@Yale.edu

it may have, it is dispensable in some multicellular organisms or in yeast, respectively.

Many functions other than inhibition of macrophage migration have been described for MIF. These include the counter-regulatory activity of glucocorticoid function,[5] sustained ERK-1/2 activation,[6] p53 suppression,[7] increased TNF, IL-1β, and PGE_2 production,[8,9] and the regulation of Jab1 transcriptional activity.[10] The crystallographic structure of MIF provides insight into an unexpected activity that may be associated with some of the above effects.

THE THREE-DIMENSIONAL STRUCTURE OF MIF

The three-dimensional structure of MIF was solved independently by three groups in 1996 and defined a new $\alpha + \beta$ protein class named the tautomerase/MIF fold and superfamily (http://scop.mrc-lmb.cam.ac.uk/scop/).[11–14] The structure was surprising in that it was a trimer of identical subunits (Fig. 1A and B). Previous biophysical studies involving analytical ultracentrifugation indicated that MIF was a dimer.[15] The oligomeric structure continued to be studied after the three-dimensional structure was determined. Cross-linking, mutational analysis, and size exclusion chromatography led to the conclusion that monomers, dimers, and trimers

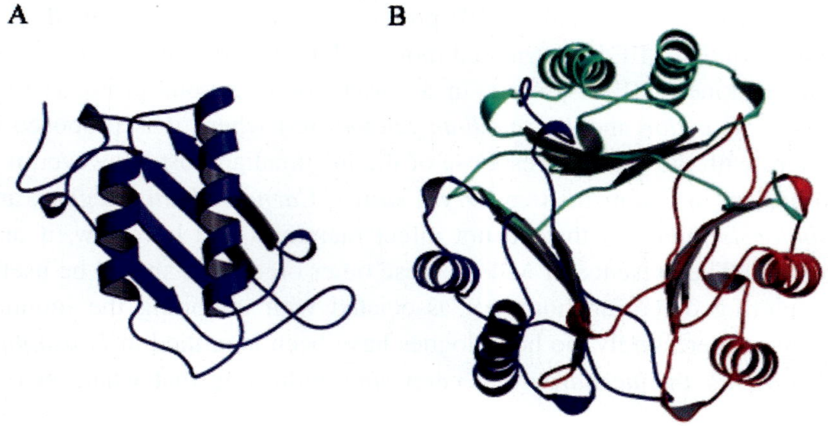

Fig. 1 Three-dimensional structure of (**A**) the human MIF monomer; and (**B**) trimer drawn from PDB entry 1MIF.[11] Figures 1–3 and 5 were produced using Molscript[44] and Raster3D.[45]

existed in solution, with monomers and dimers representing the major species in physiological concentration.[16] A more recent re-examination of the oligomeric form using sedimentation velocity and equilibrium experiments indicates that MIF is homogeneous in solution, forms a strongly-associated trimeric structure, and possesses an unusual partial specific volume.[17] This agreed with all other MIF crystal structures from different species that have been determined. Table 1 lists all known crystal structures of the tautomerase/MIF family that are currently available online (http://www.rcsb.org/pdb/home/home.do).[18] Despite sequence identities of less than 10% to human MIF, in no case has MIF been crystallized in any other oligomeric form. (However, there is one sub-family in the tautomerase/MIF superfamily that has a different oligomeric form — see below).

The secondary structure of the MIF monomer folds in such a way as to give the appearance of a 2-fold symmetry (Fig. 1A). In the orientation shown, the two anti-parallel α-helices are placed above a four stranded β-sheet with a pseudo 2-fold axis perpendicular to the plane of the figure, passing in between the two helices and through the center of the β-sheet. The four-stranded β-sheet arises from a β-strand — α-helix — β-strand motif (β-α-β) with two β-strands (β_1 and β_2) interacting in a parallel arrangement. The second set of β-strands (β_4 and β_5) is also parallel to each other but antiparallel to the first set of β-strands strands. The two remaining β-strands (β_3 and β_6) are part of intertwining loops and form interactions with β-sheets of adjacent subunits, contributing to the stabilization of the trimer. The global structure of the trimer is of three β-sheets that completely wrap around to form a barrel surrounded by six α-helices (Fig. 1B).

The oligomeric structure of MIF is important because it is directly related to an unexpected MIF activity. MIF has the same monomeric and trimeric structure as the microbial enzymes 5-carboxymethyl-2-hydroxymuconate isomerase (CHMI) (Fig. 2A and B) and 4-oxalocrotonate tautomerase (4-OT).[19] The 4-OT polypeptide is roughly half the size of MIF but dimerizes (Fig. 2C) in a manner consistent with the pseudo 2-fold symmetry of MIF; trimerization of the 4-OT dimer produces a hexamer similar to the global structure of MIF (Fig. 2D). These three proteins form families within the tautomerase/MIF superfamily. Five major subfamilies are present within the 4-OT family.[20] Proteins in four

Table 1 The coordinates, complexed ligands, PDB codes, and crystallographic parameters of proteins present in the PDB database (http:// www.rcsb.org/pdb/home/home.do)[18] that are members of the tautomerase/MIF superfamily (http://scop.mrc-lmb.cam.ac.uk/scop/).[14]

Protein	Ligand	PDB code	Resolution (Å)	R/R_{free} (%)
E. coli 5-caboxymethyl-2-hydroxymuconate isomerase	—	1OTG	2.10	18/NA
Pseudomonas sp CF600 strain 4-oxalocrotonate tautomerase	—	1OTF	1.90	20/NA
Human MIF	—	1MIF	2.60	20/NA
Rat MIF	—	1FIM	2.20	20/24
Human GIF	—	1GIF	1.90	17/NA
Pseudomonas putida 4-oxalocrotonate tautomerase	2-oxo-3-pentynoate	1BJP	2.40	21/24
Mouse MIF	(E)-2-fluoro-p-hydroxycinnamate	1MFI	1.80	18/23
Pseudomonas putida 4-oxalocrotonate tautomerase	—	4OTA	2.75	23/27
Pseudomonas putida 4-oxalocrotonate tautomerase	—	4OTB	2.50	22/26
Pseudomonas putida 4-oxalocrotonate tautomerase	—	4OTC	2.28	23/27
MIF Y95F mutant	—	1MFF	2.00	19/23
Human MIF	Hydroxyphenyl pyruvate	1CA7	2.50	20/26
Human P1G active site mutant	—	1P1G	2.50	23/27
Human inactivating mutant insert (PAM)	—	1CGQ	2.00	23/28
Human MIF	Citric acid (allosteric ligand)	1GD0	1.50	19/21
Human MIF	7-hydroxy-2-oxo-chromene-3-carboxylic acid ethyl ester inhibitor	1GCZ	1.90	20/24
Trichinella Spiralis MIF	—	1HFO	1.65	23/28
Haemophilus Influenza Hypothetical Protein	—	1MWW	2.08	20/27
E. coli 4-oxalocrotonate tautomerase	—	1GYY	1.35	15/17
Xenopus laevis MIF	—	1UIZ	2.50	24/27
Pseudomonas putida 4-oxalocrotonate tautomerase L8R mutant	—	2FM7	2.8	23/30
Mouse MIF	Covalently modified 2-oxo-4-phenyl-3-butynoate	2GDG	1.45	15/18
Human MIF	7-hydroxy-2-oxo-chromene-3-carboxylic acid ethyl ester inhibitor	1GCZ	1.90	20/24

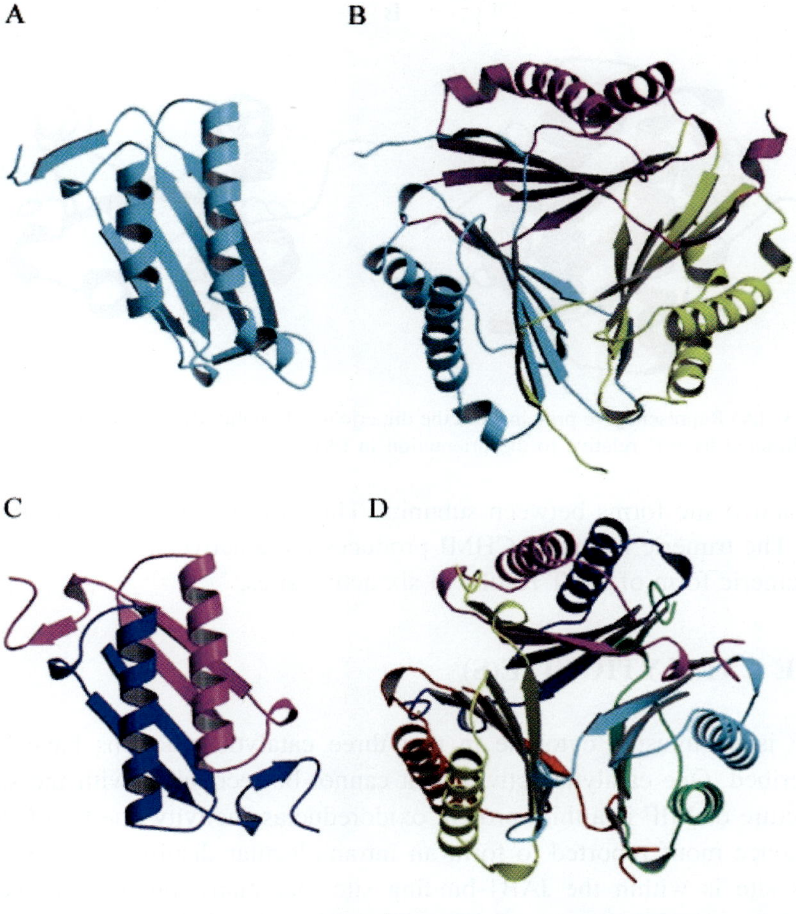

Fig. 2 (**A**) Monomeric; and (**B**) trimeric forms of 5-carboxymethyl-2-hydroxymuconate isomerase (CHMI). Drawn from 1OTG.[19] (**C**) View of the dimer of 4-oxalocrotonate tautomerase (4-OT) showing resemblance to monomers of MIF and CHMI, and (**D**) view of 4-OT hexamer. Drawn from PDB entry 1OTF.[19]

of the five subfamilies are predicted to form hexamers (trimers of dimers), but a protein in the fifth subfamily forms only a dimer with the active site formed by residues from each subunit.[20] This protein is different from 4-OT in that there are two α-helices between the second β-strand and the C-terminus. In the dimeric structure, the β-sheet separates the first α-helix from the two C-terminal α-helices (Fig. 3). This example shows the diversity of the oligomeric forms within the superfamily and highlights that

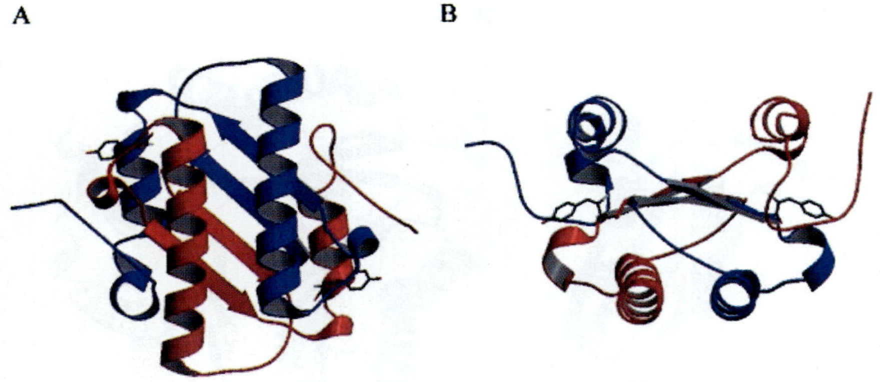

Fig. 3 (**A**) Representative protein from the dimeric 4-OT subfamily (PDB entry 1GYY).[20] (**B**) Rotated by 90° relative to the orientation in (**A**).

the active site forms between subunits. This is also true of CHMI and 4-OT. The trimeric nature of CHMI produces three active sites, whereas the hexameric form of 4-OT results in six active sites.

THE CATALYTIC SITE(S)

MIF is an unusual cytokine in that three catalytic reactions have been described. One catalytic activity that cannot be reconciled with the static structure of MIF is a thiol-protein oxidoreductase activity due to a CALC sequence motif reported to form an intramolecular disulfide bridge.[21–23] This site is within the JAB1-binding site that many, but not all, vertebrate homologues possess.[24] The three-dimensional structure of human MIF indicates that none of the three cysteine residues are within disulfide-bond distance and that a conformational change would be required to form a disulfide bond between any two of the three cysteines.

The other two activities for MIF are tautomerase activities. Although there is no sequence homology among MIF, CHMI, and 4-OT, the amino-terminal residue (after post-translational removal of the initiating methionine) for each protein is a proline. Pro-1 for CHMI and 4-OT is the catalytic base deep within a solvent accessible enzymatic site responsible for an isomerase or tautomerase activity, respectively.[19] Similarly, Pro-1 for MIF is at the bottom of a solvent-exposed pocket that appears to be a catalytic site for a small molecule substrate. This raised the question of whether the pro-inflammatory role of MIF is based on an uncharacterized enzymatic activity or direct receptor binding and activation.

A catalytic activity for a non-physiological substrate, the *D*-isomer of 2-carboxy-2,3-dihydroindole-5,6-quinone (dopachrome) converted by MIF into 5,6-dihydroxyindole-2-carboxylic acid (DHICA), was fortuitously identified.[25] In a subsequent study, the *D*-dopachrome tautomerase activity of MIF was used to screen physiological molecules that resulted in the discovery of the enol-keto tautomerization of phenylpyruvate and hydroxyphenylpyruvate.[26] However, phenylpyruvate tautomerase is not believed to be the physiological enzymatic activity of MIF.[24] Nonetheless, the identification of [pseudo]substrates for MIF allowed inhibitory analogues to be synthesized and the MIF catalytic reactions to be used in high throughput screening of potential small molecules inhibitors. Various crystal structures of small molecule inhibitors of the MIF catalytic site have been determined and are being used for the development of small molecule therapeutics.[27–29] In addition, a crystal structure of the complex between MIF and *p*-hydroxyphenylpyruvate (HPP) was used to investigate the mechanism of the catalytic activity as a model for the uncharacterized physiological enzyme activity.[30] Figure 4 shows the distances of HPP with some of the active site residues of MIF. The nitrogen of Pro-1 is within 3.3 Å of the C3 carbon of HPP, which would allow for facile proton transfer between these atoms. For Pro-1 to have a role as a catalytic base, the nitrogen atom of the proline would need to have a pKa significantly decreased from a value of over 9.0 for a proline amide.[31] NMR studies at different pH values were carried out to determine a pKa value of 5.6 for the nitrogen of Pro-1, almost four pH units below the expected pKa for the nitrogen in a proline amide.[32] This value was consistent with pKa values for the kcat/Km for the enolization of phenylpyruvate (pKa of 6.0) and the pH-dependent inactivation of MIF by 3-bromopyruvate k(inact)/Ki (pKa of 5.7).[33]

The relationship between the MIF catalytic site and inflammation has been under active investigation. Thus far, only one cellular study has demonstrated that the catalytic site is not important for a pro-inflammatory activity. In this study, there is no difference between catalytically inactive mutant and wild-type control in the MIF inhibition of macrophage

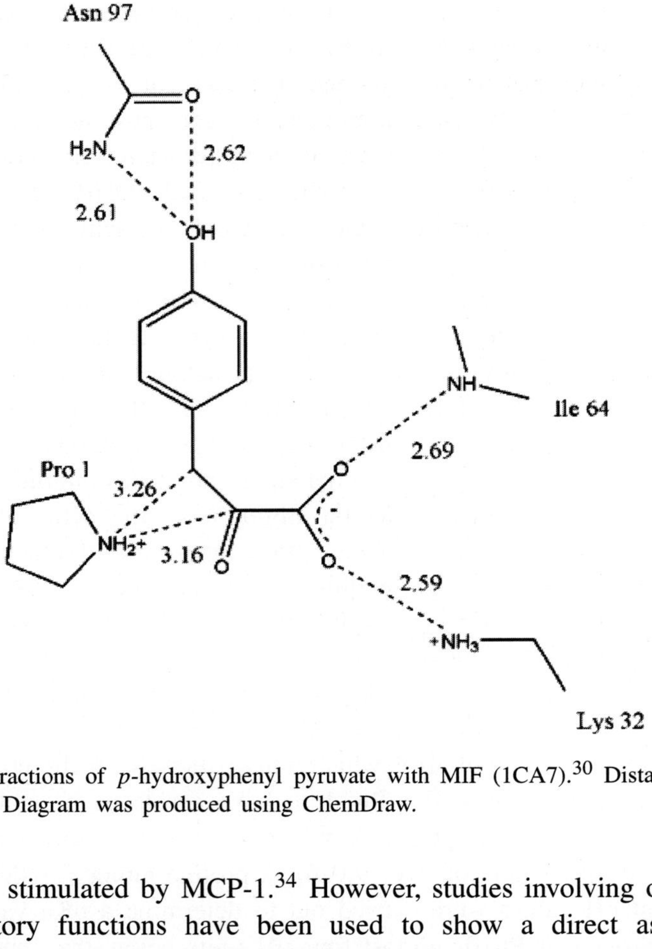

Fig. 4 Interactions of *p*-hydroxyphenyl pyruvate with MIF (1CA7).[30] Distances are in Ångstroms. Diagram was produced using ChemDraw.

migration stimulated by MCP-1.[34] However, studies involving other pro-inflammatory functions have been used to show a direct association between the catalytic site of MIF and functional activity. The results from mutants of Pro-1 led to the conclusion that the catalytic site is important for the MIF counter-regulatory activity of glucocorticoids[29] and for neutrophil priming.[32] Another mutant of Pro-1 does not up-regulate metalloproteinases-1 and -3 in rheumatoid arthritis synovial fibroblasts, in contrast to wild-type MIF.[35] In more recent experiments, the small molecule (S,R)-3-(4-hydroxyphenyl)-4,5-dihydro-5-isoxazole acetic acid methyl ester (ISO-1) was characterized as a catalytic inhibitor of the MIF active and shown to inhibit the glucocorticoid counter-regulation activity of MIF and MIF-induced production of TNF-α.[29] *In vivo* administration of different amounts of ISO-1 demonstrated a significant dose-response effect on protection against death in a mouse model of septic shock.[36] It therefore appears that the catalytic site is important for pro-inflammatory activity, but the details that link this site to inflammation are not known. One possibility is that this site actually catalyzes a chemical reaction, converting a substrate into a product that is pro-inflammatory. The other possibility is that the catalytic site affects binding and/or activation to the MIF receptor.[37]

THE MIF RECEPTOR

CD74 is a type II receptor that is well known for its function as the invariant chain (Ii) that processes and transports MHC class II proteins from the endoplasmic reticulum to compartments where peptide loading of MHC class II takes place.[38] The invariant chain, like MIF, is a homotrimer (Fig. 5).[39] In addition to the intracellular role of Ii, 2–5% of the total cellular invariant chain is found at the cell surface. CD74 also is widely expressed in cells that do not present antigen through the MHC class II mechanism, including T cells, B cells, and epithelial cells.[40,41] The idea that proteins may have multiple and, sometimes, unrelated functions has recently gained wider acceptance.[42] The presence of CD74 in many cells that do not express class II MHC suggests another function for this

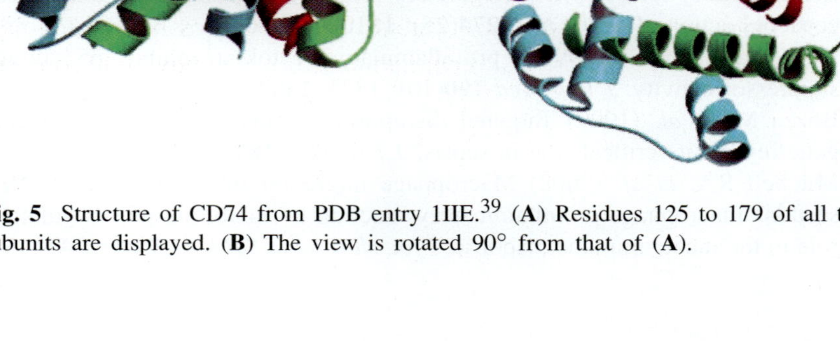

Fig. 5 Structure of CD74 from PDB entry 1IIE.[39] (**A**) Residues 125 to 179 of all three subunits are displayed. (**B**) The view is rotated 90° from that of (**A**).

polypeptide. The binding of MIF to CD74 in the presence of CD44 leads to activation of the MAP kinase pathway and phosphorylation of ERK-1/2 as well as the production of prostaglandin E_2. MIF has no apparent interaction with CD44 nor does CD44 increase the affinity of MIF for CD74.[43] This bears a similarity to other multi-subunit cytokine receptors. For example, the IL-2 receptor has α, β, and γ subunits. The α and β subunits bind to IL-2 but the γ subunit, which is required for signaling, does not. In the case of MIF, the relationship among the catalytic site, binding to the receptor, and activating CD74 and CD44 is under active investigation. Whether additional receptor subunits are required for the pro-inflammatory activities of MIF are also unknown. The MIF-receptor complex remains fertile ground for biochemical, signaling, and structural studies to determine how MIF induces its pro-inflammatory activities.

REFERENCES

1. George M, Vaughan JH. (1962) *In vitro* cell migration as a model for delayed hypersensitivity. *Proc Soc Exptl Biol Med* **111**: 514–521.
2. Bloom BR, Bennett B. (1966) Mechanism of a reaction *in vitro* associated with delayed-type hypersensitivity. *Science* **153**(731): 80–82.
3. David JR. (1966) Delayed hypersensitivity *in vitro*: its mediation by cell-free substances formed by lymphoid cell-antigen interaction. *Proc Natl Acad Sci USA* **56**(1): 72–77.
4. Weiser WY, *et al.* (1989) Molecular cloning of a cDNA encoding a human macrophage migration inhibitory factor. *Proc Natl Acad Sci USA* **86**(19): 7522–7526.
5. Calandra T, *et al.* (1995) MIF as a glucocorticoid-induced modulator of cytokine production. *Nature* **377**(6544): 68–71.
6. Mitchell RA, *et al.* (1999) Sustained mitogen-activated protein kinase (MAPK) and cytoplasmic phospholipase A2 activation by macrophage migration inhibitory factor (MIF). Regulatory role in cell proliferation and glucocorticoid action. *J Biol Chem* **274**(25): 18100–18106.
7. Hudson JD, *et al.* (1999) A proinflammatory cytokine inhibits p53 tumor suppressor activity. *J Exp Med* **190**(10): 1375–1382.
8. Bozza M, *et al.* (1999) Targeted disruption of migration inhibitory factor gene reveals its critical role in sepsis. *J Exp Med* **189**(2): 341–346.
9. Mitchell RA, *et al.* (2002) Macrophage migration inhibitory factor (MIF) sustains macrophage proinflammatory function by inhibiting p53: regulatory role in the innate immune response. *Proc Natl Acad Sci USA* **99**(1): 345–350.

10. Kleemann R, *et al.* (2000) Intracellular action of the cytokine MIF to modulate AP-1 activity and the cell cycle through Jab1. *Nature* **408**(6809): 211–216.
11. Sun HW, *et al.* (1996) Crystal structure at 2.6-A resolution of human macrophage migration inhibitory factor. *Proc Natl Acad Sci USA* **93**(11): 5191–5196.
12. Kato Y, *et al.* (1996) The crystal structure of human glycosylation-inhibiting factor is a trimeric barrel with three 6-stranded beta-sheets. *Proc Natl Acad Sci USA* **93**(7): 3007–3010.
13. Suzuki M, *et al.* (1996) Crystal structure of the macrophage migration inhibitory factor from rat liver. *Nat Struct Biol* **3**(3): 259–266.
14. Murzin AG, *et al.* (1995) SCOP: a structural classification of proteins database for the investigation of sequences and structures. *J Mol Biol* **247**(4): 536–540.
15. Nishihira J, *et al.* (1995) The structure and physicochemical properties of rat liver macrophage migration inhibitory factor. *Biochim Biophys Acta* **1247**(1): 159–162.
16. Mischke R, *et al.* (1998) Cross-linking and mutational analysis of the oligomerization state of the cytokine macrophage migration inhibitory factor (MIF). *FEBS Lett* **427**(1): 85–90.
17. Philo JS, Yang TH, LaBarre M. (2004) Re-examining the oligomerization state of macrophage migration inhibitory factor (MIF) in solution. *Biophys Chem* **108**(1–3): 77–87.
18. Berman HM, *et al.* (2000) The Protein Data Bank. *Nucleic Acids Res* **28**(1): 235–242.
19. Subramanya HS, *et al.* (1996) Enzymatic ketonization of 2-hydroxymuconate: specificity and mechanism investigated by the crystal structures of two isomerases. *Biochemistry* **35**(3): 792–802.
20. Almrud JJ, *et al.* (2002) The crystal structure of YdcE, a 4-Oxalocrotonate tautomerase homologue from *Escherichia coli*, confirms the structural basis for oligomer diversity. *Biochemistry* **41**(40): 12010–12024.
21. Kleemann R, *et al.* (1998) Disulfide analysis reveals a role for macrophage migration inhibitory factor (MIF) as thiol-protein oxidoreductase. *J Mol Biol* **280**(1): 85–102.
22. Kleemann R, *et al.* (1998) Specific reduction of insulin disulfides by macrophage migration inhibitory factor (MIF) with glutathione and dihydrolipoamide: potential role in cellular redox processes. *FEBS Lett* **430**(3): 191–196.
23. Kleemann R, *et al.* (2000) Dissection of the enzymatic and immunologic functions of macrophage migration inhibitory factor. Full immunologic activity of N-terminally truncated mutants. *Eur J Biochem* **267**(24): 7183–7193.

24. Sato A, *et al.* (2003) Macrophage migration inhibitory factor (MIF) of jawed and jawless fishes: implications for its evolutionary origin. *Dev Comp Immunol* **27**(5): 401–412.
25. Rosengren E, *et al.* (1996) The immunoregulatory mediator macrophage migration inhibitory factor (MIF) catalyzes a tautomerization reaction. *Mol Med* **2**(1): 143–149.
26. Rosengren E, *et al.* (1997) The macrophage migration inhibitory factor MIF is a phenylpyruvate tautomerase. *FEBS Lett* **417**(1): 85–88.
27. Taylor AB, *et al.* (1999) Crystal structure of macrophage migration inhibitory factor complexed with (E)-2-fluoro-p-hydroxycinnamate at 1.8 A resolution: implications for enzymatic catalysis and inhibition. *Biochemistry* **38**(23): 7444–7452.
28. Orita M, *et al.* (2001) Coumarin and chromen-4-one analogues as tautomerase inhibitors of macrophage migration inhibitory factor: discovery and X-ray crystallography. *J Med Chem* **44**(4): 540–547.
29. Lubetsky JB, *et al.* (2002) The tautomerase active site of macrophage migration inhibitory factor is a potential target for discovery of novel anti-inflammatory agents. *J Biol Chem* **277**(28): 24976–24982.
30. Lubetsky JB, *et al.* (1999) Pro-1 of macrophage migration inhibitory factor functions as a catalytic base in the phenylpyruvate tautomerase activity. *Biochemistry* **38**(22): 7346–7354.
31. Stivers JT, *et al.* (1996) 4-Oxalocrotonate tautomerase: pH dependence of catalysis and pKa values of active site residues. *Biochemistry* **35**(3): 814–823.
32. Swope M, *et al.* (1998) Direct link between cytokine activity and a catalytic site for macrophage migration inhibitory factor. *Embo J* **17**(13): 3534–3541.
33. Stamps SL, Fitzgerald MC, Whitman CP. (1998) Characterization of the role of the amino-terminal proline in the enzymatic activity catalyzed by macrophage migration inhibitory factor. *Biochemistry* **37**(28): 10195–10202.
34. Hermanowski-Vosatka A, *et al.* (1999) Enzymatically inactive macrophage migration inhibitory factor inhibits monocyte chemotaxis and random migration. *Biochemistry* **38**(39): 12841–12849.
35. Onodera S, *et al.* (2000) Macrophage migration inhibitory factor up-regulates expression of matrix metalloproteinases in synovial fibroblasts of rheumatoid arthritis. *J Biol Chem* **275**(1): 444–450.
36. Al-Abed Y, *et al.* (2005) ISO-1 Binding to the tautomerase active site of MIF inhibits its pro-inflammatory activity and increases survival in severe sepsis. *J Biol Chem* **280**(44): 36541–36544.
37. Leng L, *et al.* (2003) MIF signal transduction initiated by binding to CD74. *J Experi Medicine* **197**(11): 1467–1476.
38. Cresswell P. (1994) Assembly, transport, and function of MHC class II molecules. *Annu Rev Immunol* **12**: 259–293.

39. Jasanoff A, Wagner G, Wiley DC. (1998) Structure of a trimeric domain of the MHC class II-associated chaperonin and targeting protein Ii. *Embo J* **17**(23): 6812–6818.
40. Beswick EJ, *et al.* (2005) Helicobacter pylori binds to CD74 on gastric epithelial cells and stimulates interleukin-8 production. *Infect Immun* **73**(5): 2736–2743.
41. Ong GL, *et al.* (1999) Cell surface expression and metabolism of major histocompatibility complex class II invariant chain (CD74) by diverse cell lines. *Immunology* **98**(2): 296–302.
42. Jeffery CJ. (1999) Moonlighting proteins. *Trends Biochem Sci* **24**(1): 8–11.
43. Shi X, *et al.* (2006) CD44 is the signaling component of the macrophage migration inhibitory factor-CD74 receptor complex. *Immunity* **25**(4): 595–606.
44. Kraulis PJ. (1991) MOLSCRIPT: a program to produce both detailed and schematic plots of protein structures. *J Appl Cryst* **24**: 946–950.
45. Merritt EA, Bacon DJ. (1997) Raster3D Photorealistic Molecular Graphics. *Methods in Enzymology* **277**: 505–524.

5

MIF Redox Activity

*Jürgen Bernhagen**

SYNOPSIS

MIF is an evolutionarily conserved inflammatory cytokine with unique structural properties that does not belong to any of the known cytokine protein families. Instead, MIF has two catalytically competent, conserved structural motives that otherwise occur in bacterial and several other organisms. One such sequence is the so-called Cys-Xaa-Xaa-Cys (CXXC) thiol-protein oxidoreductase (TPOR) motif located between residues 57 and 60 of the MIF sequence, which can undergo reversible intramolecular disulfide formation. MIF shares this motif with a variety of enzymes of the TPOR family, including thioredoxin, protein disulfide isomerase and glutaredoxin. The CXXC motif is a sensitive structural site within the MIF molecule as MIF mutants with residues exchanged in this site exhibit a dramatically altered folding behavior. More importantly, the CXXC motif of MIF, in particular cysteine 60, contributes to MIF receptor action and several of its molecular and cellular activities such as the inhibition of apoptosis or glucocorticoid overriding. The precise pathways of MIF-dependent redox signaling are currently under intensive investigations. Physiologically, MIF

*Department of Biochemistry and Molecular Cell Biology, Institute of Biochemistry, University Hospital RWTH Aachen, Pauwelsstrasse 30, 52074 Aachen, Germany. Tel.: +49-241-80 88840; Fax: +49-241-80 82427; E-mail: jbernhagen@ukaachen.de

appears to be critically involved in the regulation of cellular homeostasis redox, where it can contribute to protecting cells from pro-oxidative stress. In this chapter, the structural, biochemical and biological evidence for a role of the TPOR activity for MIF function is summarized and discussed.

INTRODUCTION

Identified over 40 years ago as a lymphocyte-derived immune activity inhibiting the random migration of macrophages, MIF was among the very first soluble immune mediators to be discovered. In those early studies, the migration inhibitory properties of MIF on macrophages and its role in cellular immunity, in particular delayed-type hypersensitivity (DTH) reactions, were investigated.[1,2] The molecular entity responsible for the discovered MIF activities, let alone the structural properties of that entity, remained obscure for almost another quarter of a century until MIF was eventually cloned in 1989 from T cells.[3]

As detailed elsewhere in this book and in a number of recent review articles, it is known today that MIF is a fairly ubiquitously expressed mediator that serves as a pleiotropic inflammatory cytokine and endocrine factor with critical functions in innate and acquired immunity. As is known for several other inflammatory cytokines, MIF can aggravate inflammatory processes to promote pathophysiological derailments. MIF was consequently recognized to be a pivotal mediator of acute and chronic inflammatory conditions such as septic shock, colitis, rheumatoid arthritis, acute respiratory distress syndrome, atherosclerosis, pancreatitis, uveitis as well as tumorigenesis.[4–8]

The structure of human MIF was resolved in 1996 by Lolis and coworkers[9] (see Chapter 4 of this book). According to the crystal structure elucidation, MIF is a homotrimer with a barrel-like structure, with the monomers featuring a $\beta\alpha\beta\beta\beta\alpha\beta\beta$ motif. The three-fold symmetry of the barrel structure formed by the MIF trimer somewhat resembles the three-fold symmetry of tumor necrosis factor (TNF) and the structure of the trefoil cytokine interleukin-1β (IL-1β). The MIF monomer has an interesting architectural resemblance to the interleukin-8 (IL-8) dimer and the peptide binding domain of the major histocompatibility complex (MHC)

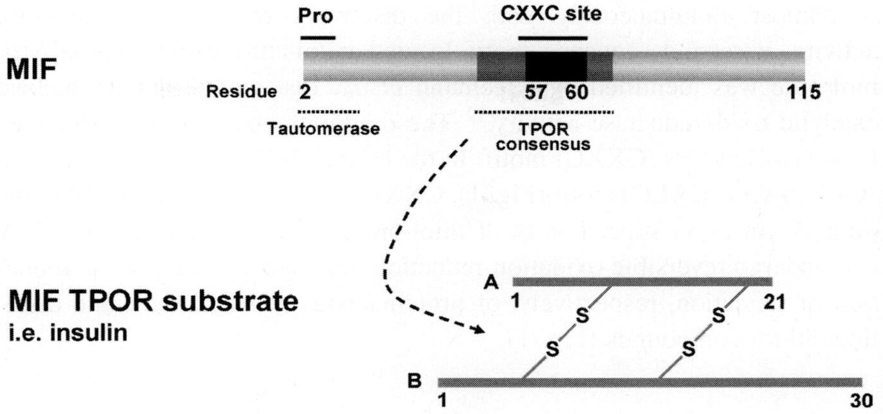

Fig. 1 Scheme of human MIF and localization of its CXXC and extended TPOR motif. The CXXC motif is located between residues 57 and 60 (black). The thiol-protein oxidoreductase (TPOR) consensus region is also indicated (dark grey). The N-terminal proline residue (Pro) within the tautomerase site of MIF is depicted. Numbering of amino acids refers to the translated cDNA sequence, with the N-terminal and processed Met residue numbered as "1." The disulfide A and B chains of insulin are shown as an example TPOR substrate of MIF.

MIF, macrophage migration inhibitory factor; CXXC, Cys-Xaa-Xaa-Cys (Cys-Ala-Leu-Cys in the case of MIF).

class I. However, overall MIF does not have any apparent domains, nor could a tight structural homology with any of the known cytokine protein families be established.[9,10]

Given that MIF is a relatively small protein mediator with a tight fold, it is probably not surprising that it exhibits no typical domain structure. However, two conserved sequence motives have been identified. The motives have been found to represent local catalytic centers within the MIF sequence that are responsible for two distinct catalytic activities, a tautomerase/isomerase and a thiol-protein oxidoreductase (TPOR) activity. The tautomerase site of MIF was discovered in 1996 by Rorsman and colleagues,[11] and is dealt with in detail in Chapter 4. With respect to the overall 3D structure of MIF, it is noteworthy that MIF has a striking architectural homology with certain bacterial tautomerases, as well as a human D-dopachrome tautomerase, that also share the catalytic tautomerase site with MIF. However, there is only a marginal sequence homology between MIF and these proteins.

Almost simultaneously with the discovery of MIF's tautomerase activity, a second sequence motif located within the center of the MIF molecule was identified by Kleemann *et al.*, that was found to mediate catalytic oxidoreductase activity.[12] The oxidoreductase site is a so-called Cys-Xaa-Xaa-Cys (CXXC) motif. In the case of MIF, the CXXC is a Cys-Ala-Leu-Cys (CALC) motif (Fig. 1). CXXC, which is typically found in the thioredoxin (Trx) super-family of thiol-protein oxidoreductases (TPORs), can undergo reversible oxidation-reduction reactions to catalyze the reduction or oxidation, respectively, of proteinaceous or small molecule disulfide/dithiol compounds (Fig. 1).

THIOL-PROTEIN OXIDOREDUCTASES AND THIOREDOXIN

Thioredoxin is the prototypical member of the TPOR family of proteins. In addition to human Trx and various thioredoxins in other species, redox-regulating proteins such as the glutaredoxins Grx1, Grx2 and Grx3, the peroxiredoxins (Prxs), protein disulfide isomerase (PDI), and disulfide bond proteins (Dsb) belong to the TPOR family.[13–17] While the residues between the vicinal cysteines vary between the TPOR family members, these proteins all share additional, conserved residues upstream of CXXC. These are a phenylalanine 5–7 residues upstream of the N-terminal cysteine of CXXC and a leucine or valine 2–3 residues further upstream.[18] MIF also fulfils the "extended" TPOR motif requirements in having a Phe at position −7 and a Leu residue at position −10 of its CXXC motif.[12]

TPORs are enzymes involved in disulfide-mediated oxidation-reduction and folding reactions and their catalytic activity is based on the formation or reduction of a catalytic disulfide bridge between the two vicinal cysteines of the CXXC region,[16] these cysteines being placed in a so-called thioredoxin/glutaredoxin fold.[19] Exceptionally, a one-cysteine mechanism may be involved. The Grx proteins, by using their glutathione binding site, uniquely reduce mixed disulfides with glutathione via a monothiol mechanism, where only an N-terminal low pKa Cys residue within their CXXC active site sequence is required.[13] Similarly, the 1-Cys subclass of Prxs, i.e. Prx6, uses GSH as an electron donor to reduce hydroperoxide, whereby the oxidation of the active site Cys residue to a sulfenic acid during catalysis requires piGST-catalyzed glutathionylation and reduction with GSH to complete the enzymatic cycle.[20]

Depending on the overall 3D structure of the protein and the residues in the vicinity of the CXXC region, TPORs can have a more reducing or oxidizing redox potential. For example, Trx has a strongly reducing potential of $\sim -270\,\text{mV}$, thus it acts primarily as a disulfide reductase.[21]

Functionally, TPORs are involved in a variety of cellular effects, including operating as dithiol reductants for small molecule disulfides (such as glutathione or cystine) and/or protein disulfides. TPORs regulate the formation of deoxyribonucleotides for DNA synthesis (by reducing the essential enzyme ribonucleotide reductase) or the generation of reduced sulfur (via 3′-phosphoadenylylsulfate reductase). In addition, TPORs can be involved in protein-protein interactions to modulate target protein function. For example, Trx interacts directly with key signaling molecules such as apoptosis signal-regulating kinase-1 (ASK-1), c-Jun activation domain binding protein-1 (JAB1), or Ref-1 to control cellular signaling processes. At least some of these signaling processes regulated by the interaction of the signaling proteins with Trx are independent of the redox activity of Trx. The redox effects of the TPORs on the cellular redox state often require the sequential action of more than one TPOR. In the "glutaredoxin system," glutaredoxins utilize the reducing power of glutathione to catalyze disulfide reductions in the presence of NADPH and glutathione reductase (GR). Similarly, the "thioredoxin system" consists of Trx, thioredoxin reductase and NADPH.

Through these various effects, TPORs participate in the regulation of cellular reduction/oxidation (redox) homeostasis, protein folding, and the defense against oxidative and reactive oxygen species (ROS) stress.

More recently, it was recognized that the modifications of protein thiols by oxidation, glutathionylation, cysteinylation, or S-nitrosylation are not only "undesired" and irreversible oxidative stress-induced damage products. Rather, it appears that the oxidoreduction of protein thiols/disulfides constitutes a reversible regulatory mechanism, representing a previously unrecognized cellular signaling mechanism,[22] which is now known as an important feature of "cellular redox signaling pathways".[23,24] Also, low doses of peroxides, glutathione, and ROS have been recognized to convey signaling responses.[23,25,26] In this context, TPORs act as "redox sensors" as they can undergo protein-thiol modifications themselves, and can act to form, modulate, or reverse thiol modifications of critical signaling proteins.[13,27,28]

A number of years ago, Trx was surprisingly demonstrated to be an important inhibitor of immune cell apoptosis and apoptosis inhibition by Trx was found to be mediated by its TPOR activity.[29–32] These findings were based on earlier work showing that the oxidation of cellular sulfhydryl groups induces apoptosis in T lymphocytes,[33] and that human Trx is identical with the adult T cell leukemia-derived factor (ADF) that induces IL-2 receptor and T cell proliferation.[34] Numerous publications have since confirmed that Trx plays a key role in inhibiting apoptosis in a variety of cells and cell stress situations, and that this function is mainly dependent on its TPOR activity.[28,35–38] Interestingly, the catalytic redox activity of Trx seems to be dispensible for its inhibitory effect on apoptosis through signal-regulating kinase-1 (ASK-1) regulation.[39] However, as has been elegantly shown by the use of active site Cys mutants, Trx binding to ASK-1 is dependent on the presence of at least one of the CXXC cysteines, while the Trx-mediated inhibition of ASK-1 via the ubiquitination and degradation of ASK-1 is independent of the redox activity of Trx.[40,41] Thus, in addition to the redox signaling processes discussed above, in which Trx is involved and through which Trx regulates cell functions such as apoptosis, one mechanism by which Trx participates in cellular apoptosis inhibition is through the inhibition of ASK-1.

Interestingly, human Trx, which is a 12 kDa cytosolic protein, acts as a co-cytokine upon leaderless secretion in addition to functioning as a protein disulfide reductase and as a key component of cellular redox regulation. Accordingly, extracellular Trx exhibits a variety of immune and inflammatory activities. Notably, a 10 kDa C-terminally truncated Trx (Trx80) comprising the 80 or 84 N-terminal amino acids is also secreted and present in plasma, where it was originally purified and identified as an "eosinophilic cytotoxicity enhancing factor." Trx80 is a potent mitogenic cytokine that stimulates the growth of resting human peripheral blood mononuclear cells (PBMC), an effect that Trx lacks. In terms of its structure function profile, Trx80 is different from Trx, because it is a dimer lacking in reductase activity, and the cytokine activity is not dependent on the Cys residues of the Trx active-site motif. The primary targets of Trx80 in PBMC are monocytes that are activated to proliferate and increase the expression of a variety of immune mediators and receptors. In essence, Trx80 is a potent cytokine for monocytes directing the immune system to a Th_1 response via IL-12 production.[42–44]

Thus, although Trx is the prototypical member of the Trx superfamily of TPOR proteins, it exhibits a much greater variety of functions than most of the other TPORs, including activities that are not related to the CXXC active site-mediated redox activity that is characteristic of the TPORs. This is noteworthy in the context of this book, as MIF shares with Trx several functional characteristics, including its intra- and extracellular activities, its leaderless secretion, and its potent anti-apoptotic effect (see below).

THE MIF CXXC MOTIF, CXXC DISULFIDE BOND FORMATION AND CATALYTIC THIOL-PROTEIN OXIDOREDUCTASE ACTIVITY OF MIF

The CALC motif of MIF is well conserved through all mammalian MIF orthologues. The parasite proteins *B. malayi* MIF and *W. bancrofti* MIF also have a similar motif, with CAVC and CPVC sequences, respectively. Moreover, while not identical, the residues between the CXXC cysteines in MIF are somewhat similar in size and hydrophobicity to those in other TPORs.[12] Of note is the finding that the various MIF orthologues exhibit almost identical β-turn propensities (around 9%) in their CXXC regions when compared with TPORs such as Grx or DsbA.[12,45] It is believed that the disulfide formation between the CXXC cysteines of the catalytic site of a TPOR is accompanied by the corresponding formation of a β-turn. Together, this suggests that disulfide formation occurs at the CXXC site of MIF, and that MIF might be able to catalyze CXXC-mediated oxidation-reduction reactions in the cell. This notion is strongly supported by S-alkylation experiments in combination with the mass spectrometric analysis of natively folded, recombinant human MIF. In these studies, it was demonstrated that under oxidizing conditions, a significant portion of MIF can form a disulfide structure involving Cys residues 57 and 60, i.e those cysteines constituting the CXXC. In addition, near UV circular dichroism (CD) spectropolarimetry revealed a characteristic maximum at ~280 nm, which provided additional evidence that intramolecular disulfide formation occurs upon oxidation of MIF.[12]

As discussed above, MIF forms oligomeric structures, but it has remained unclear whether the monomer, dimer or trimer is the physiologically relevant species, or whether there is an equilibrium between these species in solution. The X-ray crystallographic analyses unanimously show that all three Cys residues of MIF exist in their reduced thiol forms in the crystallized trimer.[9,46] Biochemical and crosslinking studies of MIF further indicate that none of the Cys residues is involved in an intermolecular disulfide bond.[3,47,48] Yet, the biochemical solution methods argue for the formation of an intramolecular disulfide bond between Cys57 and 60 under appropriate conditions.[12] This apparent contradiction may be resolved if one considers the different preparation procedures and MIF concentrations used. For the X-ray crystallographic analyses, MIF was prepared from *E. coli* overexpression systems under reducing conditions, and crystallization was achieved from a solution of ~10 mg/ml.[9,46] In contrast, the biochemical solution analyses involve a renaturation procedure during which *E. coli*-derived MIF is refolded under controlled oxidizing conditions. Renaturation is also performed at more physiological MIF concentrations of around 300 μg/ml, and biologically active recombinant MIF (rMIF) at concentrations between 1 and 100 ng/ml is derived from such preparations by simple dilution.[12,48,49] Thus, it appears that MIF can form an intramolecular CXXC disulfide bond, when it is refolded under oxidizing conditions. These oxidizing conditions are similar to those prevailing in the circulation or within distinct subcellular compartments. The enzyme(s) responsible for reducing/re-reducing MIF's CXXC thiol groups in the cell is unknown.

The CXXC consensus motif of typical TPORs such as Trx lies in a so-called Trx-fold, which is located at the N-terminus of an α-helix. Although the overall 3D structure of the MIF monomer is not only architecturally similar to the IL-8 dimer but also shows a remote resemblance to the Trx monomer, MIF is not structurally homologous to the TPOR proteins. Also, the CALC redox motif of MIF lies at the N-terminus of a β-strand element with Cys-57 located in the preceding loop rather than in a Trx-like fold. Since MIF exhibits CXXC-dependent TPOR activity, is able to catalyze the reduction of insulin and small molecular weight compound disulfides similar to Trx family proteins, and interacts with insulin, the peroxiredoxin PAG and possibly hepatopoietin/ALR (activator of liver regeneration) through a disulfide mechanism (see below for details), it appears that during catalysis, the structure of MIF may be partially unfolded or changed.[50] In this respect, it should be remembered that mono- or dimeric MIF could represent the relevant structure in solution, and that those species may have slightly different conformational properties.[47] Interestingly, thioredoxin reductase (Trr), another member of the TPOR family, does not contain an apparent Trx fold structure, but otherwise shares similar redox properties with the TPOR proteins, indicating that it may undergo a marked conformational change during catalysis.[51] The active conformation could be a "relaxed" conformation, as substrates for the redox activity of Trx have been found to bind to the active site cleft in an extended strand structure.[52] The fact that TPOR-derived small peptide sequences have TPOR protein-like catalytic properties as shown in Trx, Grx and PDI peptides,[21,53–56] further indicates that partially unfolded, altered or relaxed conformational elements are involved in CXXC redox catalysis. Of note is the identification of MIF-derived 10 and 16 amino acid peptides that undergo disulfide formation of their CXXC cysteines, and that exhibit catalytic TPOR and MIF-like biological activity.[12,57–59] Thus, upon interaction with its redox substrates, MIF may undergo a conformational change that may allow for disulfide bond formation within its CXXC site and the concomitant reduction of the substrate.

MIF exhibits catalytic TPOR activity *in vitro*, apparently acting primarily as a disulfide reductase. It can reduce protein and peptide disulfides as shown for insulin and a cysteinylated insulin peptide.[12,60,61] Glutathione (GSH) and dihydrolipoamide act as co-substrates for the insulin-reducing activity.[60] MIF can also reduce small molecular weight disulfides such as 2-hydroxyethyldisulfide (HED).[12] The co-substrate profile utilized by MIF suggests that MIF's redox activity properties could be more similar to those of Grx than Trx. In support of this notion is the finding that MIF-derived redox-active peptides can form mixed disulfides with cysteine,[57] as well as the observation that the catalytic TPOR activity is fully dependent on the presence of Cys60, while the Cys57 mutant shows partial rest activity. Glutaredoxins can engage in disulfide/dithiol reactions through a monothiol mechanism, in which the critical active site Cys residue reacts with a mixed disulfide between the target protein and glutathione ("R-S-S-G" mixed disulfide).[13,62] In the MIF mutant C60SMIF, the critical active site Cys residue (see also below for the structure activity profile of the CXXC motif of MIF) exhibits essentially no catalytic activity in the 2-HED and insulin reduction assays (5 and 14% of wildtype (wt) MIF, respectively). Similarly, the double mutant C57SC60SMIF has only 20% of the insulin reduction activity of wtMIF. In contrast, C57SMIF retains more than 50% of the activity of wtMIF in both assays. The latter is indicative of a monothiol mechanism as it has been established for Grx and suggested for Prx.[13,62,63]

The actual redox potential of MIF's redox site is unknown, but the redox potential of the MIF-derived CXXC-spanning peptide MIF(50–65) with an $E'0$ value of –258 mV suggests that the redox potential of full-length MIF, similar to that of the peptide, is more reducing than oxidizing.[57] In line with this notion, MIF is able to reduce 2-HED (see above). The TPOR-derived small peptide sequences encompassing the CXXC motifs of the parent proteins have been studied extensively, and the data on their redox potentials have been in part useful in predicting the redox potentials of the corresponding full-length proteins.[19,21,53,55] In addition to its CXXC cysteine residues at positions 57 and 60, MIF has a third cysteine at position 81, but the extensive structure activity studies of C81SMIF in comparison with wtMIF and the catalytic center mutants C60SMIF and C57SMIF suggest that this residue is not involved in the oxidoreductase activity of MIF and hence is not critical for the biological and immunological activities of MIF.[64] Recent cell signaling studies, in which the effects of C81SMIF on extracellular-regulated signal kinase (ERK)-mitogen-activated protein kinase (MAPK) and c-Jun-N-terminal kinase (JNK) activity were compared with those of wtMIF (Lue H, Schütz A, Bernhagen J; unpublished data), confirm this contention. It should be noted that PPT2 (also alternatively termed DDT, according to its functions as phenylpyruvate tautomerase 2 and D-dopachrome tautomerase, respectively), which shares a 27% sequence identity with MIF and a highly similar 3D architecture,[65] but has no CXXC motif due to the lack of a cysteine residue at position 60, is devoid of any catalytic TPOR and TPOR-dependent biological activity.[58,64]

Compared to Trx and Grx, MIF has a relatively low catalytic activity in the insulin disulfide reduction and 2-HED assays, respectively. This could mean that MIF is a low-efficiency redox catalyst such as the peroxiredoxins (see also below). Alternatively, MIF may rather act as a donor of proteinaceous reducing equivalents through its reduced CXXC thiol groups. Such a mechanism has also been suggested for the antiapoptotic activity of Trx.[31] Thus, MIF could be both a low-efficiency redox catalyst and a thiol group donor.

ROLE FOR MIF IN CELLULAR REDOX HOMEOSTASIS, REDOX SIGNALING, AND MIF-DRIVEN CELLULAR RESPONSES

It is increasingly becoming evident that MIF plays a role in cellular redox regulation. As MIF preferentially utilizes the physiological co-substrates glutathione (GSH) and dihydrolipoamide as reducing agents in insulin reduction, and because GSH and lipoamide are potent antioxidants and regulators of cellular oxidative stress,[13,27,66,67] it was postulated that MIF could be involved in cellular redox regulation.

In fact, cellular MIF levels are induced by hydrogen peroxide (H_2O_2), and MIF is secreted upon H_2O_2 stimulation from epithelial cells, macrophages, fibroblasts, neonatal cardiac myocytes,[60,68–70] and ovary cells (Finkelmeier D, Bernhagen J; unpublished observations). The specific link between MIF expression/secretion and peroxide stimulation could be verified in experiments performed in the presence of a catalase, which completely abolished MIF production.[70] Moreover, in cardiac myocytes MIF was identified as a key "factor induced by oxidative stress" ("FSO").[70] That the production of MIF is controlled by the redox state of the cell was also demonstrated *in vivo*. Peritoneal macrophages from rodents produce significantly less MIF upon LPS or phorbol ester stimulation, when the animals are previously injected with vitamin E, a treatment that results in a marked increase of the cellular α-tocopherol content and a decrease of cellular oxidative stress.[71]

MIF is not only an indicator of the oxidative stress situations of the cell, which cause it to be produced and secreted at increased rates, but it also seems to directly participate in modulating the cellular response to redox stress. H_2O_2 treatment of cells leads to changes in the cellular GSH/GSSG ratio with a resultant decrease in the GSH pool. It was thus of interest whether MIF could modulate cellular GSH levels. Applying an inducible Tet-off cell system that allows for the defined enhancement of intracellular MIF levels in HeLa cells subjected to oxidative stress, it was demonstrated that MIF leads to a significant increase in the cellular GSH concentration and thus promotes the elevation of cellular antioxidants in stress situations.[72]

The mechanism by which MIF regulates cellular GSH levels is unknown. However, Jung and colleagues have recently identified one possible mechanism by which this kind of regulation may occur. They identified specific protein complexes between MIF and the peroxiredoxin PAG, a thiol-specific cellular antioxidant protein.[73] As indicated above, peroxiredoxins (Prxs) are a sub-family of low-efficiency peroxidases that lack prosthetic groups or catalytically active heteroatoms, but use thiols as reductants. Their peroxidative activity is due to a conserved cysteine. In mammals, Prxs appear to be responsible for the redox regulation of diverse metabolic processes.[17,20,74–77] As there are substantial differences in the co-substrate requirements of the Prxs of pathogenic microorganisms and their mammalian host, mammalian Prxs prominently inhibit the antioxidant defense of pathogens, thereby contributing to the innate immune response of the host.[76,78,79] The interaction between PAG and MIF involves a mixed disulfide, with Cys173 of PAG involved in the disulfide bond. The Cys residue of MIF participating in this complex formation has not been identified, but it may be speculated that it is Cys60. Cys173 of PAG and the disulfide bond-mediated interaction between MIF and PAG also mutually influence the enzymatic activities of both proteins.

The identified disulfide-dependent PAG-MIF interaction confirms that MIF can interact with cellular proteins carrying susceptible disulfide bonds or thiols. In the case of insulin (see above), MIF interacts with a disulfide bond-carrying protein and reduces its disulfide structure. It is unknown whether mixed protein disulfides are an intermediate in this reaction. Functionally, the reduction of insulin disulfides by MIF could play a role in the regulation of insulin secretion, as MIF colocalizes with insulin in the secretory vesicles of pancreatic β cells (see below and Ref. 80). Alternatively, it has been suggested that MIF could be involved in the reduction of the disulfide structures of insulin-derived (or other class II-presented peptide antigens) antigenic peptides to improve peptide binding to the MHC class II cleft. As the optimum of MIF's observed reducing capacity is in the neutral pH range, it has been postulated that this kind of "MIF-dependent antigen processing" would occur in the extracellular milieu during the presentation of externally generated antigens.[61]

PAG is one of 10 proteins currently known to bind to MIF (for an overview see Table 1). In addition to PAG and insulin, at least one other MIF binding protein, viz. the hepatopoeitin (HPO) or the augmenter of liver regeneration (ALR), is associated with MIF's redox-modulating activities. HPO is a flavin-linked sulfhydryl oxidase, and the invariant CXXC motif in HPO, which forms an intramolecular disulfide, is essential for its catalytic activity. HPO exhibits both extracellular and intracellular functions, resembling Trx and MIF in this regard. Extracellular HPO triggers the mitogen-activated protein kinase (MAPK) pathway by binding to its specific cell surface receptor. Intracellular HPO potentiates the AP-1 pathway in a MAPK-independent fashion.[81]

Table 1 MIF Binding Proteins and Their Association with the MIF TPOR Activity

Protein (including abbreviations)	Association with MIF's TPOR Activity	References
JAB1/CSN5	suggested	(82, 84, 108, 109)
CSN6	not known	(108)
Insulin	yes	(61)
CD74/invariant chain	not known	(110, 111)
MHC class II	yes	(61)
CD44	not known	(111, 112)
Hepatopoietin/HPO/ALR	yes	(82)
Peroxiredoxin PAG	yes	(73)
BNIPL	not known	(113)
MLCK	not known	(114)

Legend: Only proteins and references for/in which a protein-protein interaction with MIF has been reported are listed. The list contains only MIF-interacting proteins that were explicitly reported as such, and may not be complete. We apologize if an interaction partner has been left out. Abbreviations used: JAB1, c-Jun activation domain binding protein-1; CSN, constitutive photomorphogenesis (COP9) signalosome; HPO, hepatopoietin; ALR, augmenter of liver regeneration; BNIPL, Bcl-2/adenovirus E1B 19 kDa interacting protein 2-like, BNIP-2-like; MLCK, myosin light chain kinase.

While insulin may be regarded as a substrate of MIF's disulfide reductase activity, PAG and HPO could be the regulatory targets of MIF's redox-modulating functions. Moreover, the binding of PAG and HPO to MIF is associated with the function of the CXXC-active site of these proteins.

The MIF-HPO interaction was first identified by using a yeast two-hybrid screen.[82] Their interaction was further verified by a His-lag-based pull-down assay *in vitro* and coimmunoprecipitation *in vivo* in mammalian cells. MIF and HPO colocalize in the cytoplasm.

Interestingly, both HPO and MIF bind to the coactivator and COP9 signalosome component JAB1/CSN5 and modulate the AP-1 pathway.[82] The MIF-HPO binding is stronger than that between HPO and JAB1 or MIF and JAB1. Potentiation of the AP-1 pathway by intracellular HPO occurs through JAB1; HPO colocalizes with the CSN in the nucleus of hepatic cells. HPO-mediated potentiation of AP-1 activity is inhibited by curcumin, a potent inhibitor of a CSN-associated kinase, indicating that HPO-JAB1-regulated AP-1 is controlled by cell signaling events.[81,83]

The MIF-HPO interaction that leads to the heterodimerization of these proteins is associated with a reduced binding between MIF and JAB1. While it is unclear whether MIF-HPO heterodimerization actually occurs by a disulfide mechanism, and whether the CXXC cysteines contribute indirectly to the binding process, it is evident that these two redox factors mutually modulate their effects on cell signaling. Interestingly, HPO, as a liver-specific regeneration augmenter, was found to be induced by the expression of MIF in hepatoma cells.[82]

MIF's role in cellular redox regulation appears to be connected with cell signaling processes. Direct evidence for this notion comes from the recent identification of the intracellular interaction between MIF and JAB1/CSN5.[84] Briefly, JAB1 is both a transcriptional coactivator of the activator protein-1 (AP-1) transcriptional pathway and a component of the COP9 signalosome. The CSN is a multiprotein complex that modulates ubiquitin-26 S proteasome-dependent protein degradation via deneddylation of cullins.[85–88] MIF inhibits JAB1-mediated AP-1 activation and counter-regulates JAB1-dependent p27 degradation and G1 cell cycle arrest. Precisely how MIF's redox activity may impose on JAB1 and the CSN is unclear. The binding of MIF to JAB1 is dependent on the sequence region 50–65 of MIF, but there is no requirement for the presence of an sequence region 50–65 of MIF, but there is no requirement for the presence of an intact CXXC motif for successful binding as both a bis-serine variant of MIF(50–65) and the redox-dead mutant C60SMIF can bind to JAB1.[84] Yet, the JAB1-antagonistic effects of MIF appear to be CXXC-dependent as C60SMIF loses this activity almost completely. Interestingly, JAB1 was recently identified to also bind to Trx, with Trx antagonizing the same JAB1-activated signaling pathways as MIF.[89] JAB1 contains several Cys residues, but it is unknown if they are involved in the interactions with MIF or Trx. Thus, further structure activity studies with MIF CXXC mutants and the examination of such mutants in additional MIF-JAB1-dependent cellular processes are needed to clarify whether MIF's redox activity is directly involved in JAB1 modulation, and whether the inability of C60SMIF to modulate the JAB1 function is due to a conformational effect.

However, the possibility that JAB1 and/or the CSN could in fact be regulated by a redox mechanism is underpinned by the notion that JAB1 not only binds to MIF and Trx, but also to HPO, and as discussed above, MIF binds to HPO too.[81–83,90] Neither HPO dimerization nor its binding to JAB1 is CXXC-dependent. However, similar to MIF, HPO variants with the Cys residues at the active site substituted for Ser do not have sulfhydryl oxidase activity; they do not retain the c-Jun phosphorylation-enhancing activity, and fail to potentiate JAB1-mediated AP-1 activation. This means that the JAB1-mediated potentiation of AP-1 by HPO is dependent on its sulfhydryl oxidase activity. Thus, the HPO-JAB1 link provides strong evidence that JAB1 could in fact be regulated by a redox mechanism.

Additional evidence for a connection between CXXC-dependent TPOR effects of MIF and cell signaling processes comes from a study by Nguyen and coworkers.[57] MIF(50–65) was found to be sufficient in triggering mitogen-activated kinase (MAPK) activity in fibroblasts, suggesting that the CXXC-based TPOR activity of MIF is either sufficient for MAPK activation and/or that receptor activation involves the CXXC motif or one of the critical Cys residues therein (see also below).

A special kind of MIF-mediated redox-regulatory process has been observed in rat sperm maturation. Eickhoff and colleagues showed that the incubation of rat sperm cells with MIF leads to an increase of detectable free thiol groups in the sperm flagellum. At the same time, a decrease in the cellular zinc content was measured.[91] Since this effect is observed with MIF concentrations in the range of 25–50 ng/ml, it is likely that the MIF-mediated increase in sperm protein thiol groups occurs by a true catalytic process. The release of zinc could be a subsequent step. These data confirm that MIF can act as a redox catalyst *in vivo*. The experiments by Eickhoff and colleagues were performed with exogenously applied recombinant MIF and vesicular MIF. While this affirms a role for extracellular MIF in sperm redox regulation, intracellular MIF likely plays a critical role in these processes too. In fact, vesicular MIF is probably efficiently shuttled into the sperm cell cytosol.

In vivo evidence for a role of MIF's TPOR activity in cellular redox regulation was also obtained in a neuronal cell model. As also discussed in more detail below ("role of MIF in ROS quenching"), MIF's TPOR activity is critically involved in angiotensin II-induced neuronal firing.[59] MIF acts as an inhibitor of the chronotropic actions of angiotensin II in hypothalamal neurons. Of note is the finding that this inhibitory function is mediated by the TPOR activity of MIF in a CXXC-dependent manner, as inoculated wtMIF and MIF peptide fragment MIF(50–65), but not inoculated MIF variant C60SMIF or the bis-serine variant peptide MIF peptide C57SC60SMIF(50-65), fully exhibit an inhibitory action on angiotensin II-stimulated neuronal firing.[58,59] The target proteins and the precise mechanism by which MIF redox regulation of neuronal firing occurs are unknown.

MIF is a pivotal mediator of the innate immune response of the host.[5] MIF's oxidoreductase activity is likely to play a role in MIF-mediated immune cell functions. In contrast to wtMIF, the redox-dead mutant, C60SMIF, is unable to activate macrophages to kill leishmanial parasites.[12] With respect to the same macrophage-activatory effect, C57SMIF exhibits less than 70% of the activity of wtMIF. Similarly, the redox center mutants, C57SMIF and C60SMIF, exhibit only ~65% and 37%, respectively, of the glucocorticoid overriding activity[64] that wtMIF exerts on monocytes/macrophages.[92] While the contribution of MIF's TPOR activity is thus obviously not at 100%, it is nevertheless suggested from these data that MIF-mediated redox effects play a role in regulating inflammatory monocyte/macrophage functions. The glucocorticoid overriding data[92] in conjunction with structure activity studies of the 16-meric CXXC-spanning peptide MIF(50–65), of which cysteine mutants and lactam bridge-cyclized variants were also investigated,[57] suggests that a CXXC disulfide-dependent conformation could be involved in the receptor docking and activation process. Since such a receptor-active conformation might also be induced by a receptor-dependent induced-fit mechanism, redox-independent receptor activation may occur to a certain extent; this could explain the rest activity of the C60SMIF mutant regarding its glucocorticoid overriding capacity. Of note are the findings that the MIF redox peptide MIF(50–65) can bind to the MIF receptor CD74 and C60SMIF exhibits a reduced binding potency to CD74 (Kamir D, Bucala R, Bernhagen J; unpublished observations).

MIF is not only linked to hyper- or pro-oxidative stress situations of the cell, but may also be a determinant of hypoxia-induced cell responses, viz. apoptosis. Tumor cells are often resistant to hypoxia-induced apoptosis; MIF was identified as a genetic determinant that may mediate this phenotype in certain colon cancer cell lines.[93] Thus, MIF also appears to be regulated by, and linked to, the hypoxic states of the cell.

ROS, APOPTOSIS AND INHIBITION OF CELLULAR ROS STRESS BY MIF

Reactive oxygen species (ROS) are the major cause of pro-oxidative cellular stress. ROS are molecular oxygen species that are either reactive *per se* or are readily convertible into reactive species. ROS encompass peroxides such as hydrogen peroxide (H_2O_2) or superoxide anion radical ($\bullet O_2^-$). Endogenously, ROS are generated during the process of oxidative phosphorylation in the mitochondrial respiratory chain. Endogenous ROS generation is due to erroneous electron transfer reactions, and it is assumed that 1–2% of the oxygen (O_2) metabolized in the respiratory chain is falsely converted to $\bullet O_2^-$. ROS can lead to membrane damage by lipid peroxidation, DNA strand breaks, and protein oxidation at several amino acids. ROS-damaged lipids contain alkoxyl (•RO) and peroxiradicals (•ROO), which can lead, among other effects, to destabilization of the mitochondrial membrane.[94]

In proteins, the thiol-containing amino acids are predominantly prone to modification by ROS. While sulfur is usually considered as part of the cellular antioxidant systems, there is mounting evidence that reactive sulfur species (RSS) with stressor properties similar to those found in ROS are formed under conditions of oxidative stress. Thiols, as well as disulfides, are easily oxidized to sulfur species with sulfur in higher oxidation states. Such agents include thiyl radicals, disulfides, sulfenic acids and disulfide-S-oxides.[95,96] A generated cysteine-derived thiyl radical can react with another thiyl radical to form an oxidized disulfide structure, which, in the case of a cytosolic protein, can result in conformational change and/or inactivation. ROS-mediated damage of DNA can affect both the bases and the sugar-phosphate backbone. As to the former, oxidation can lead to base conversion (guanine $\rightarrow$ 8-hydroxyguanine) and subsequent mispairing, while radical-induced oxidation of the backbone affords DNA strand breaks.[97]

ROS quenching can occur by various mechanisms. The classical components of the ROS detoxification machinery are the enzymes catalase, superoxide dismutase (SOD), and the Grx and Trx systems. For example, $\bullet O_2^-$ generated from O_2 by cytochrom c in the inner mitochondrial membrane is converted to H_2O_2 by the mitochondrial SOD. Subsequently, H_2O_2 is reduced and detoxified to H_2O by glutathione peroxidase (GPR) or a catalase. Catalase, SOD and GPR are localized in both the cytosol and mitochondrium as well as in other organells. Mitochondrial $\bullet O_2^-$ generation oxidizes critical cysteine residues within the adenine nucleotide translocator (ANT), leading to unspecific channel formation, destabilization of the mitochondrial membrane ("mild uncoupling"), cytochrome c release, apoptosome formation, and initiation of apoptosis.[98,99] Glutathione (GSH) is probably the most important cellular antioxidant. Located in the cytosol, nucleus and mitochondrium, it is capable of reacting with, and detoxifying, ROS such as $\bullet O_2^-$, $\bullet OH$ (hydroxyl radical), H_2O_2, organic peroxides (ROOH), as well as singlet oxygen ($^1O_2$). The re-reduction of the formed GSSG is then catalyzed by glutathione reductase in a NADPH-dependent manner. Additional ROS quenching may be based on defense mechanisms by less specific processes involving iron-copper-binding proteins (albumin, ferritin, metallothionin) and heat shock proteins. Exogenous ROS detoxification is afforded by antioxidants such as ascorbate, GSH, tocopherole or carotinoids.

As to the ROS-induced mitochondrial apoptosis pathway, the mitochondrial Trx system (Trx2) and Prxs constitute additional and potent ROS quenching machineries.[35] At the gene transcription level, protective pathways such as the JNK/AP-1-mediated expression of manganese SOD (MnSOD) and HO-1 are induced. Under "mild" oxidative conditions, the NF_kB pathway is activated, leading to a proinflammatory and proproliferative response that has been recognized more recently as a redox signaling pathway. Only when the cellular ROS levels exceed a certain threshold value, is apoptosis triggered. The threshold value, as well as the effects of ROS on the proliferative properties of cells, differs between differentiated, mitotic-competent and tumorigenic cells. Differentiated and tumor cells exhibit a relatively low responsiveness, with tumor cells reacting somewhat to an increasing oxidized state and increasing ROS levels of the cell during the proliferation and apoptosis phase. In contrast, mitotic-competent cells strongly react to increasing ROS levels in the mid-oxidized state of a cell. When ROS increase further, such cells are able to respond by a strong pro-apoptotic response.

Interestingly, MIF has potent antiapoptotic activities.[72,100–102] and has been reported to be able to quench cellular ROS levels.[58,59] In a neuronal cell model of angiotensin II-elicited neuronal firing through the Ang II type 1 receptor (AT1-R), MIF was identified as an intracellular inhibitor of the actions of Ang II. Firstly, Ang II increased the intracellular levels of MIF in neurons cultured from rat hypothalamus and brainstem. Secondly, the elevation of intracellular MIF by Ang II prevented further chronotropic actions of this peptide. Thirdly, the intracellular application of exogenous recombinant MIF abolished the Ang II-induced chronotropic action in neurons. Finally, the intracellular application of the MIF(50–65), which harbors the thiol oxidoreductase property of MIF, mimicked the inhibitory actions of MIF on Ang II-stimulated neuronal firing, indicating that the MIF TPOR activity is involved. At the same time, MIF and MIF(50–65) were capable of reducing the cellular ROS burden in those cells. Thus, this study demonstrated a link between MIF's role as an intracellular negative regulator of Ang II-induced actions in neurons and an MIF-dependent ROS-protective effect.[59] A link between the inhibition of neuronal firing by MIF, and MIF's TPOR and ROS quenching properties was confirmed by Matsuura and coworkers.[58] An intracellular perfusion of physiological concentrations of MIF into neuronal cultures caused a significant increase in the delayed rectifier K+ current (I(Kv)), one of the component currents whose activity contributes to neuronal firing. This effect was apparent after 3 min, and was maximal after 20–30 min. MIF that had been inactivated by boiling did not alter I(Kv), and MIF-neutralizing antibodies abolished the action of rMIF. The stimulatory effect of MIF on I(Kv) current density was mimicked by the intracellular application of peptide MIF(50–65), corroborating the involvement of the TPOR activity of the MIF molecule. Conversely, neither C60SMIF nor the MIF homologue D-dopachrome tautomerase, both of which lack TPOR activity, altered I(Kv). As the increase in I(Kv) produced by rMIF was abolished by the $\bullet O_2^-$ scavenger Tiron, these studies confirmed that the neuronal action of MIF includes a stimulatory action on I(Kv) that may be mediated by an MIF-dependent TPOR/$\bullet O_2^-$-scavenging mechanism.[58]

A connection between ROS quenching by MIF and certain of its cellular activities is suggested by yet another study done by Sakamoto and coworkers.[103] Differing from the above findings in the neuronal cell system, this study indicates that secreted MIF may also be able to function as an antioxidant and ROS scavenger in the extracellular milieu or serum. ROS may contribute to aging and osteoporosis resulting from decreases in plasma antioxidants in aged osteoporotic women, whereas vitamin K2 supplementation has been reported to decrease osteoporotic fractures in postmenopausal women. In searching for the mechanism by which vitamin K2 prevents osteoporosis, Sakamoto *et al.* identified MIF as an involved serum antioxidant. Specifically, these authors investigated the effect of vitamin K2 on the osteoporosis of aged rats by evaluating the relationship between serum antioxidant levels and bone metabolism. In their study, the aged female rats not only exhibited lower serum alkaline phosphatase activity and osteocalcin levels, but also had lower serum levels of the antioxidants 17β-estradiol, GPx activity, and interestingly MIF, as compared with the young female rats. Of note is the correlation of between vitamin K2 supplementation and increased serum alkaline phosphatase activity and elevated MIF levels, while a classical antioxidant such as GPx was unaltered.[103]

In terms of MIF's antiapoptotic effects, an association with protection from ROS stress is also emerging. ROS are primarily involved in triggering the mitochondrial apoptosis pathway. Subsequently, p53-mediated apoptosis is induced via the activation of the stress kinases JNK and ATM.

MIF inhibits p53-dependent apoptotic processes,[100–102] and blocks the stress-induced JNK pathway.[72,84] Of note is a direct link between MIF-mediated ROS quenching and the inhibition of the p53 pathway that could be established recently in mouse embryonal fibroblasts for both exogenously applied ROS and endogenous ROS generation (Thiele M, Bernhagen J; unpublished observation).

CONNECTION BETWEEN MIF, THIOREDOXIN AND APOPTOSIS INHIBITION

MIF inhibits p53-dependent apoptotic processes, and endogenous MIF may be involved in ROS scavenging and ROS-mediated apoptosis (see above). In addition, studies by Nguyen *et al.* suggest that redox effects could play a role in MIF's antiapoptotic activity and that MIF shares with Trx properties in the suppression of pro-oxidative stress-induced apoptosis.[72] Similar to Trx and low molecular weight thiol compounds such as β-mercaptoethanol, micromolar concentrations of extracellular MIF are able to inhibit oxidative stress-induced apoptosis, but the overexpression of intracellular MIF also leads to a marked anti-apoptotic effect.[72] This suggests that like Trx, MIF may inhibit apoptosis by a catalytic mechanism or as a donor of proteinaceous thiol groups.

Another potential link between the apoptosis-modulating properties of the redox regulators Trx and MIF was demonstrated by Kondo and coworkers in CD4+ T cells.[104] Accordingly, MIF and Trx reciprocally control their expression levels upon pro-oxidative stress-induced apoptosis. MIF levels are reduced upon the overexpression of Trx, and Trx levels are enhanced in T cells from MIF knock-out mice. This could mean that the anti-apoptotic activities of MIF and Trx are at least in part coordinated by an interdependent regulation of the expression levels of these cellular players. It is currently unclear, whether MIF and Trx may nevertheless act synergistically during oxidative stress-induced apoptosis, and whether other means of regulation such as post-translational modifications or interaction with JAB1 or p53 could play a role herein. It should be noted that GIF, which was shown to be essentially identical to MIF in its sequence and 3D structure, is cysteinylated under certain conditions.[105]

MIF and Trx not only share striking similarities with respect to their redox-related antiapoptotic activity and modulation of JAB1 functions (see also Chapter 17), but are also alike in that they can both be viewed as secretable cellular enzymes with distinct extracellular functions. It has therefore been proposed to term them "redoxkines" (redox-acting cytokines) or "cytozymes" (enzymes with cytokine functions).[50] In fact, Trx, which was originally defined as a classical intracellular enzyme is now recognized as a co-cytokine.[44] Interestingly, both Trx and MIF, and also truncated Trx, are secreted by a leaderless, non-classical export pathway.[106,107] Once secreted, both MIF and Trx exert a broad spectrum of cytokine activities to modulate the immune and inflammatory response of the host. As discussed above, this includes the enhancement of lymphocyte proliferation, activation of macrophages, induction of cytokines and stress-related gene products, and the modulation of cell migration.

CONCLUSION

MIF is a potent inflammatory cytokine that fulfils key activities through receptor interaction in the extracellular milieu, thereby mediating both physiological and pathophysiological processes of the host immune and inflammatory response. In addition, MIF is intracellularly expressed at appreciable concentrations in a variety of cells. It may be argued that MIF is an intracellular protein that serves as an archaic regulator of the critical aspects of cell behavior and innate immunity, having "acquired" an additional extracellular function through evolution.

There is accumulating evidence that the intracellular function of MIF could be due to its activity as TPOR. Owing to its conserved CXXC site, MIF is capable of modulating the cellular redox state and protecting the cell from redox stress. This involves an enhancement of reduced thiol equivalents in the cell such as GSH and the quenching of both extracellularly and endogenously generated ROS. The modulation of cellular redox homeostasis by MIF is most likely related to the control of cell behavior such as apoptosis. These activities are dependent on MIF's TPOR center and its critical Cys60 residue. MIF interacts with the cellular redox regulators, Prx and HPO, and possibly others as well, but the precise molecular link between MIF-mediated TPOR activity, its interaction with other cellular

TPORs, and MIF's potent effects on key cellular responses such as apoptosis, proliferation or migration is still unresolved.

REFERENCES

1. David JR. (1966) Delayed hypersensitivity *in vitro*: its mediation by cell-free substances formed by lymphoid cell-antigen interaction. *Proc Natl Acad Sci USA* **56**: 72–77.
2. George M, Vaughan JH. (1962) Migration of mononuclear cells in capillary tubes. *Proc Soc Exp Biol Med* **111**: 514–521.
3. Weiser WY, Temple DM, Witek-Gianotti JS, *et al.* (1989) Molecular cloning of cDNA encoding a human macrophage migration inhibtion factor. *Proc Natl Acad Sci USA* **86**: 7522–7526.
4. Bucala R. (1996) MIF rediscovered — cytokine, pituitary hormone, and glucocorticoid-induced regulator of the immune response. *FASEB J* **10**: 1607–1613.
5. Calandra T, Roger T. (2003) Macrophage migration inhibitory factor: a regulator of innate immunity. *Nat Rev Immunol* **3**: 791–800.
6. Morand EF, Leech M, Bernhagen J. (2006) MIF: a new cytokine link between rheumatoid arthritis and atherosclerosis. *Nat Rev Drug Discov* Advance online publication.
7. Mitchell RA. (2004) Mechanisms and effectors of MIF-dependent promotion of tumourigenesis. *Cell Signal* **16**: 13–19.
8. Donnelly SC, Bucala R, Metz CN, *et al.* (1999) Macrophage migration inhibitory factor and acute lung injury. *Chest* **116**: 111S.
9. Sun H, Bernhagen J, Bucala R, Lolis E. (1996) Crystal structure at 2.6 Å resolution of human macrophage migration inhibitory factor. *Proc Natl Acad Sci USA* **93**: 5191–5196.
10. Kato Y, Muto T, Tomura T, *et al.* (1996) The crystal structure of human glycosylation-inhibiting factor is a trimeric barrel with three 6-stranded β-sheets. *Proc Natl Acad Sci USA* **93**: 3007–3010.
11. Rosengren E, Bucala R, Åman P, *et al.* (1996) The immunoregulatory mediator macrophage migration inhibitory factor (MIF) catalyzes a tautomerization reaction. *Mol Med* **2**: 143–149.
12. Kleemann R, Kapurniotu A, Frank RW, *et al.* (1998) Disulfide analysis reveals a role for macrophage migration inhibitory factor (MIF) as a thiol-protein oxidoreductase. *J Mol Biol* **280**: 85–102.
13. Fernandes AP, Holmgren A. (2004) Glutaredoxins: glutathione-dependent redox enzymes with functions far beyond a simple thioredoxin backup system. *Antioxid Redox Signal* **6**: 63–74.

14. Freedman RB, Klappa P, Ruddock LW. (2002) Protein disulfide isomerases exploit synergy between catalytic and specific binding domains. *EMBO Rep* **3**: 136–140.
15. Holmgren A. (1985) Thioredoxin. *Ann Rev Biochem* **54**: 237–271.
16. Nakamura H. (2005) Thioredoxin and its related molecules: update 2005. *Antioxid Redox Signal* **7**: 823–828.
17. Wood ZA, Schroder E, Robin Harris J, Poole LB. (2003) Structure, mechanism and regulation of peroxiredoxins. *Trends Biochem Sci* **28**: 32–40.
18. Ellis LBM, Saurugger P, Woodward C. (1992) Identification of the three-dimensional thioredoxin motif: related structure in the ORF3 protein of the *Staphylococcus aureus mer* operon. *Biochemistry* **31**: 4882–4891.
19. Martin JL. (1995) Thioredoxin — a fold for all reasons. *Structure* **3**: 245–250.
20. Manevich Y, Fisher AB. (2005) Peroxiredoxin 6, a 1-Cys peroxiredoxin, functions in antioxidant defense and lung phospholipid metabolism. *Free Radic Biol Med* **38**: 1422–1432.
21. Siedler F, Rudolph-Böhner S, Doi M, *et al.* (1993) Redox potentials of active-site bis(cysteinyl) fragments of thiol-protein oxidoreductases. *Biochemistry* **32**: 7488–7495.
22. Ghezzi P. (2005) Oxidoreduction of protein thiols in redox regulation. *Biochem Soc Trans* **33**: 1378–1381.
23. Fratelli M, Goodwin LO, Orom UA, *et al.* (2005) Gene expression profiling reveals a signaling role of glutathione in redox regulation. *Proc Natl Acad Sci USA* **102**: 13998–14003.
24. Ghezzi P, Bonetto V, Fratelli M. (2005) Thiol-disulfide balance: from the concept of oxidative stress to that of redox regulation. *Antioxid Redox Signal* **7**: 964–972.
25. Haendeler J, Tischler V, Hoffmann J, *et al.* (2004) Low doses of reactive oxygen species protect endothelial cells from apoptosis by increasing thioredoxin-1 expression. *FEBS Lett* **577**: 427–433.
26. Jones DP, Go YM, Anderson CL, *et al.* (2004) Cysteine/cystine couple is a newly recognized node in the circuitry for biologic redox signaling and control. *FASEB J* **18**: 1246–1248.
27. Ghezzi P. (2005) Regulation of protein function by glutathionylation. *Free Radic Res* **39**: 573–580.
28. Haendeler J, Hoffmann J, Tischler V, *et al.* (2002) Redox regulatory and anti-apoptotic functions of thioredoxin depend on S-nitrosylation at cysteine 69. *Nat Cell Biol* **4**: 743–749.
29. Andoh T, Chock PB, Chiueh CC. (2002) The roles of thioredoxin in protection against oxidative stress-induced apoptosis in SH-SY5Y cells. *J Biol Chem* **277**: 9655–9660.

30. Baker A, Payne CM, Briehl MM, Powis G. (1997) Thioredoxin, a gene found overexpressed in human cancer, inhibits apoptosis *in vitro* and *in vivo*. *Cancer Res* **57**: 5162–5167.
31. Iwata S, Hori T, Sato N, *et al.* (1997) Adult T cell leukemia (ATL)-derived factor/human thioredoxin prevents apoptosis of lymphoid cells induced by L-cystine and glutathione depletion: possible involvement of thiol-mediated redox regulation in apoptosis caused by pro-oxidant state. *J Immunol* **158**: 3108–3117.
32. Powis G, Mustacich D, Coon A. (2000) The role of the redox protein thioredoxin in cell growth and cancer. *Free Radic Biol Med* **29**: 312–322.
33. Sato N, Iwata S, Nakamura K, *et al.* (1995) Thiol-mediated redox regulation of apoptosis. Possible roles of cellular thiols other than glutathione in T cell apoptosis. *J Immunol* **154**: 3194–3203.
34. Tagaya Y, Maeda Y, Mitsui A, *et al.* (1989) ATL-derived factor (ADF), an IL-2 receptor/Tac inducer homologous to thioredoxin; possible involvement of dithiol-reduction in the IL-2 receptor induction. *EMBO J* **8**: 757–764.
35. Tanaka T, Hosoi F, Yamaguchi-Iwai Y, *et al.* (2002) Thioredoxin-2 (TRX-2) is an essential gene regulating mitochondria-dependent apoptosis. *EMBO J* **21**: 1695–1703.
36. Nilsson J, Soderberg O, Nilsson K, Rosen A. (2000) Thioredoxin prolongs survival of B-type chronic lymphocytic leukemia cells. *Blood* **95**: 1420–1426.
37. Kondo N, Ishii Y, Kwon YW, *et al.* (2004) Redox-sensing release of human thioredoxin from T lymphocytes with negative feedback loops. *J Immunol* **172**: 442–448.
38. Freemerman AJ, Powis G. (2000) A redox-inactive thioredoxin reduces growth and enhances apoptosis in WEHI7.2 cells. *Biochem Biophys Res Commun* **274**: 136–141.
39. Saitoh M, Nishitoh H, Fujii M, *et al.* (1998) Mammalian thioredoxin is a direct inhibitor of apoptosis signal-regulating kinase (ASK) 1. *EMBO J* **17**: 2596–2606.
40. Bishopric NH, Webster KA. (2002) Preventing apoptosis with thioredoxin: ASK me how. *Circ Res* **90**: 1237–1239.
41. Liu Y, Min W. (2002) Thioredoxin promotes ASK1 ubiquitination and degradation to inhibit ASK1-mediated apoptosis in a redox activity-independent manner. *Circ Res* **90**: 1259–1266.
42. Pekkari K, Gurunath R, Arner ES, Holmgren A. (2000) Truncated thioredoxin is a mitogenic cytokine for resting human peripheral blood mononuclear cells and is present in human plasma. *J Biol Chem* **275**: 37474–37480.
43. Pekkari K, Avila-Carino J, Bengtsson A, *et al.* (2001) Truncated thioredoxin (Trx80) induces production of interleukin-12 and enhances CD14 expression in human monocytes. *Blood* **97**: 3184–3190.

44. Pekkari K, Holmgren A. (2004) Truncated thioredoxin: physiological functions and mechanism. *Antioxid Redox Signal* **6**: 53–61.
45. Smith JA, Pease LG. (1980) Reverse turns in peptides and proteins. *CRC Crit Rev Biochem* **8**: 315–399.
46. Suzuki M, Sugimoto H, Nakagawa A, *et al.* (1996) Crystal structure of the macrophage migration inibitory factor from rat liver. *Nat Struct Biol* **3**: 259–266.
47. Mischke R, Kleemann R, Brunner H, Bernhagen J. (1998) Crosslinking and mutational analysis of the oligomerization state of the cytokine macrophage migration inhibitory factor (MIF). *FEBS Lett* **427**: 85–90.
48. Bernhagen J, Mitchell RA, Calandra T, *et al.* (1994) Purification, bioactivity, and secondary structure analysis of mouse and human macrophage migration inhibitory factor (MIF). *Biochemistry* **33**: 14144–14155.
49. Calandra T, Bernhagen J, Mitchell RA, Bucala R. (1994) The macrophage is an important and previously unrecognized source of macrophage migration inhibitory factor. *J Exp Med* **179**: 1985–1902.
50. Thiele M, Bernhagen J. (2005) Link between macrophage migration inhibitory factor (MIF) and cellular redox regulation. *Antioxid Redox Signal* **7**: 1234–1248.
51. Waksman G, Krishna TS, Williams CHJ, Kuriyan J. (1994) Crystal structure of *Escherichia coli* thioredoxin reductase refined at 2 A resolution. Implications for a large conformational change during catalysis. *J Mol Biol* **236**: 800–816.
52. Eguchi M, McMillan M, Nguyen C, *et al.* (2003) Chemogenomics with peptide secondary structure mimetics. *Comb Chem High Throughput Screen* **6**: 611–621.
53. Cabrele C, Flori S, Pegoraro S, Moroder L. (2002) Redox-active cyclic bis(cysteinyl)peptides as catalysts for *in vitro* oxidative protein folding. *Chem Biol* **9**: 731–740.
54. Shi T, Spain SM, Rabenstein DL. (2004) Unexpectedly fast cis/trans isomerization of Xaa-Pro peptide bonds in disulfide-constrained cyclic peptides. *J Am Chem Soc* **126**: 790–796.
55. Siedler F, Quarzago D, Rudolph-Bohner S, Moroder L. (1994) Redox-active bis-cysteinyl peptides. II. Comparative study on the sequence-dependent tendency for disulfide loop formation. *Biopolymers* **34**: 1563–1572.
56. Zhang R, Snyder GH. (1989) Dependence of formation of small disulfide loops in two-cysteine peptides on the number and types of intervening amino acids. *J Biol Chem* **264**: 18472–18479.
57. Nguyen MT, Beck J, Lue H, *et al.* (2003) A sixteen residue peptide fragment of macrophage migration inhibitory factor, MIF(50–65), exhibits redox activity and has MIF-like biological functions. *J Biol Chem* **278**: 33654–33671.

58. Matsuura T, Sun C, Leng L, *et al.* (2006) Macrophage migration inhibitory factor increases neuronal delayed rectifier K+ current. *J Neurophysiol* **95**: 1042–1048.
59. Sun C, Li H, Leng L, *et al.* (2004) Macrophage migration inhibitory factor: an intracellular inhibitor of angiotensin II-induced increases in neuronal activity. *J Neurosci* **24**: 9944–9952.
60. Kleemann R, Mischke R, Kapurniotu A, *et al.* (1998) Specific reduction of insulin disulfides by macrophage migration inhibitory factor (MIF) with glutathione and dihydrolipoamide: potential role in cellular redox processes. *FEBS Lett* **430**: 191–196.
61. Potolicchio I, Santambrogio L, Strominger JL. (2003) Molecular interaction and enzymatic activity of macrophage migration inhibitory factor (MIF) with immunorelevant peptides. *J Biol Chem* **278**: 30889–30895.
62. Lundström-Ljung J, Holmgren A. (1995) Glutaredoxin accelerates glutathione-dependent folding of reduced ribonuclease A together with protein disulfide-isomerase. *J Biol Chem* **270**: 7822–7828.
63. Rouhier N, Gelhaye E, Jacquot JP. (2002) Glutaredoxin-dependent peroxiredoxin from poplar: protein-protein interaction and catalytic mechanism. *J Biol Chem* **277**: 13609–13614.
64. Kleemann R, Kapurniotu A, Mischke R, *et al.* (1999) Characterization of catalytic center mutants of macrophage migration inhibitory factor (MIF) and comparison with C81S MIF. *Eur J Biochem* **261**: 753–766.
65. Sugimoto H, Taniguchi M, Nakagawa A, *et al.* (1999) Crystal structure of human D-dopachrome tautomerase, a homologue of macrophage migration inhibitory factor, at 1.54 angstrom resolution. *Biochemistry* **38**: 3268–3279.
66. Bast A, Haenen GR. (1988) Interplay between lipoic acid and glutathione in the protection against microsomal lipid peroxidation. *Biochim Biophys Acta* **963**: 558–561.
67. Kagan VE, Shvedova A, Serbinova E, *et al.* (1992) Dihydrolipoic acid — a universal antioxidant both in the membrane and in the aqueous phase. Reduction of peroxyl, ascorbyl and chromanoxyl radicals. *Biochem Pharmacol* **44**: 1637–1649.
68. Fukuzawa J, Nishihira J, Hasebe N, *et al.* (2002) Contribution of macrophage migration inhibitory factor to extracellular signal-regulated kinase activation by oxidative stress in cardiomyocytes. *J Biol Chem* **277**: 24889–24895.
69. Sakamoto W, Fujie K, Isomura S, *et al.* (2002) Secretion of macrophage migration inhibitory factor differs from interleukin-6 in hydrogen peroxide- and LPS-stimulated human fibroblasts. *Int Immunopharmacol* **2**: 1123–1131.
70. Takahashi M, Nishihira J, Shimpo M, *et al.* (2001) Macrophage migration inhibitory factor as a redox-sensitive cytokine in cardiac myocytes. *Cardiovasc Res* **52**: 438–445.

71. Sakamoto W, Nishihira J, Fujie K, *et al.* (1998) Inhibition of macrophage migration inhibitory factor secretion from macrophages by vitamin E. *Biochim Biophys Acta* **1404**: 427–434.
72. Nguyen M, Lue H, Kleemann R, *et al.* (2003) The cytokine macrophage migration inhibitory factor (MIF) reduces pro-oxidative stress-induced apoptosis. *J Immunol* **170**: 3337–3347.
73. Jung H, Kim T, Chae HZ, *et al.* (2001) Regulation of macrophage migration inhibitory factor and thiol-specific antioxidant protein PAG by direct interaction. *J Biol Chem* **276**: 15504–15510.
74. Rhee SG, Chae HZ, Kim K. (2005) Peroxiredoxins: a historical overview and speculative preview of novel mechanisms and emerging concepts in cell signaling. *Free Radic Biol Med* **38**: 1543–1552.
75. Immenschuh S, Baumgart-Vogt E. (2005) Peroxiredoxins, oxidative stress, and cell proliferation. *Antioxid Redox Signal* **7**: 768–777.
76. Flohe L, Budde H, Hofmann B. (2003) Peroxiredoxins in antioxidant defense and redox regulation. *Biofactors* **19**: 3–10.
77. Rhee SG, Kang SW, Chang TS, *et al.* (2001) Peroxiredoxin, a novel family of peroxidases. *IUBMB Life* **52**: 35–41.
78. Rhee SG, Kang SW, Netto LE, *et al.* (1999) A family of novel peroxidases, peroxiredoxins. *Biofactors* **10**: 207–209.
79. Hofmann B, Hecht HJ, Flohe L. (2002) Peroxiredoxins. *Biol Chem* **383**: 347–364.
80. Waeber G, Calandra T, Roduit R, *et al.* (1997) Insulin secretion is regulated by the glucose-dependent production of islet beta cell macrophage migration inhibitory factor. *Proc Natl Acad Sci USA* **94**: 4782–4787.
81. Lu C, Li Y, Zhao Y, *et al.* (2002) Intracrine hepatopoietin potentiates AP-1 activity through JAB1 independent of MAPK pathway. *FASEB J* **16**: 90–92.
82. Li Y, Lu C, Xing G, *et al.* (2004) Macrophage migration inhibitory factor directly interacts with hepatopoietin and regulates the proliferation of hepatoma cell. *Exp Cell Res* **300**: 379–387.
83. Wang Y, Lu C, Wei H, *et al.* (2004) Hepatopoietin interacts directly with COP9 signalosome and regulates AP-1 activity. *FEBS Lett* **572**: 85–91.
84. Kleemann R, Hausser A, Geiger G, *et al.* (2000) Intracellular action of the cytokine MIF to modulate AP-1 activity and the cell cycle through Jab1. *Nature* **408**: 211–216.
85. Chamovitz DA, Segal D. (2001) JAB1/CSN5 and the COP9 signalosome. A complex situation. *EMBO Rep* **2**: 96–101.
86. Wei N, Deng XW. (2003) The COP9 signalosome. *Annu Rev Cell Dev Biol* **19**: 261–286.
87. Wolf DA, Zhou C, Wee S. (2003) The COP9 signalosome: an assembly and maintenance platform for cullin ubiquitin ligases? *Nat Cell Biol* **5**: 1029–1033.

88. Bech-Otschir D, Seeger M, Dubiel W. (2002) The COP9 signalosome: at the interface between signal transduction and ubiquitin-dependent proteolysis. *J Cell Sci* **115**: 467–473.
89. Hwang CY, Ryu YS, Chung MS, *et al.* (2004) Thioredoxin modulates activator protein 1 (AP-1) activity and p27Kip1 degradation through direct interaction with Jab1. *Oncogene* **23**: 8868–8875.
90. Chen X, Li Y, Wei K, *et al.* (2003) The potentiation role of hepatopoietin on activator protein-1 is dependent on its sulfhydryl oxidase activity. *J Biol Chem* **278**: 49022–49030.
91. Eickhoff R, Baldauf C, Koyro HW, *et al.* (2004) Influence of macrophage migration inhibitory factor (MIF) on the zinc content and redox state of protein-bound sulphydryl groups in rat sperm: indications for a new role of MIF in sperm maturation. *Mol Hum Reprod* **10**: 605–611.
92. Calandra T, Bernhagen J, Metz CN, *et al.* (1995) MIF as a glucocorticoid-induced modulator of cytokine production. *Nature* **377**: 68–71.
93. Yao K, Shida S, Selvakumaran M, *et al.* (2005) Macrophage migration inhibitory factor is a determinant of hypoxia-induced apoptosis in colon cancer cell lines. *Clin Cancer Res* **11**: 7264–7272.
94. Le Bras M, Clement MV, Pervaiz S, Brenner C. (2005) Reactive oxygen species and the mitochondrial signaling pathway of cell death. *Histol Histopathol* **20**: 205–219.
95. Giles GI, Jacob C. (2002) Reactive sulfur species: an emerging concept in oxidative stress. *Biol Chem* **383**: 375–388.
96. Giles GI, Tasker KM, Jacob C. (2001) Hypothesis: the role of reactive sulfur species in oxidative stress. *Free Radic Biol Med* **31**: 1279–1283.
97. Barzilai A, Yamamoto K. (2004) DNA damage responses to oxidative stress. *DNA Repair (Amst)* **3**: 1109–1115.
98. Zoratti M, Szabo I, De Marchi U. (2005) Mitochondrial permeability transitions: how many doors to the house? *Biochim Biophys Acta* **1706**: 40–52.
99. Zoratti M, Szabo I. (1995) The mitochondrial permeability transition. *Biochim Biophys Acta* **1241**: 139–176.
100. Fingerle-Rowson G, Petrenko O, Metz CN, *et al.* (2003) The p53-dependent effects of macrophage migration inhibitory factor revealed by gene targeting. *Proc Natl Acad Sci USA* **100**: 9354–9359.
101. Hudson JD, Shoaibi MA, Maestro R, *et al.* (1999) A proinflammatory cytokine inhibits p53 tumor suppressor activity. *J Exp Med* **190**: 1375–1382.
102. Mitchell RA, Liao H, Chesney J, *et al.* (2002) Macrophage migration inhibitory factor (MIF) sustains macrophage proinflammatory function by inhibiting p53: regulatory role in the innate immune response. *Proc Natl Acad Sci USA* **99**: 345–350.
103. Sakamoto W, Isomura H, Fujie K, *et al.* (2005) The effect of vitamin K2 on bone metabolism in aged female rats. *Osteoporos Int* **16**: 1604–1610.

104. Kondo N, Ishii Y, Son A, *et al.* (2004) Cysteine-dependent immune regulation by TRX and MIF/GIF family proteins. *Immunol Lett* **92**: 143–147.
105. Watarai H, Nozawa R, Tokunaga A, *et al.* (2000) Post-translational modification of the glycosylation inhibiting factor (GIF) gene product generates bioactive GIF. *Proc Natl Acad Sci USA* **97**: 13251–13256.
106. Nickel W. (2003) The mystery of nonclassical protein secretion. A current view on cargo proteins and potential export routes. *Eur J Biochem* **270**: 2109–2119.
107. Flieger O, Engling A, Bucala R, *et al.* (2003) Regulated secretion of macrophage migration inhibitory factor is mediated by a non-classical pathway involving an ABC transporter. *FEBS Lett* **551**: 78–86.
108. Burger-Kentischer A, Finkelmeier D, Thiele M, *et al.* (2005) Binding of JAB1/CSN5 to MIF is mediated by the MPN domain but is independent of the JAMM motif. *FEBS Lett* **579**: 1693–1701.
109. Pyle ME, Korbonits M, Gueorguiev M, *et al.* (2003) Macrophage migration inhibitory factor expression is increased in pituitary adenoma cell nuclei. *J Endocrinol* **176**: 103–110.
110. Leng L, Metz CN, Fang YXJ, *et al.* (2003) MIF signal transduction initiated by binding to CD74. *J Exp Med* **197**: 1467–1476.
111. Meyer-Siegler KL, Leifheit EC, Vera PL. (2004) Inhibition of macrophage migration inhibitory factor decreases proliferation and cytokine expression in bladder cancer cells. *BMC Cancer* **4**: 34.
112. Meyer-Siegler KL, Vera PL. (2005) Substance P induced changes in CD74 and CD44 in the rat bladder. *J Urol* **173**: 615–620.
113. Shen L, Hu J, Lu H, *et al.* (2003) The apoptosis-associated protein BNIPL interacts with two cell proliferation-related proteins, MIF and GFER. *FEBS Lett* **540**: 86–90.
114. Wadgaonkar R, Dudek SM, Zaiman AL, *et al.* (2005) Intracellular interaction of myosin light chain kinase with macrophage migration inhibition factor (MIF) in endothelium. *J Cell Biochem* **95**: 849–858.

6

MIF Receptor-mediated Signal Transduction

*Lin Leng**

INTRODUCTION

While MIF is one of the first cytokine activities to be described, with its eponymic activity first reported in the late 1950s,[1] the identity of its binding receptor and its molecular mechanism of action escaped understanding until very recently. This delay in knowledge can be attributed first to difficulties in the isolation, molecular characterization, and cloning of MIF, which was not achieved until 1989.[2] The independent cloning of a homologous mouse MIF sequence in 1993 allowed for the preparation of pure, recombinant protein that became widely available for *in vivo* studies.[3,4] The tertiary structure of MIF continued to remain unsettled for some time, however, and it was not until the MIF crystal structure was solved that this information also became known.[5–7]

The three-dimensional structure of MIF showed a new protein fold, but concurrent observations suggested that MIF also possessed an enzymatic activity. This evidence was based first on the serendipitous finding that MIF catalyzed the tautomerization of non-physiologic isomers

*Department of Medicine, Yale University School of Medicine, The Anlyan Center, S525, 300 Cedar Street, New Haven, CT 06520, Lin.Leng@Yale.edu.

of dopachrome,[8] and second, on the prediction that the highly conserved -C-X-X-C- sequence motif could mediate redox activity.[9] These observations prompted interest in the biologic relevance of MIF's enzymatic function(s), which together with the lack of information about a cell surface receptor, suggested to some investigators that MIF mediated its biologic action via a substrate.[10] There was significant doubt cast on the requirement for a discrete, cell surface receptor, and controversy ensued about the relative role(s) of the receptor versus non-receptor (i.e. enzymatic) modes of action of MIF.[11–15]

The powerful neutralizing action of anti-MIF antibody together with the retention of MIF biologic activity in most (but not all) studies of recombinantly-engineered MIF muteins that lacked the catalytic N-terminal proline prompted some groups to continue investigation into a specific, cell surface receptor for MIF. Strong support for a receptor was also provided by Mitchell *et al.*,[16] who examined the intracellular signaling pathways activated by MIF stimulation. To obviate concerns about contaminating LPS, which has the property of broadly activating diverse signaling pathways in monocytes/macrophages, this initial work was performed in fibroblasts, which are insensitive to LPS. MIF-induced increases in the phosphorylation and activation of the p44/p42 extracellular-signal-regulated kinases (ERK-1/2) family of MAP kinase proteins were observed.[16] The ERK1/2 pathway remains the best studied action of MIF on MAPKs, but there is also evidence that the p38 subfamily of MAPK, which is strongly proinflammatory, may be activated. This may occur by a cell-specific or secondary action of MIF.[17]

In most physiological contexts, the activation of ERK-1/2 is transient so that phosphorylation decays in less than 90 minutes. An unusual feature of MIF stimulation, which is also shared by the stimuli provided by oncogenic mutations in Ras or by integrin co-ligation,[18–20] is that the ensuing activation of ERK-1/2 is sustained for many hours.[21,22] This notable property of MIF has significant bearing on its downstream signaling actions, which may vary in a cell-specific fashion. An important concept in this regard is that the temporal pattern of ERK-1/2 activation (transient versus sustained) critically determines the fate of the ensuing cellular response.[22] Emerging evidence also supports a role for the MIF-induced activation of intracellular pathways that include P-I-3 kinase, and Akt.[23]

ERK-1/2 activates several downstream effector proteins that are involved in the inflammatory response, such as transcription factors (*c-myc*, NF-κB, and Ets), cytoskeletal proteins mediating membrane activation and phagocytosis, and other protein kinases.[22,24] MIF activation leads to the phosphorylation of ELK-1,[16] which is a member of the Ets family of transcription factors that have a role in mediating the expression of Toll-like receptor 4,[25] which is the receptor for endotoxin. Among the ERK-1/2 effector proteins, cytoplasmic phospholipase A_2 ($cPLA_2$) is an important component of the inflammatory cascade that is necessary for the production of arachidonic acid, which is the precursor of prostaglandins and leukotrienes.[26,27] Arachidonic acid also activates the *c-jun* N-terminal kinase, which is required for the efficient translation of TNF-α mRNA.[28] $cPLA_2$ is an important target for the anti-inflammatory action of glucocortiocids,[29] and experiments showed that MIF fully "overrides" glucocorticoid inhibition of $cPLA_2$ activation.[16] This observation provided the first insight into a molecular mechanism by which MIF could counter-regulate the glucocorticoid effects on macrophages. MIF also decreases the expression of MAPK phosphatase (MPK-1), which normally acts to dephosphorylate ERK-1/2 and downregulate the MAPK activation response.[25] This action may contribute further to the glucocorticoid insensitivity of MIF-stimulated cells.

MOLECULAR CLONING OF THE MIF RECEPTOR

The specific ability of MIF to induce cellular ERK-1/2 activation.[16] strongly affirmed the hypothesis that MIF interacts with a specific cell surface receptor. Initial attempts by our group to quantify high-affinity cell surface binding sites for MIF were frustrated by the loss of MIF biologic activity upon radioiodination. MIF that was metabolically labeled with ^{35}S was observed to bind to monocytes in a saturable and competitive manner, but the specific activity of binding was too low for the further purification or selection of high-affinity binding sites (Metz C, Bucala R, *unpublished data*). Our group ultimately succeeded in creating a fluorescent-labeled MIF species that showed the retention of biologic activity (i.e. proliferation, inhibition of apoptosis) by linking MIF to the dye Alexa-488 using gentle reaction conditions that were also optimized to yield an average ligand/MIF (homotrimer) ratio of 1:1, which was established by MALDI-MS.[15]

Confocal microscopy and direct visualization of human THP-1 monocytes at 4°C established the cell surface binding of Alexa-MIF, and cell-bound Alexa-MIF was internalized upon shifting the temperature to 37°C (Fig. 1).[30] Alexa-MIF bound specifically to a subpopulation of human THP-1 monocytes by flow cytometry,[30] and high affinity binding sites increased in number after stimulation with the known macrophage activating cytokine, interferon-γ. Quantitative binding studies performed with increasing concentrations of MIF and analyzed by flow cytometry provided data consistent with two classes of cell surface receptors on THP-1 monocytes. The higher affinity binding activity showed a Kd of 3.7×10^{-8} M, and the lower affinity binding showed a Kd of 3.5×10^{-7} M[30,31] (Fig. 2).

Given the prior difficulties encountered in different biochemical approaches, the author and colleagues chose to pursue an expression cloning strategy to identify the MIF receptor. As outlined in Fig. 3, this approach relied on a mammalian expression library prepared from the cDNA of IFNγ-stimulated, human THP-1 monocytes, high-speed fluorescence-activated cell sorting, and the selection of the positively-staining clones for reamplification. Library aliquots were transfected into COS-7 cells, and the cells which stained with Alexa-MIF were purified by fluorescence cell sorting. The cDNA clones were then amplified and retransfected for additional rounds of cell sorting. After four rounds of

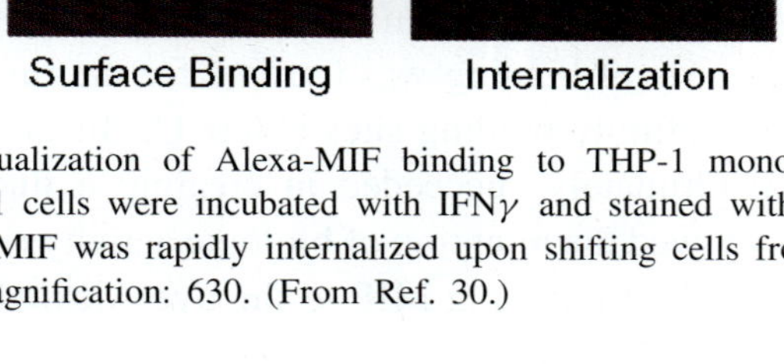

Fig. 1 Direct visualization of Alexa-MIF binding to THP-1 monocytes by confocal microscopy. THP-1 cells were incubated with IFNγ and stained with Alexa-MIF (*left*). Cell-bound Alexa-MIF was rapidly internalized upon shifting cells from 4° to 37°C for 15 min (*right*). Magnification: 630. (From Ref. 30.)

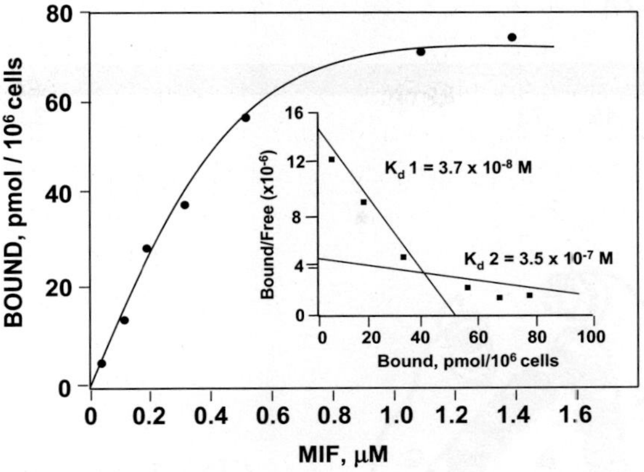

Fig. 2 Binding kinetics for the interaction between Alexa-MIF and THP-1 monocytes generated by using CellQuest Flow Cytometry Software. (Orginal data from Refs. 30 and 31.)

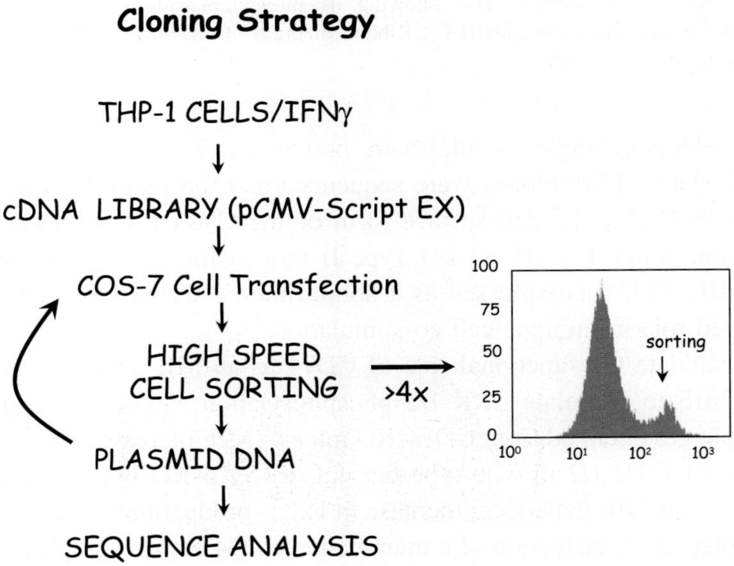

Fig. 3 Expression cloning and FACS selection strategy used to isolate the MIF receptor. (From description in Ref. 30.)

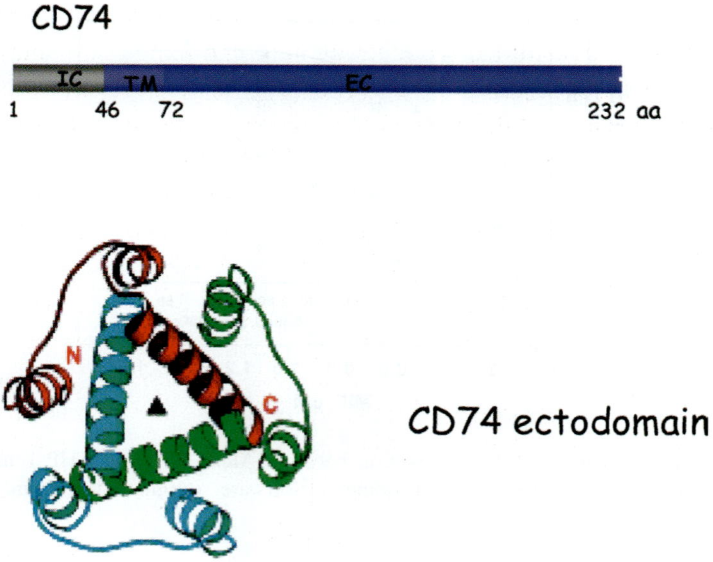

Fig. 4 Diagram of human CD74 showing its intracytoplasmic (IC), transmembrane (TM), and extracelluelr domains (EC). Ribbon diagram of CD74 derived from its three-dimensional crystal structure.[32]

FACS selection, single colonies were prepared in *E. coli* and the plasmid DNA isolated. Fifty clones were sequenced and the majority were found to encode CD74, the cell surface form of the Class II-associated invariant chain, which is a 31–41 kD Type II transmembrane protein (Fig. 4). Like MIF, CD74 is expressed as a homotrimer.[32] CD74 has a previously described role in immune cell co-stimulation.[33,34]

To validate the functional role of CD74 as an MIF receptor, the ability of MIF to stimulate ERK-1/2 phosphorylation was tested in primary macrophages obtained from CD74-KO mice.[35] MIF increased the phosphorylation of ERK-1/2 in wild-type but not in CD74-KO macrophages, and there was no MIF-dependent increase in PGE_2 production in the $CD74^{-/-}$ macrophages. Transfection of a mammalian CD74 expression plasmid into the COS-7 cell line (which does not express $CD74^{-}$) also reconstituted MIF binding and the induction of ERK-1/2 phosphorylation.[30]

BIAcore analysis of the binding interaction between MIF and $sCD74^{73-232}$ revealed an equilibrium Kd of 9.0×10^{-9}–2.3×10^{-10}. This Kd value may be somewhat lower than the dissociation constant *in vivo* because native CD74 is a trimer and the removal of the CD74 transmembrane domain (to facilitate cloning and expression) may inhibit native CD74 oligomerization.[36] Nevertheless, this Kd value is well within the range expected for the nanomolar concentrations of MIF that exist in the extracellular milieu.

These studies provided strong evidence for CD74 as the cell surface receptor for MIF. Notably, it had previously been known that CD74 expression occurs independently of class II in diverse cell types including monocytes, B cells, activated T cells, and fibroblasts.[37,38] CD74 surface expression is also known to be regulated by the length of the protein's N-terminal, intracellular domain, which may vary depending on which of the two in-phase initiation codons is utilized.[39] Whether this differential translation of CD74 mRNA mediates cellular sensitivity to MIF is worthy of further investigation.

HOW DOES MIF MEDIATE SIGNAL TRANSDUCTION THROUGH ITS RECEPTOR?

The mechanism(s) for MIF signaling are not immediately apparent from the structure of CD74. The cytosolic domain of CD74 is only 29–46 amino acids in length — depending on which of the two alternative initiation codons is translated.[39] It lacks homology with the catalytic domains of tyrosine or serine/threonine kinases, or with domains that might interact with non-receptor tyrosine kianses, or nucleotide binding proteins. The intracytoplasmic portion of CD74 is known to undergo phosphorylation in a protein-kinase dependent manner, but evidence supporting ligand activation in this process has yet to be reported.[40] Nevertheless it is noteworthy that recent studies have identified an accessory role for CD74 in T cell stimulation, by a means that requires a chondroitin-sulphate-dependent interaction between CD74 and CD44.[33,34] CD44 is a widely-expressed, polymorphic transmembrane protein with known tyrosine kinase activation properties.[41] Indeed, recent evidence in support of a molecular complex between MIF, CD74 and CD44 has been provided by Meyer-Siegler *et al.*,[42] who utilized an immunoprecipitation approach in cultured carcinoma cells.

Figure 5 schematizes current hypotheses with respect to the likely mechanisms for MIF signal transduction via CD74 based on the known

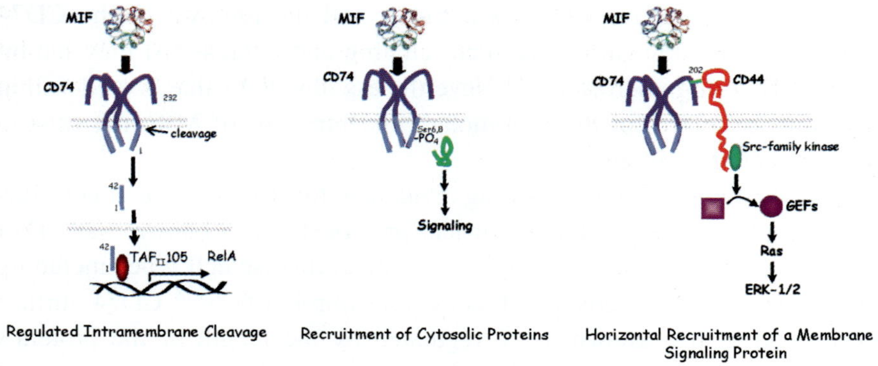

Fig. 5 Proposed pathways for MIF signaling based on the known molecular biology of the MIF receptor (CD74).[31]

molecular biology of these proteins. These mechanisms are not mutually exclusive. CD74-regulated intramembrane cleavage is an interesting and well worked out mechanism for signaling in B lymphocytes. Activation of the RelA/NF-κB family of transcription factors has yet to be reported as a feature of MIF action; however, and MIF action in B lymphocytes is only now beginning to be investigated. Of note, the effect of antibody ligation of CD74 in inducing BCL-X_L transcription and B cell survival was reported recently to be mimicked — in a CD74-dependent fashion — by MIF stimulation.[43]

The known phosphorylation of the CD74 intracytoplasmic domain is strongly suggestive of a signal transduction function requiring contact with second messengers. The recruitment of a second transmembrane protein, such as CD44, is yet another mechanism for MIF signaling, particulary since CD44 activates members of the Src-family tyrosine kinases leading to downstream ERK phosphorylation. Receptor activation by horizontal recruitment is typical of proteins that span the membrane only once, and it is the structural basis for signal transduction by several known cytokine receptors.[44] The IL-6 family of receptors, for instance, does not contain a functional cytoplasmic or intracellular domain, and is incapable of transducing a signal directly. Rather, the IL-6/IL-6 receptor complex interacts with a second membrane protein, gp130, which then transduces its signal.[45] The highly polymorphic nature of the CD44 ectodomain, which also functions in cell adhesion, also lends itself to cell-specific pleiotropism.[46] Such a pathway may lie within the expectation of MIF's broad regulatory role in cell survival and apoptosis,[47–49] and suggests that MIF actions, which are considered to be broad for a classical "proinflammatory" cytokine, may be regulated productively by the interaction of a cell surface binding protein (i.e. CD74) with a polymorphic, signal transduction molecule (CD44).

ACKNOWLEDGMENTS

Studies in the author's laboratory are supported by grants from the NIH.

Abbreviations

Alexa-MIF:	Alexa-488-modified MIF,
ERK-1/2:	extracellular signal-related kinase,
GEFs:	guanine nucleotide exchange factors,
KO:	gene knock-out,
LPS:	lipopolysaccharide,
MAPK:	mitogen-activated protein kinase,
MIF:	macrophage migration inhibitory factor,
RIP:	regulated intramembrane proteolysis,
TLR-4:	Toll-like receptor 4.

REFERENCES

1. George M, Vaughn JH. (1962) *In vitro* cell migration as a model for delayed hypersensitivity. *Proc Soc Exptl Biol Med* **111**: 514–521.
2. Weiser WY, Temple PA, Witek-Giannotti JS, *et al.* (1989) Molecular cloning of a cDNA encoding a human macrophage migration inhibitory factor. *Proc Natl Acad Sci USA* **86**: 7522–7526.
3. Bernhagen J, Calandra T, Mitchell RA, *et al.* (1993) MIF is a pituitary-derived cytokine that potentiates lethal endotoxaemia. *Nature* **365**: 756–759.
4. Bernhagen J, Mitchell RA, Calandra T, *et al.* (1994) Purification, bioactivity, and secondary structure analysis of mouse and human macrophage migration inhibitory factor (MIF). *Biochemistry* **33**: 14144–14155.
5. Sugimoto H, Suzuki M, Nakagawa A, *et al.* (1996) Crystal structure of macrophage migration inhibitory factor from human lymphocyte at 2.1 A resolution. *FEBS Letters* **389**: 145–148.

6. Sun HW, Bernhagen J, Bucala R, Lolis E. (1996) Crystal structure at 2.6-A resolution of human macrophage migration inhibitory factor. *Proc Natl Acad Sci USA* **93**: 5191–5196.
7. Kato Y, Muto T, Tomura T, *et al.* (1996) The crystal structure of human glycosylation-inhibiting factor is a trimeric barrel with three 6-stranded beta-sheets. *Proc Natl Acad Sci USA* **93**: 3007–3010.
8. Rosengren E, Bucala R, Aman P, *et al.* (1996) The immunoregulatory mediator macrophage migration inhibitory factor (MIF) catalyzes a tautomerization reaction. *Mol Med* **2**: 143–149.
9. Kleemann R, Kapurniotu A, Frank RW, *et al.* (1998) Disulfide analysis reveals a role for macrophage migration inhibitory factor (MIF) as thiol-protein oxidoreductase. *J Mol Biol* **280**: 85–102.
10. Swope MD, Lolis E. (1999) Macrophage migration inhibitory factor: cytokine, hormone, or enzyme. *Rev Physiol Biochem Pharmacol* **139**: 1–32.
11. Swope M, Sun HW, Blake PR, Lolis E. (1998) Direct link between cytokine activity and a catalytic site for macrophage migration inhibitory factor. *EMBO J* **17**: 3534–3541.
12. Onodera S, Kaneda K, Mizue Y, *et al.* (2000) Macrophage migration inhibitory factor upregulates expression of matrix metalloproteinases in synovial fibroblasts of rheumatoid arthritis. *J Biol Chem* **275**: 444–450.
13. Bendrat K, Al-Abed Y, Callaway DJ, *et al.* (1997) Biochemical and mutational investigations of the enzymatic activity of macrophage migration inhibitory factor. *Biochemistry* **36**: 15356–15362.
14. Hermanowski-Vosatka A, Mundt SS, Ayala JM, *et al.* (1999) Enzymatically inactive macrophage migration inhibitory factor inhibits monocyte chemotaxis and random migration. *Biochemistry* **38**: 12841–12949.
15. Senter PD, Al-Abed Y, Metz CN, *et al.* (2002) Inhibition of macrophage migration inhibitory factor (MIF) tautomerase and biological activities by acetaminophen metabolites. *Proc Natl Acad Sci USA* **99**: 144–149.
16. Mitchell RA, Metz CN, Peng T, Bucala R. (1999) Sustained mitogen-activated protein kinase (MAPK) and cytoplasmic phospholipase A2 activation by macrophage migration inhibitory factor (MIF). Regulatory role in cell proliferation and glucocorticoid action. *J Biol Chem* **274**: 18100–18106.
17. Santos LL, Lacey D, Yang YH, *et al.* (2004) Activation of synovial cell p38 MAP kinase by macrophage migration inhibitory factor. *J Rheumatol* **31**: 1038–1043.
18. Howe AK, Juliano RL. (1998) Distinct mechanisms mediate the initial and sustained phases of integrin-mediated activation of the Raf/MEK/mitogen-activated protein kinase cascade. *J Biol Chem* **273**: 27268–27274.
19. Eliceiri BPKR, Stromblad S, Cheresh DA. (1998) Integrin alphavbeta3 requirement for sustained mitogenactivated protein kinase activity during angiogenesis. *J Cell Biol* **140**: 1255–1263.

20. Roovers K, Davey G, Zhu X, *et al.* (1999) Alpha5beta1 integrin controls cyclin D1 expression by sustaining mitogen-activated protein kinase activity in growth factor-treated cells. *Mol Cell Biol* **10**: 3197–3204.
21. Marshall C. (1995) Specificity of receptor tyrosine kinase signaling: transient versus sustained extracellular signal regulated kinase activation. *Cell* **80**: 179–185.
22. Vojtek AB, Der CJ. (1998) Increasing complexity of the Ras signaling pathway. *J Biol Chem* **273**: 19925–19928.
23. Amin MA, Volpert OV, Woods JM, *et al.* (2003). Migration inhibitory factor mediates angiogenesis via mitogen-activated protein kinase and phosphatidylinositol kinase. *Circ Res* **93**; 321–329.
24. Robinson MJ, Cobb MH. (1997) Mitogen-activated protein kinase pathways. *Curr Opin Cell Biol* **9**: 180–186.
25. Roger T, Miconnet I, Schiesser AL, *et al.* (2005). Critical role for Ets, AP-1 and GATA-like transcription factors in regulating mouse Toll-like receptor 4 (Tlr4) gene expression. *Biochem J* **387**: 355–365.
26. Exton JH. (1994) Phosphatidylcholine breakdown and signal transduction. *Biochim Biophys Acta* **1212**: 26–42.
27. Murakami M, Nakatani Y, Atsumi G, *et al.* (1997) Regulatory functions of phospholipase A2. *Crit Rev Immunol* **17**: 225–283.
28. Swantek JL, Cobb MH, Geppert TD. (1997) Jun N-terminal kinase/stress activated protein kinase (JNK/SAPK) is required for lipopolysaccharide stimulation of tumor necrosis factor alpha (TNF-α) translation: glucocorticoids inhibit TNF-α translation by blocking JNK/SAPK. *Mol Cell Biol* **17**: 6274–6282.
29. Goppelt-Struebe M, Rehfeldt W. (1992) Glucocorticoids inhibit TNF alpha-induced cytosolic phospholipase A2 activity. *Biochim Biophys Acta* **1127**: 163–167.
30. Leng L, Metz C, Fang Y, *et al.* (2003) MIF signal transduction initiated by binding to CD74. *J Exp Med* **197**: 1467–1476.
31. Leng L, Bucala R. (2006) Insight into the biology of macrophage migration inhibitory factor (MIF) revealed by the cloning of its cell surface receptor. *Cell Res* **16**: 162–168.
32. Jasanoff A, Wagner G, Wiley DC. (1998) Structure of a trimeric domain of the MHC class II-associated chaperonin and targeting protein Ii. *EMBO J* **17**: 6812–6818.
33. Naujokas MF, Morin M, Anderson MS, *et al.* (1993) The chondroitin sulfate form of invariant chain can enhance stimulation of T cell responses through interaction with CD44. *Cell* **74**: 257–268.
34. Naujokas MF, Arneson LS, Fineschi B, *et al.* (1995). Potent effects of low levels of MHC class II associated invariant chain on $CD4^+$ T cell development. *Immunity* **3**: 359–372.

35. Shachar I, Flavell RA. (1996) Requirement for invariant chain in B cell maturation and function. *Science* **274**: 106–108.
36. Ashman JB, Miller J. (1999) A role for the transmembrane domain in the trimerization of the MHC Class II-associated invariant chain. *J Immunol* **163**: 2704–2712.
37. Koch H, Harris AW. (1984) Differential expression of the invariant chain in mouse tumor cells: relationship to B lymphoid development. *J Immunol* **132**: 12–15.
38. Henne C, Schwenk F, Koch N, Möller P. (1995) Surface expression of the invariant chain (CD74) is independent of concomitant expression of major histocompatibility complex class II antigens. *Immunol* **84**: 177–182.
39. Arunachalam B, Lamb C, Cresswell P. (1993) Transport properties of free and MHC class II-associated oligomers containing different isoforms of human invariant chain. *Int Immunol* **6**: 451.
40. Anderson HA, Bergstralh DT, Kawamura T, *et al.* (1999) Phosphorylation of the invariant chain by protein kinase C regulates MHC class II trafficking to antigen-processing compartments. *J Immunol* **163**: 5435–5443.
41. Lesley J, Hyman R, Kincade P. (1993) CD44 and its interaction with extracellular matrix. *Adv Immunol* **54**: 271–335.
42. Meyer-Siegler KL, Leifheit EC, Vera PL. (2004) Inhibition of macrophage migration inhibitory factor decreases proliferation and cytokine expression in bladder cancer cells. *Bmc Cancer* **4**.
43. Starlet D, Gore Y, Binsky I, *et al.* (2006). Cell surface CD74 initiates a signaling cascade leading to cell proliferation and survival. *Blood* Feb 16; [Epub ahead of print].
44. Stroud RM, LaPorte S, Wells JA. (2001) Cytokine-receptor signaling at the structural level. In: Oppenheim JJ, Feldmann M (eds) *Cytokine Reference Vol 1: Ligands*. Academic Press, San Diego, pp. 21–34.
45. Taga T, Kishimoto T. (1997) Gp130 and the interleukin-6 family of cytokines. *Annu Rev Immunol* **15**: 797–819.
46. Turley EA, Noble PW, Bourguignon LYW. (2002). Signaling properties of hyaluronan receptors. *J Biol Chem* **277**: 4589–4592.
47. Hudson JD, Shoaibi MA, Maestro R, *et al.* (1999) A proinflammatory cytokine inhibits p53 tumor suppressor activity. *J Exp Med* **190**: 1375–1382.
48. Mitchell RA, Liao H, Chesney J, *et al.* (2002) Macrophage migration inhibitory factor (MIF) sustains macrophage proinflammatory function by inhibiting p53: regulatory role in the innate immune response. *Proc Natl Acad Sci USA* **99**: 345–350.
49. Fingerle-Rowson G, Petrenko O, Netz CN, *et al.* (2003) The p53-dependent effects of macrophage migration inhibitory factor revealed by gene targeting. *Proc Natl Acad Sci USA* **100**: 9354–9359.

7

MIF in Innate Immunity and Infectious Diseases

Chloé Kaech, Thierry Roger and Thierry Calandra

INNATE IMMUNITY

The capacity to eliminate invading microorganisms is essential for the survival of vertebrates and invertebrates. The host antimicrobial defense response is orchestrated by the innate immune system.[1–3] Sensing of invasive microbial pathogens is carried out by the sentinel cells of the innate immune system (i.e. monocytes/macrophages, dendritic cells, natural killer cell and granulocytes). The components of the microbial cell wall (such as endotoxin/lipopolysaccharide (LPS), peptidoglycan, lipopeptides, glucans and mannans), nucleic acids (unmethylated CpG DNA motifs, single-stranded and double-stranded RNA) and flagellin, collectively designated as microbe-associated molecular motifs, are recognized by specific classes of germline-encoded microbial sensors known as pattern-recognition molecules. Pattern-recognition molecules comprise humoral mediators, such as the LPS-binding protein (LBP), the mannan-binding lectin and components of the complement system, and cell-associated molecules, including CD14, Toll-like receptors

Infectious Diseases Service, Department of Medicine, Centre Hospitalier Universitaire Vaudois, CH-1011 Lausanne, Switzerland. E-mail: Thierry.Calandra@chuv.ch

(TLR), nucleotide-binding oligomerization domain-like receptors (NLR) (also known as nucleotide-binding site and leucin-rich repeat proteins or NBS-LRR), peptidoglycan-recognition proteins, the mannose receptor and dectin-1. Binding of ligands to pattern-recognition molecules activates signal transduction pathways and transcription of immune genes. Upregulation of major histocompatibility complex (MHC) class II molecules and expression of co-stimulatory molecules at the surface of antigen-presenting cells and the release of effector molecules into the extracellular milieu are critical early events of the host defense innate immune response.

The cytokine family of effector molecules initiates the host inflammatory response and coordinates cellular and humoral mechanisms aimed at the eradication or the walling-off of invading pathogens.[4,5] Proinflammatory cytokines recruit and activate cells of the innate immune system and induce the production of proinflammatory mediators, including other cytokines, eicosanoids, nitric oxide, platelet activating factor and free radicals. Cytokine-induced upregulated expression of MHC class I and class II molecules contributes to the activation and proliferation of B and T lymphocytes. Conversely, anti-inflammatory cytokines downregulate inflammatory responses, thereby assuming important counter-regulatory activities.

Control of the host inflammatory response is a delicate balancing act, and dysregulated innate immune responses may have dramatic consequences for the host. On the one hand, a "cytokine storm" caused by an overwhelming systemic proinflammatory response, which may be induced by highly virulent pathogens or due to gain-of-function mutations of innate immune genes, can cause septic shock and trigger multiple organ dysfunctions and death. On the other hand, qualitative or quantitative innate immune defects due to inherited loss-of-function mutations or acquired immunodeficiencies may facilitate microbial proliferation and invasion, which in turn can lead to septic shock and multi-organ failure. This concept is supported by experimental data showing increased susceptibility to infection in animals lacking specific pattern-recognition molecules, cytokines, or cytokine receptors.[6] Similarly, human studies have linked the presence of polymorphisms of promoter or coding regions of cytokines, cytokine receptors and pattern-recognition receptors with susceptibility to and outcome of infectious diseases.[7,8] Tight regulation of the expression of molecules implicated in antimicrobial host defense is thus essential for proper resolution of microbial attacks.

MACROPHAGE MIGRATION INHIBITORY FACTOR (MIF)

Shortly after its rediscovery as a peptide secreted in a hormone-like fashion from the anterior pituitary gland (see Chapter 1), MIF was rapidly found to be a pivotal proinflammatory cytokine of innate immunity (reviewed in Ref. 9). The aim of this chapter is to review the role played by MIF in host defenses against infectious diseases (except for parasitic infections covered in Chapter 3).

Cellular Sources and Expression Patterns

MIF has a broad cellular and tissue distribution and is most notably expressed in compartments that are in direct contact with the host's natural environment. MIF has been detected in endothelial cells, fibroblasts, lungs, gastrointestinal tract, kidneys, liver, pancreas, prostate, brain, skin, eyes, fat tissue, breasts, pituitary gland, adrenal glands, testes, ovaries, bones and joints.[10,11] Within the immune system, monocytes, macrophages, dendritic cells, T and B cells, neutrophils, eosinophils, basophils, and mast cells have been found to express MIF (reviewed in Refs. 12 and 13). How much each of these cells and tissues contribute to the circulating baseline levels of MIF is unknown. However, one can reasonably speculate that both immune and endocrine cells are important contributors of systemic MIF.[14] In contrast to most cytokines, MIF is constitutively expressed and stored in intracellular pools and does therefore not require *de novo* protein synthesis before secretion into the extracellular milieu.

At first thought to be only a target of MIF action, cells of the monocyte/macrophage lineage were shown to be an important source of MIF.[14] MIF is released by macrophages stimulated with proinflammatory stimuli, including bacterial endotoxin and exotoxins and the cytokines tumor necrosis factor (TNF) and interferon (IFN)γ (Table 1). Many of these stimuli induce a typical "bell-shaped" MIF dose-response curve with minimal production at both ends of the curve, which might serve to protect the host from the detrimental effects of excessive production of this proinflammatory cytokine. Although recent work has demonstrated that MIF secretion may occur via a non-classical pathway involving an ABCA1 transporter, the molecular basis of MIF release remains largely unknown.[15]

Table 1 Microbial Products and Immunomodulatory Stimuli Up-regulating MIF Expression *In Vitro*

Stimulus	Cell Type	References
Endotoxin	Monocytes/macrophages	(14)
	Neutrophils	(81)
	Dendritic cells	(82)
	Peripheral blood mononuclear cells	(47)
Exotoxins (SPEA, TSST-1)	Macrophages, splenocytes	(64)
Peptidoglycan	Peripheral blood mononuclear cells	(47)
Heat-killed *E. coli*	Peripheral blood mononuclear cells	(47)
Heat-killed *S. aureus*	Peripheral blood mononuclear cells	(47)
M. tuberculosis	Macrophages	(73)
Influenza virus	Pulmonary epithelial cells	(67)
TNF, IFNγ	Macrophages	(14)
C5a	Eosinophils, neutrophils	(81; 83)
IL-5	Eosinophils	(83)
IL-9	Mast cells	(84)
Nitric oxide	Fetal membranes	(85)
Anti-CD3 antibodies	T cells	(17)
Low-dose glucocorticoids	Monocytes/macrophages, T cells	(16)

As illustrated in Table 1, numerous microbial products and immunomodulatory stimuli have been reported to induce MIF production *in vitro* by a variety of cell types. A tissue distribution study in the rat showed that preformed MIF protein is released from the pituitary, adrenal glands, lungs, liver, spleen, kidneys and skin within six hours of LPS injection and contributes to increasing the circulating MIF concentrations.[10] Simultaneously, MIF mRNA expression is upregulated, so that supranormal intracellular MIF levels are restored within 24 hours of LPS stimulation. When released, MIF acts as a typical proinflammatory cytokine promoting inflammatory responses. The molecules induced directly or indirectly by MIF include cytokines (TNF, IFNγ, interleukin (IL)-1β, IL-2, IL-6, IL-8, IL-12, CXCL-2),[14,16–19] nitric oxide,[20,21] products of the arachidonic acid pathway[22,23] and matrix metalloproteases and their inhibitors.[24,25]

Molecular Basis of MIF-mediated Effects

The modes of action of MIF relevant to antimicrobial host defenses are summarized in Fig. 1. The molecular mechanisms involved in other biological functions of MIF are described in Chapters 6, 15 and 16. The molecular

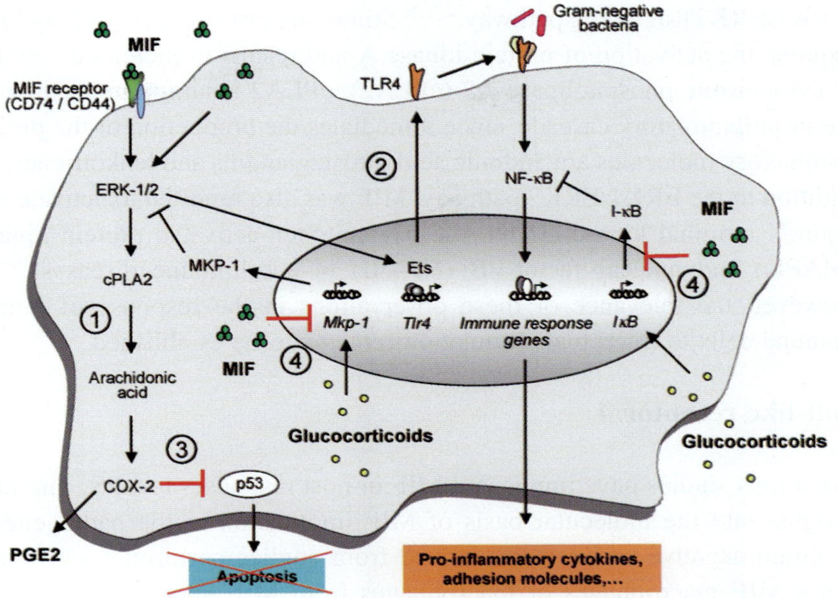

Fig. 1 Biological activities of MIF implicated in antimicrobial host defenses. Via activation of the ERK1/ERK2 signaling pathway, MIF enhances the production of arachidonic acid and its derivates (1); upregulates TLR4 expression (2); and represses LPS-induced p53-mediated apoptosis of macrophages (3). Furthermore, MIF counteracts anti-inflammatory effects of glucocorticoids (4).

bases of MIF biological activity have been the focus of extensive research and may not yet be completely unravelled. Recent work has shown that MIF can be taken up by cells by endocytosis (in a process that may be either receptor-dependent or receptor-independent).[26] MIF was also shown to be a ligand for CD74, the cell surface form of the MHC class II-associated invariant chain.[27] Whereas signal transduction has been previously associated with CD74, more recent work seems to suggest that CD44 is the signal transducing molecule of the MIF receptor complex (reviewed in Ref. 28 and in Chapter 16).

Intracellular signaling pathways

Studies of MIF proliferative activity in fibroblasts have indicated that MIF induces the activation of the extracellular-signal-regulated kinase

(ERK1/ERK2) signaling pathway.[22,29] Stimulation of ERK1/ERK2 by MIF requires the activation of protein kinase A and results in increased activity of cytoplasmic phospholipase A2 (cPLA2). cPLA2 is an important link in the proinflammatory cascade, since it mediates the production of the proinflammatory molecules arachidonic acid, prostaglandins and leukotrienes. In addition to the ERK1/ERK2 pathway, MIF was also reported to activate the c-jun-N terminal kinase (JNK), the p38 mitogen-activated protein kinase (MAPK) and nuclear factor-κB (NF-κB) in autoimmune diseases.[30–32] However, the relevance of these observations in the response of innate immune cells to microbial stimulation remains to be established.

Toll-like receptor 4

Numerous studies have implicated MIF in host responses to LPS. The first insights into the molecular basis of MIF implication in the pathogenesis of Gram-negative sepsis were derived from studies performed with antisense MIF macrophages or macrophages from MIF knockout (MIF$^{-/-}$) mice.[33,34] MIF-deficient macrophages were found to be hyporesponsive to stimulation with LPS and Gram-negative bacteria, as measured by decreased NF-κB activity and reduced production of TNF, IL-1β, IL-6 and IL-12. The impaired response of MIF-deficient macrophages to LPS and Gram-negative bacteria was due to downregulation of TLR4 — the signal-transducing molecule of the LPS receptor complex. MIF-dependent TLR4 upregulation is mediated by the Ets family of transcription factors that are activated by the ERK1/ERK2 signaling pathway and are essential for *TLR4* gene transcription.[35] By increasing TLR4 expression, MIF facilitates the detection of endotoxin-containing bacteria and the production of proinflammatory cytokines essential for a rapid antimicrobial host response. This effect seems to be especially important for the elimination of intracellular Gram-negative pathogens requiring Th1 cytokine-based macrophage activation. Indeed, MIF$^{-/-}$ mice were shown to be highly susceptible to infection with the intracellular bacterium *Salmonella typhimurium*.[36]

p53

Both p53-dependent and p53–independent apoptosis of lymphocytes and endothelial cells occurs in patients with sepsis.[37,38] Interestingly, in contrast to lymphocytes and dendritic cells, there is no depletion of tissue macrophages in septic patients.[39] Infection-induced apoptosis has been proposed as a regulatory mechanism which limits the lifespan of activated cells and therefore the inflammatory response. In functional screens aimed at the identification of genes which bypass p53-mediated growth arrest or apoptosis, MIF was identified as a negative regulator of p53 activity.[40] Consistent with these findings, the proinflammatory functions and viability of MIF-deficient macrophages were observed to be severely impaired after LPS-challenge due to unchecked activation of p53-mediated apoptotic signals.[23] Inhibition of p53 activity by MIF requires serial activation of ERK1/ERK2, cPLA2 and cyclooxygenase 2. MIF thus promotes proinflammatory and innate immune responses via an upregulation of TLR4 expression and prolongation of the lifespan of macrophages.

Glucocorticoids

Activation of the hypothalamic-pituitary-adrenal (HPA) axis with ensuing release of glucocorticoid hormones is an integral component of the host response to infection (reviewed in Ref. 41). In addition to their important role in metabolic homeostasis, glucocorticoids are powerful anti-inflammatory mediators. It was therefore very surprising to find that macrophage and T cell MIF production was induced rather than inhibited by glucocorticoids in rodents.[16,17] The dose-response curve of macrophage MIF production induced by cortisol or dexamethasone was bell-shaped, with a peak of MIF release at glucocorticoid concentrations in the range of 10^{-12} to 10^{-14} M. Moreover, intravenous injection of dexamethasone or stress-induced activation of the HPA axis increased circulating MIF concentrations in the rat.[16] Similarly, administration of dexamethasone increased MIF protein expression in organs considered to be sensitive to the growth-inhibitory effects of glucocorticoids (immune and endocrine tissues, skin, muscles).[42] At first glance, these results appear paradoxical and difficult to reconcile with the proinflammatory properties of MIF. However, MIF was subsequently found to override the anti-inflammatory and immunosuppressive effects of glucocorticoids. *In vitro*, MIF reversed glucocorticoid-induced inhibition of: 1) TNF, IL-1, IL-6 and IL-8 synthesis by LPS-stimulated human monocytes;[16] 2) TNF-induced arachidonic acid release by murine fibroblasts;[22] 3) proliferation of human T cells; and 4) IL-2 and IFNγ synthesis by human T cells stimulated with concanavalin A or tetanus toxoid.[17] Likewise, *in vivo*, MIF reversed the protective effects mediated by glucocorticoids in a mouse model of lethal endotoxemia.[16] Hence the concept that MIF is a physiological antagonist of the immuno-modulatory effects of glucocorticoids. MIF and glucocorticoids thus appear to function as a counter-regulatory dyad that modulates pro- and anti-inflammatory host responses.

In humans, serum MIF levels were reported to follow a circadian rhythm similar to that of cortisol, with peak levels in the morning.[43] However, in healthy volunteers, stimulation of the HPA axis by hypoglycemia, corticotropin-releasing hormone or Synacthen test, or suppression by dexamethasone was not associated with consistent changes in serum MIF concentration.[43,44] Serum MIF levels increased after oral administration of low-dose cortisone, but decreased after intravenous administration of high-dose dexamethasone.[43] At a cellular level, MIF production in human synoviocytes was induced by dexamethasone at concentrations in the range of 10^{-10} to 10^{-12} M, but was suppressed at higher (10^{-7} M) dexamethasone concentrations.[45] In human T lymphoblasts and in human lung epithelial cells, concentrations of dexamethasone between 10^{-6} and 10^{-12} M suppressed MIF production.[46] No induction was observed at lower dexamethasone concentrations. In a recent study, *ex vivo* production of MIF by peripheral blood mononuclear cells (PBMCs) in response to stimulation with LPS, peptidoglycan or heat-killed *Escherichia coli* was higher in cells obtained from septic shock patients than from healthy subjects.[47] However, upregulated MIF production was not observed in PBMCs from septic shock patients treated with cortisol. Overall, these data suggest that glucocorticoid-induced regulation of MIF production may depend on species, cell type and glucocorticoid type and dose.

Several mechanisms whereby MIF can override the anti-inflammatory effects of glucocorticoids have been identified. First, MIF antagonized the inhibitory effect of cortisol on NF-κB activation in human PBMCs by counteracting the steroid-induced upregulation of cytosolic IκBα, an inhibitor of the NF-κB pathway.[48] Second, in human and mouse monocytic cells, MIF suppressed glucocorticoid-induced expression of MAPK phosphatase 1 (MKP-1), a dual specificity phosphatase known to inactivate ERK1/ERK2, p38 and JNK MAPK activities and cytokine production induced by proinflammatory and microbial stimuli.[49,50] The identification of MKP-1 as a target of MIF-glucocorticoid crosstalk also provides a molecular basis for the reported opposing effects of MIF and dexamethasone on cPLA2 activity.[22]

MIF IN EXPERIMENTAL ANIMAL MODELS OF TOXIC SHOCK SYNDROMES AND SEPSIS

Over the last years, MIF has been ascribed important functions in the innate immune system and was found to be an important mediator of toxic shock syndromes and live bacterial sepsis in several experimental animal models (summarized in Table 2).

Table 2 Role of MIF in Experimental Animal Models of Sepsis

Animal Model	Treatment or Condition	Effect on Mortality	References
Endotoxemia and Gram-negative sepsis			
High dose LPS	rMIF	Aggravation	(51)
High dose LPS	anti-MIF, MIF ko, ISO-1	Protection	(20; 51; 52)
LPS and Bacille Calmette-Guerin	anti-MIF, antisense MIF adenovirus	Protection	(54; 55)
LPS and pancreatitis	anti-MIF, MIF ko	Protection	(53)
E. coli peritonitis	rMIF	Aggravation	(60)
E. coli peritonitis	anti-MIF	Protection	(60)
S. typhimurium i.v./p.o.	MIF ko	Aggravation	(36)
Gram-positive toxic shock syndromes and sepsis			
Staphylococcal enterotoxin B	MIF ko	Protection	(20)
Toxic shock syndrome toxin 1	anti-MIF	Protection	(64)
S. pneumoniae pneumonia	anti-MIF	Protection	(T.C. *et al.*)
Polymicrobial sepsis			
Cecal ligation and puncture (CLP)	anti-MIF, ISO-1	Protection	(52; 60)
CLP followed 2 days later by:			
- *S. typhimurium* i.p.	rMIF	Protection	(66)
- *P. aeruginosa* i.n.	rMIF	Protection	(66)
- *L. monocytogenes* i.p.	rMIF	Protection	(66)
CLP followed 2 days later by:			
- *S. typhimurium* i.p.	anti-MIF	Aggravation	(66)
- *P. aeruginosa* i.n.	anti-MIF	Aggravation	(66)

Endotoxemia and Gram-Negative Sepsis

Endotoxemia

Endotoxin or LPS is a main virulence factor of Gram-negative bacteria. Experimental endotoxemia in mice caused a rapid increase in MIF serum levels, peaking 20 hours after LPS challenge.[51] Administration of recombinant MIF did not, like TNF, induce shock when administered systemically in mice, but was found to markedly increase lethality when co-injected with LPS. Conversely, pre-treatment with neutralizing anti-MIF antibodies reduced proinflammatory cytokine production (50% decrease in serum TNF levels) and improved survival.[51] Similarly, MIF$^{-/-}$ mice were shown to produce lower amounts of TNF, but not of IL-6 and IL-10, than wild-type mice after exposure to LPS and to exhibit increased survival.[20] Consistent with these observations, the small molecule ISO-1 ((S,R)-3-(4-hydroxyphenyl)-4,5-dihydro-5-isoxazole acetic acid methyl ester), which was developed as an inhibitor of MIF tautomerase activity, also reduced TNF production by LPS-stimulated peritoneal macrophages and protected mice from lethal endotoxemia.[52]

Investigations of the role played by high MIF levels in LPS-induced organ injury showed that neutralization of MIF activity reduced the levels of CXCL-2, a powerful neutrophil chemoattractant, and attenuated neutrophil migration into the lungs.[19] Likewise, in a model of LPS-injury in the context of acute pancreatitis, MIF$^{-/-}$ mice and wild-type mice treated with anti-MIF antibodies exhibited reduced lung injury and mortality when compared to control mice.[53] In agreement with our previous observations of a role for MIF in TLR4 expression, anti-MIF therapy suppressed pancreatitis-induced upregulation of pulmonary TLR4 expression. MIF was also shown to be involved in the pathogenesis of acute hepatic failure induced by LPS and bacille Calmette-Guerin.[54,55] Anti-MIF antibodies[54] or antisense MIF therapy[55] improved liver injury, modulated TNF upregulation and increased survival. Finally, MIF was established as an important mediator of LPS-induced myocardial dysfunction.[56] LPS causes MIF release from cardiomyocytes into the systemic circulation[56] and from the lungs into the pulmonary circulation 30 hours after LPS challenge.[57] MIF released from the lungs may contribute to the persistent cardiac dysfunction seen in late sepsis. Indeed, treatment with anti-MIF antibodies reversed LPS-induced cardiac dysfunction as evaluated by echocardiography.[56,58] Mitigation of MIF-induced activation of pro-apoptotic pathways and acute inflammation in the heart might be the mechanism whereby anti-MIF antibodies improved cardiac function and survival.[58]

Escherichia coli (*E. coli*) peritonitis

Bacterial peritonitis triggers an acute inflammatory response characterized by the rapid appearance of proinflammatory cytokines (such as TNF or IL-1β) first in the peritoneal cavity and then in the systemic circulation.[59] High concentrations of MIF were also detected in the peritoneal exudate and in the blood of mice with acute *E. coli* peritonitis.[60] As in experimental endotoxemia, co-administration of recombinant MIF and *E. coli* increased mortality, while prophylactic treatment with anti-MIF antibodies[60] or with the MIF inhibitor ISO-1[52] improved survival. Protection afforded by neutralization of MIF activity was associated with lower bacterial counts and TNF levels in the systemic circulation 12 hours after the onset of *E. coli* peritonitis.[60]

Although anti-cytokine therapy can protect the host from a harmful "cytokine storm," neutralization of MIF activity might also impair innate immune defenses and transform a mild localized infection into a rapidly progressive systemic infection. To test this possibility, wild-type mice were treated with anti-MIF antibodies before i.p. injection of a low inoculum (LD_{10}) of *E. coli*.[60] Interestingly, MIF neutralization did not increase mortality. Moreover, mice treated with anti-MIF antibodies had lower bacterial blood counts and lower serum concentrations of TNF and IL-6 than mice treated with control antibodies. Therefore, immunoneutralization of MIF does not compromise the host's capacity to combat a mild infection caused by an extracellular Gram-negative bacterium such as *E. coli*.

Salmonella typhimurium (*S. typhimurium*) sepsis

Given that *S. typhimurium* infects, replicates and persists in the macrophage, it is often used in experimental animal models to cause an infection mimicking human typhoid fever.[61] In contrast to infections caused by extracellular bacteria, MIF seems to be indispensable for controlling infections caused by the facultatively intracellular Gram-negative bacterium *S. typhimurium*.[36] When compared to wild-type mice, MIF$^{-/-}$ mice infected with *S. typhimurium* per os (p.o.) or intravenously (i.v.) displayed reduced serum levels of TNF, IFNγ and IL-12, increased bacterial loads in the liver and spleen and higher mortality. These findings suggested that MIF deficiency is associated with the typical hallmarks of an impaired Th1 response and inadequate macrophage activation, leading to impaired killing of intracellular bacteria. Consistent with this hypothesis, administration of exogenous IFNγ rescued *S. typhimurium*-infected MIF$^{-/-}$ mice, whose mortality returned to a level similar to that of wild-type mice.[36] For infections with organisms residing in macrophages, the disadvantages of an inadequate macrophage activation in the absence of MIF thus seem to outweigh the advantages of an attenuated systemic proinflammatory response. This hypothesis is in agreement with results obtained in two experimental models of intracellular infections caused by *Mycobacterium tuberculosis* (see below) and *Leishmania* (see Chapter 3).

***Pseudomonas aeruginosa* (*P. aeruginosa*) pneumonia**

To explore the role of MIF in the pathogenesis of *P. aeruginosa* pneumonia, wild-type mice and MIF$^{-/-}$ mice were instilled intratracheally with *P. aeruginosa*.[20] Compared to wild-type mice, MIF$^{-/-}$ mice had fewer neutrophils in the bronchoalveolar lavage six hours after infection and cleared bacteria more rapidly from the lungs. Unfortunately, no survival experiments were reported in this article.

Gram-Positive Toxic Shock Syndromes and Sepsis

Toxic shock syndromes

Gram-positive exotoxins are microbial superantigens that can induce shock by stimulating macrophages and large pools of T cells to release abundant amounts of proinflammatory mediators.[62,63] Beside Gram-negative endotoxin, MIF has also been implicated in immune responses induced by Gram-positive staphylococcal and streptococcal exotoxins. As mentioned earlier in this chapter, small quantities of toxic-shock syndrome toxin 1 (TSST-1) and streptococcal pyrogenic exotoxin A (SPEA) were shown to induce MIF release by macrophages. Furthermore, in a model of lethal TSST-1-induced shock in D-galactosamine-sensitized mice, anti-MIF antibody therapy given before exotoxin challenge significantly reduced mortality.[64] Likewise, $MIF^{-/-}$ mice were resistant to lethal shock induced by staphylococcal enterotoxin B (SEB) in mice sensitized with D-galactosamine. Protection was associated with reduced serum TNF levels compared to wild-type mice.[20]

Streptococcus pneumoniae (*S. pneumoniae*) pneumonia

The role of MIF in Gram-positive sepsis was studied in a mouse model of pneumonia induced by intranasal (i.n.) administration of *S. pneumoniae*. Compared to treatment with control antibodies, treatment with anti-MIF antibodies increased survival and reduced bacterial counts and TNF and IL-6 concentrations in blood, bronchoalveolar lavage and lung homogenates (T.C. *et al.*, unpublished observations).

Polymicrobial Sepsis

The role of MIF in the pathogenesis of sepsis was also investigated in the cecal ligation and puncture (CLP) peritonitis model, a very stringent experimental model of polymicrobial sepsis caused by a mixture of aerobic and anaerobic Gram-negative and Gram-positive bacteria. CLP is considered by many investigators to be a critical preclinical test for any new sepsis therapy. Treatment with anti-MIF antibodies[60] or with ISO-1[52] protected mice from lethal CLP peritonitis. Of note, mice were still protected when anti-MIF antibodies were administered as late as 8 hours after induction of peritonitis.[60] The fact that anti-MIF antibodies had the ability to protect animals from sepsis when given therapeutically rather than prophylactically is crucial, as any anti-sepsis therapy will be given after the onset of infection in humans. Importantly, anti-MIF therapy was also found to be protective in TNF-deficient mice, showing that the pathogenic role of MIF in the CLP model is independent of TNF.[60] Taken together, the results obtained in models of toxic shock syndromes and of Gram-negative, Gram-positive

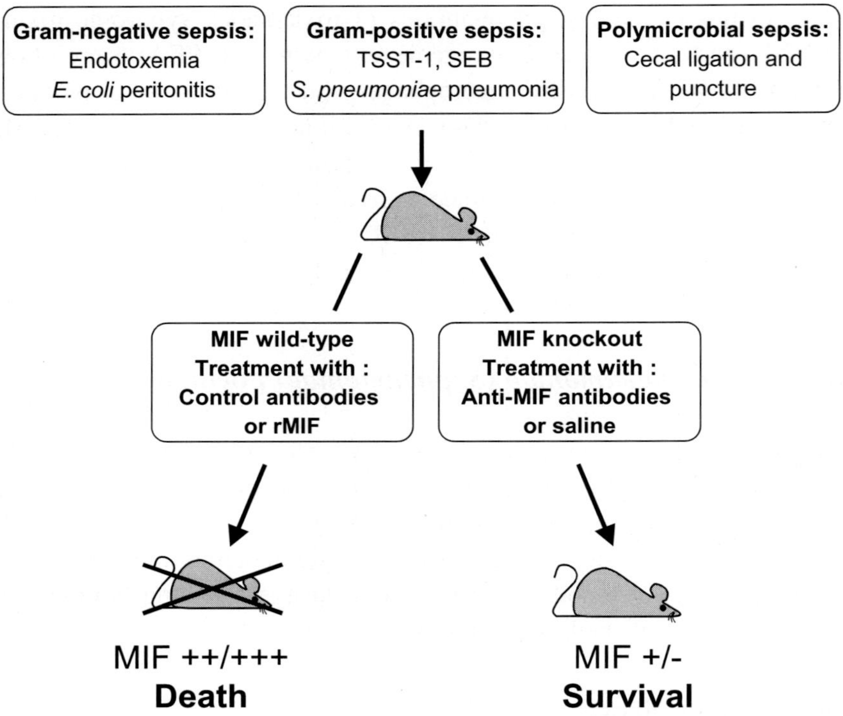

Fig. 2 Role of MIF in early sepsis. In the early phase of sepsis, high MIF levels (endogenous or recombinant exogenous MIF) are harmful, and preventive neutralization of MIF activity with specific antibodies or gene knockout improves the outcome of Gram-negative, Gram-positive and polymicrobial sepsis.

and polymicrobial sepsis indicate that the release of large quantities of MIF into the systemic circulation contributes to the lethal cascade of events triggered by severe bacterial infection and that preventive neutralization of MIF activity reduces inflammation and improves outcome (Fig. 2).

During sepsis, the initial proinflammatory immune response is followed by a state of immunosuppression that may leave the host highly vulnerable to secondary infections.[65] A recent study investigated the role played by MIF in sepsis-induced immunosuppression.[66] A few days after sublethal CLP, mice developed an immunosuppressed state characterized by low serum MIF levels and a reduced capacity to produce TNF after LPS-challenge and they exhibited an increased susceptibility to bacterial superinfections. Indeed, these mice succumbed to infection when challenged with an otherwise non-lethal inoculum of *S. typhimurium* (i.p.), *P. aeruginosa* (i.n.) or *Listeria monocytogenes* (*L. monocytogenes*) (i.p.) two days after CLP. Administration of recombinant MIF partially restored their capacity to produce TNF in response to LPS and improved survival when given shortly before superinfection with *S. typhimurium*, *P. aeruginosa* or *L. monocytogenes.* Conversely, immunoneutralization of MIF shortly before bacterial superinfection significantly enhanced mortality.[66] Although excessive production of the proinflammatory cytokine MIF is harmful in the acute hyperinflammatory phase of sepsis, the results obtained in these 2-hit sepsis models clearly demonstrate that MIF is required to prevent or control superinfection in the late hypoinflammatory phase of sepsis.

Viral Infections

Very few studies focusing on the role of MIF in host responses to viral infection have been reported in the literature to date. *In vitro,* exposure of human bronchiolar and alveolar epithelial cells to influenza A virus was associated with a massive release of MIF predominantly from preformed intracellular pools in the context of cellular necrosis.[67] Yet, this passively liberated MIF was found to be biologically active, suggesting that elevated intrapulmonary MIF levels may contribute to the host immune response during influenza A virus infection. Another recent work showed that MIF contributes to liver injury induced by hepatitis B virus (HBV)-specific cytotoxic T lymphocytes (CTL) *in vivo*.[68] Treatment of HBV-infected mice with anti-MIF antibodies before CTL injection reduced IFNγ and TNF mRNA expression in the liver, infiltration of inflammatory cells, hepatocyte apoptosis and transaminase activity. No direct antiviral effect of MIF could be demonstrated *in vitro* or *in vivo*.

MIF IN PATIENTS WITH INFECTIOUS DISEASES

Severe Sepsis and Septic Shock

In healthy human subjects, circulating concentrations of MIF are in the range of 2 to 6 ng/ml.[60] Investigators noted fluctuations in serum MIF

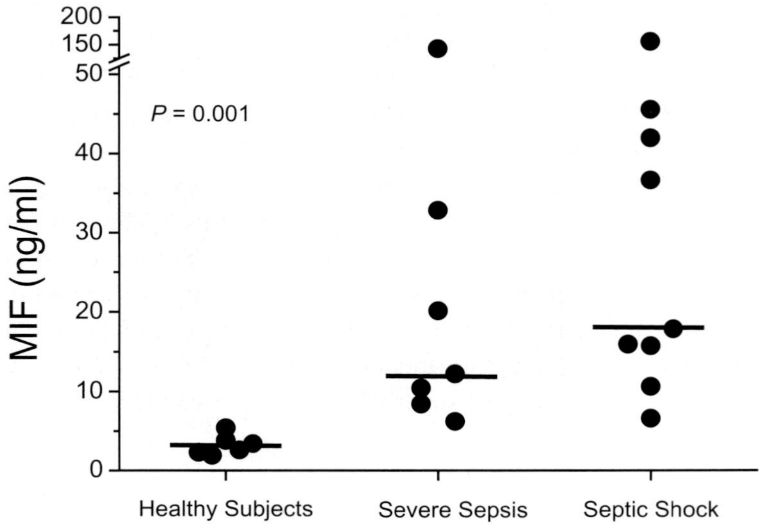

Fig. 3 Serum MIF levels in healthy subjects and in patients with severe sepsis or septic shock. Serum MIF levels are significantly higher in patients with septic shock and severe sepsis than in healthy subjects ($p = 0.001$). Solid horizontal lines indicate median values. Adapted from Ref. 60.

levels that they attributed to a possible circadian rhythm.[43] Consistent with the concept established in animal models of septic shock, serum concentrations of MIF were significantly higher in patients with septic shock and severe sepsis than in healthy controls (Fig. 3).[60] Likewise, in patients who developed a systemic inflammatory response syndrome after cardiac surgery, circulating MIF levels were significantly higher in patients with positive blood cultures than in patients without microbiologically documented sepsis.[69] Serum MIF levels were also higher in patients with septic shock than in trauma patients or control subjects during the entire course of disease.[70] In patients with septic shock, MIF concentrations were higher in non-survivors than in survivors (Fig. 4) and were correlated with levels of cortisol and IL-6. Confirming the association between MIF and survival of sepsis, a recent study demonstrated that high MIF levels were an early indicator of poor outcome in patients with septic shock.[71] The observation that serum MIF levels remained elevated for several days in patients with sepsis has important therapeutic implications. Should anti-MIF therapeutic

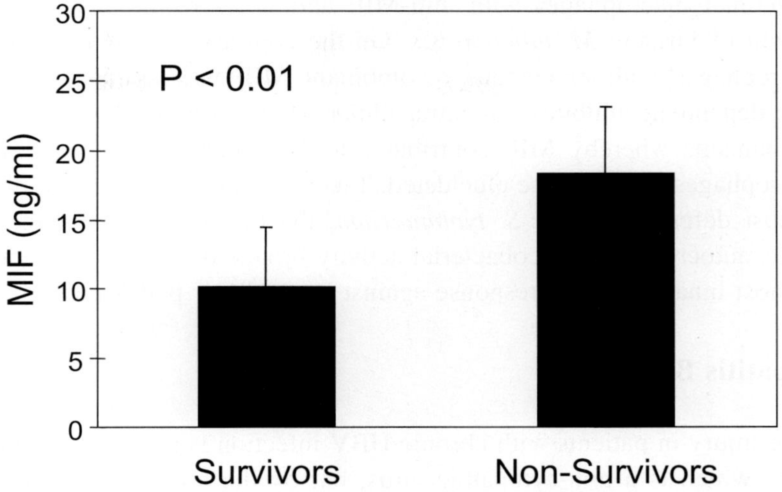

Fig. 4 Serum MIF levels in survivors and non-survivors of septic shock. In patients with septic shock, MIF concentrations are significantly higher in non-survivors than in survivors. Adapted from Ref. 70.

strategies become available in the near future, they will have a broad time window of administration. This is in sharp contrast to anti-TNF or anti-IL-1 therapies, for which an early timing of intervention is essential. Given the production kinetics of these cytokines, these treatments were often applied too late in septic patients (i.e., at a time when TNF and IL-1 had already returned to baseline levels). This fact might at least partly explain why immunoneutralization of TNF and IL-1 did not improve survival in patients with sepsis (reviewed in Ref. 72).

Tuberculosis

Beside pyogenic bacteria, MIF has recently been shown to participate in the innate immune response of macrophages against mycobacteria. Human macrophages exposed to virulent *M. tuberculosis* or stimulated with mannose-capped lipoarabinomannan derived from *M. tuberculosis* released abundant quantities of MIF.[73] Consistent with these *in vitro* observations, serum MIF levels were reported to be elevated in patients with pulmonary tuberculosis.[74] *In vitro*, neutralization of endogenous MIF released by human macrophages with anti-MIF antibodies resulted in enhanced growth of virulent *M. tuberculosis*. On the contrary, treatment of human macrophages with exogenous recombinant MIF was associated with a dose-dependent inhibition of intracellular *M. tuberculosis* growth.[73] The mechanisms whereby MIF contributes to *M. tuberculosis* killing within macrophages remain to be elucidated. Taken together with the role of MIF in host defenses against *S. typhimurium*, the fact that macrophage MIF exerts autocrine anti-mycobacterial activity broadens MIF's implication in the host innate immune response against intracellular pathogens.

Hepatitis B

Liver injury in patients with chronic HBV infection is not caused by HBV itself, which is a non-cytopathic virus, but by the host antiviral immune response.[75] Circulating blood levels and intrahepatic expression of MIF were measured in patients in the immune-tolerance and in the immune-clearance phase of perinatally acquired chronic hepatitis B.[76] Serum MIF levels were higher and intrahepatic MIF expression was stronger in patients with immune clearance than in patients with immune tolerance or than in healthy controls. Serum MIF levels correlated with serum transaminase activity, intrahepatic inflammatory infiltrates and necroinflammatory liver injury. These data support a role for MIF in the immune-mediated hepatic injury of chronic hepatitis B and are in accordance with the observation that anti-MIF antibody therapy in HBV-infected mice prior to injection of specific cytotoxic T lymphocytes attenuated liver damage.[68]

Dengue

Infection with the dengue virus can be either asymptomatic or cause dengue fever or dengue hemorrhagic fever with possible dengue shock syndrome. In a recent study, the serum concentrations of MIF were measured in adult patients with dengue virus infection.[77] Similar to what has been seen in bacterial septic shock, non-survivors of dengue hemorrhagic fever had significantly higher MIF levels than survivors and than patients with dengue fever.

Other Infections

Consistent with the concept that MIF is produced and released in response to infection, elevated MIF levels have been detected in the urine, amniotic fluid and middle ear exudate obtained from patients with urinary tract infections,[78] infectious amnionitis[79] and otitis media.[80]

CONCLUSIONS

Over the last 15 years, MIF has emerged as a pivotal mediator of the innate immune system. In contrast to most cytokines, MIF is expressed constitutively by immune cells and in tissues that are at the forefront of host antimicrobial defense. Microbial products and proinflammatory cytokines induce a rapid release of preformed MIF which acts as a proinflammatory effector molecule regulating innate immunity. Several important biological activities ascribed to MIF are likely to mediate its effects within the immune system. These include the upregulation of TLR4 expression and inhibition of p53-mediated apoptosis in the macrophage, the activation of MAPK such as ERK1/ERK2 and the counter-regulatory effect of MIF on glucocorticoid-induced immunosuppression via the inhibition of MKP-1. Several lines of evidence support a critical role for MIF in host antimicrobial defense responses. On the one hand, MIF is required for optimal host defense against infection, as shown by its capacity to limit the growth of intracellular bacterial pathogens, such as *Salmonella* and *Mycobacterium,* and to protect the host from superinfection in the late hyperinflammatory phase of sepsis. On the other hand, excessive MIF production in the acute hyperinflammatory phase of sepsis is deleterious, as shown by the fact that neutralization of MIF activity improves outcome in animal sepsis models and that serum MIF levels are increased and correlated with outcome in patients with septic shock. Given the central position of MIF in innate immunity and in the pathogenesis of infectious diseases, it is becoming increasingly evident that MIF is one of the leading candidate targets for the development of new therapeutic strategies for patients with severe sepsis and septic shock.

ACKNOWLEDGMENTS

The authors are supported by grants from the Swiss National Science Foundation (3100-066972.01), the Bristol-Myers Squibb Foundation, the Leenaards Foundation and the Santos-Suarez Foundation for Medical Research.

REFERENCES

1. Medzhitov R, Janeway CA, Jr. (1997) Innate immunity: the virtues of a nonclonal system of recognition. *Cell* **91**(3): 295–298.
2. Hoffmann JA, Kafatos FC, Janeway CA, Ezekowitz RA. (1999) Phylogenetic perspectives in innate immunity. *Science* **284**(5418): 1313–1318.
3. Janeway CA, Jr., Medzhitov R. (1998) Introduction: the role of innate immunity in the adaptive immune response. *Semin Immunol* **10**(5): 349–350.
4. Dinarello CA. (2000) Proinflammatory cytokines. *Chest* **118**(2): 503–508.
5. Calandra T, Bochud PY, Heumann D. (2002) Cytokines in septic shock. *Curr Clin Top Infect Dis* **22**: 1–23.
6. Buer J, Balling R. (2003) Mice, microbes and models of infection. *Nat Rev Genet* **4**(3): 195–205.
7. Villar J, Maca-Meyer N, Perez-Mendez L, Flores C. (2004) Bench-to-bedside review: understanding genetic predisposition to sepsis. *Crit Care* **8**(3): 180–189.
8. Hill AV. (1998) The immunogenetics of human infectious diseases. *Annu Rev Immunol* **16**: 593–617.
9. Calandra T, Roger T. (2003) Macrophage migration inhibitory factor: a regulator of innate immunity. *Nat Rev Immunol* **3**(10): 791–800.
10. Bacher M, Meinhardt A, Lan HY, *et al.* (1997) MIF expression in experimentally-induced endotoxemia. *Am J Pathol* **150**(1): 235–246.
11. Waeber G, Calandra T, Roduit R, *et al.* (1997) Insulin secretion is regulated by the glucose-dependent production of islet β-cell macrophage migration inhibitory factor. *PNAS* **94**(9): 4782–4787.
12. Baugh JA, Bucala R. (2002) Macrophage migration inhibitory factor. *Crit Care Med* **30**(1 Supp): S27–S35.
13. Lue H, Kleemann R, Calandra T, *et al.* (2002) Macrophage migration inhibitory factor (MIF): mechanisms of action and role in disease. *Microbes Infect* **4**(4): 449–460.
14. Calandra T, Bernhagen J, Mitchell RA, Bucala R. (1994) The macrophage is an important and previously unrecognized source of macrophage migration inhibitory factor. *J Exp Med* **179**: 1895–1902.

15. Flieger O, Engling A, Bucala R, *et al.* (2003) Regulated secretion of macrophage migration inhibitory factor is mediated by a non-classical pathway involving an ABC transporter. *FEBS Lett* **551**(1–3): 78–86.
16. Calandra T, Bernhagen J, Metz CN, *et al.* (1995) MIF as a glucocorticoid-induced modulator of cytokine production. *Nature* **377**: 68–71.
17. Bacher M, Metz CN, Calandra T, *et al.* (1996) An essential regulatory role for macrophage migration inhibitory factor in T-cell activation. *PNAS* **93**: 7849–7854.
18. Donnelly SC, Haslett C, Reid PT, *et al.* (1997) Regulatory role for macrophage migration inhibitory factor in acute respiratory distress syndrome. *Nature Med* **3**(3): 320–323.
19. Makita H, Nishimura M, Miyamoto K, *et al.* (1998) Effect of anti-macrophage migration inhibitory factor antibody on lipopolysaccharide-induced pulmonary neutrophil accumulation. *Am J Respir Crit Care Med* **158**: 573–579.
20. Bozza M, Satoskar AR, Lin G, *et al.* (1999) Targeted disruption of migration inhibitory factor gene reveals its critical role in sepsis. *J Exp Med* **189**(2): 341–346.
21. Bernhagen J, Mitchell RA, Calandra T, Voelter W, Cerami A, Bucala R. (1994) Purification, bioactivity, and secondary structure analysis of mouse and human macrophage migration inhibitory factor (MIF). *Biochemistry* **33**: 14144–14155.
22. Mitchell RA, Metz CN, Peng T, Bucala R. (1999) Sustained mitogen-activated protein kinase (MAPK) and cytoplasmic phopholipase A2 activation by macrophage migration inhibitory factor (MIF). *J Biol Chem* **274**(25): 18100–18106.
23. Mitchell RA, Liao H, Chesney J, *et al.* (2002) Macrophage migration inhibitory factor (MIF) sustains macrophage proinflammatory function by inhibiting p53: regulatory role in the innate immune response. *Proc Natl Acad Sci USA* **99**(1): 345–350.
24. Onodera S, Tanji H, Suzuki K, *et al.* (1999) High expression of macrophage migration inhibitory factor in the synovial tissues of rheumatoid joints. *Cytokine* **11**(2): 163–167.
25. Onodera S, Nishihira J, Iwabuchi K, *et al.* (2002) Macrophage migration inhibitory factor up-regulates matrix metalloproteinase-9 and -13 in rat osteoblasts. Relevance to intracellular signaling pathways. *J Biol Chem* **277**(10): 7865–7874.
26. Kleemann R, Hausser A, Geiger G, *et al.* (2000) Intracellular action of the cytokine MIF to modulate AP-1 activity and the cell cycle through Jab1. *Nature* **408**(6809): 211–216.
27. Leng L, Metz CN, Fang Y, *et al.* (2003) MIF signal transduction initiated by binding to CD74. *J Exp Med* **197**(11): 1467–1476.

28. Leng L, Bucala R. (2006) Insight into the biology of Macrophage Migration Inhibitory Factor (MIF) revealed by the cloning of its cell surface receptor. *Cell Res* **16**(2): 162–168.
29. Lue H, Kapurniotu A, Fingerle-Rowson G, *et al.* (2006) Rapid and transient activation of the ERK MAPK signalling pathway by macrophage migration inhibitory factor (MIF) and dependence on JAB1/CSN5 and Src kinase activity. *Cell Signal* **18**(5): 688–703.
30. Santos LL, Lacey D, Yang Y, Leech M, Morand EF. (2004) Activation of synovial cell p38 MAP kinase by macrophage migration inhibitory factor. *J Rheumatol* **31**(6): 1038–1043.
31. Onodera S, Nishihira J, Koyama Y, *et al.* (2004) Macrophage migration inhibitory factor up-regulates the expression of interleukin-8 messenger RNA in synovial fibroblasts of rheumatoid arthritis patients: common transcriptional regulatory mechanism between interleukin-8 and interleukin-1beta. *Arthritis Rheum* **50**(5): 1437–1447.
32. Amin MA, Haas CS, Zhu K, *et al.* (2006) Migration inhibitory factor up-regulates vascular cell adhesion molecule-1 and intercellular adhesion molecule-1 via Src, PI3 kinase, and NFkappaB. *Blood* **107**(6): 2252–2261.
33. Roger T, David J, Glauser MP, Calandra T. (2001) MIF regulates innate immune responses through modulation of Toll-like receptor 4. *Nature* **414**(6866): 920–924.
34. Roger T, Froidevaux C, Martin C, Calandra T. (2003) Macrophage migration inhibitory factor (MIF) regulates host responses to endotoxin through modulation of Toll-like receptor 4 (TLR4). *J Endotoxin Res* **9**(2): 119–123.
35. Roger T, Miconnet I, Schiesser AL, Kai H, Miyake K, Calandra T. (2005) Critical role for Ets, AP-1 and GATA-like transcription factors in regulating mouse Toll-like receptor 4 (Tlr4) gene expression. *Biochem J* **387**(Pt 2): 355–365.
36. Koebernick H, Grode L, David JR, *et al.* (2002) Macrophage migration inhibitory factor (MIF) plays a pivotal role in immunity against Salmonella typhimurium. *Proc Natl Acad Sci USA* **99**(21): 13681–13686.
37. Hotchkiss RS, Tinsley KW, Hui JJ, *et al.* (2000) p53-dependent and -independent pathways of apoptotic cell death in sepsis. *J Immunol* **164**(7): 3675–3680.
38. Hotchkiss RS, Karl IE. (2003) The pathophysiology and treatment of sepsis. *N Engl J Med* **348**(2): 138–150.
39. Hotchkiss RS, Tinsley KW, Swanson PE, *et al.* (2002) Depletion of dendritic cells, but not macrophages, in patients with sepsis. *J Immunol* **168**(5): 2493–2500.
40. Hudson JD, Shoaibi MA, Maestro R, Carnero A, Hannon GJ, Beach DH. (1999) A proinflammatory cytokine inhibits p53 tumor suppressor activity. *J Exp Med* **190**(10): 1375–1382.

41. Webster JI, Sternberg EM. (2004) Role of the hypothalamic-pituitary-adrenal axis, glucocorticoids and glucocorticoid receptors in toxic sequelae of exposure to bacterial and viral products. *J Endocrinol* **181**(2): 207–221.
42. Fingerle-Rowson G, Koch P, Bikoff R, *et al.* (2003) Regulation of macrophage migration inhibitory factor expression by glucocorticoids *in vivo*. *Am J Pathol* **162**(1): 47–56.
43. Petrovsky N, Socha L, Silva D, Grossman AB, Metz C, Bucala R. (2003) Macrophage migration inhibitory factor exhibits a pronounced circadian rhythm relevant to its role as a glucocorticoid counter-regulator. *Immunol Cell Biol* **81**(2): 137–143.
44. Isidori AM, Kaltsas GA, Korbonits M, *et al.* (2002) Response of serum macrophage migration inhibitory factor levels to stimulation or suppression of the hypothalamo-pituitary-adrenal axis in normal subjects and patients with Cushing's disease. *J Clin Endocrinol Metab* **87**(4): 1834–1840.
45. Leech M, Metz C, Hall P, *et al.* (1999) Macrophage migration inhibitory factor in rheumatoid arthritis: evidence of proinflammatory function and regulation by glucocorticoids. *Arthritis Rheum* **42**(8): 1601–1608.
46. Alourfi Z, Donn RP, Stevens A, Berry A, McMaster A, Ray DW. (2005) Glucocorticoids suppress macrophage migration inhibitory factor (MIF) expression in a cell-type-specific manner. *J Mol Endocrinol* **34**(2): 583–595.
47. Maxime V, Fitting C, Annane D, Cavaillon JM. (2005) Corticoids normalize leukocyte production of macrophage migration inhibitory factor in septic shock. *J Infect Dis* **191**(1): 138–144.
48. Daun JM, Cannon JG. (2000) Macrophage migration inhibitory factor antagonizes hydrocortisone-induced increases in cytosolic IkappaBalpha. *Am J Physiol Regul Integr Comp Physiol* **279**(3): R1043–R1049.
49. Roger T, Chanson AL, Knaup-Reymond M, Calandra T. (2005) Macrophage migration inhibitory factor promotes innate immune responses by suppressing glucocorticoid-induced expression of mitogen-activated protein kinase phosphatase-1. *Eur J Immunol* **35**(12): 3405–3413.
50. Aeberli D, Yang Y, Mansell A, Santos L, Leech M, Morand EF. (2006) Endogenous macrophage migration inhibitory factor modulates glucocorticoid sensitivity in macrophages via effects on MAP kinase phosphatase-1 and p38 MAP kinase. *FEBS Lett* **580**(3): 974–981.
51. Bernhagen J, Calandra T, Mitchell RA, *et al.* (1993) MIF is a pituitary-derived cytokine that potentiates lethal endotoxaemia. *Nature* **365**: 756–759.
52. Al-Abed Y, Dabideen D, Aljabari B, *et al.* (2005) ISO-1 binding to the tautomerase active site of MIF inhibits its pro-inflammatory activity and increases survival in severe sepsis. *J Biol Chem* **280**(44): 36541–36544.
53. Matsuda N, Nishihira J, Takahashi Y, Kemmotsu O, Hattori Y. (2006) Role of MIF in acute lung injury in mice with acute pancreatitis complicated by endotoxemia. *Am J Respir Cell Mol Biol.*

54. Kobayashi S, Nishihira J, Watanabe S, Todo S. (1999) Prevention of lethal acute hepatic failure by antimacrophage migration inhibitory factor antibody in mice treated with bacille Calmette-Guerin and lipopolysaccharide. *Hepathology* **29**(6): 1752–1759.
55. Iwaki T, Sugimura M, Nishihira J, Matsuura T, Kobayashi T, Kanayama N. (2003) Recombinant adenovirus vector bearing antisense macrophage migration inhibitory factor cDNA prevents acute lipopolysaccharide-induced liver failure in mice. *Lab Invest* **83**(4): 561–570.
56. Garner LB, Willis MS, Carlson DL, *et al.* (2003) Macrophage migration inhibitory factor is a cardiac-derived myocardial depressant factor. *Am J Physiol Heart Circ Physiol* **285**(6): H2500–H2509.
57. Lin X, Sakuragi T, Metz CN, *et al.* (2005) Macrophage migration inhibitory factor within the alveolar spaces induces changes in the heart during late experimental sepsis. *Shock* **24**(6): 556–563.
58. Chagnon F, Metz CN, Bucala R, Lesur O. (2005) Endotoxin-induced myocardial dysfunction: effects of macrophage migration inhibitory factor neutralization. *Circ Res* **96**(10): 1095–1102.
59. Zanetti G, Heumann D, Gerain J, *et al.* (1992) Cytokine production after intravenous or peritoneal gram-negative bacterial challenge in mice. Comparative protective efficacy of antibodies to tumor necrosis factor-α and to lipopolysaccharide. *J Immunol* **148**: 1890–1897.
60. Calandra T, Echtenacher B, Roy DL, *et al.* (2000) Protection from septic shock by neutralization of macrophage migration inhibitory factor. *Nat Med* **6**(2): 164–170.
61. Raupach B, Kaufmann SH. (2001) Immune responses to intracellular bacteria. *Curr Opin Immunol* **13**(4): 417–428.
62. Ikejima T, Dinarello CA, Gill M, Wolff SM. (1984) Induction of interleukin-1 by a product of *Staphylococcus aureus* associated with toxic shock syndrome. *J Clin Invest* **73**: 1312–1320.
63. Marrack P, Kappler J. (1990) The staphylococcal enterotoxins and their relatives. *Science* **248**: 705–711.
64. Calandra T, Spiegel LA, Metz CN, Bucala R. (1998) Macrophage migration inhibitory factor is a critical mediator of the activation of immune cells by exotoxins of Gram-positive bacteria. *Proc Natl Acad Sci USA* **95**(19): 11383–11388.
65. Kox WJ, Volk T, Kox SN, Volk HD. (2000) Immunomodulatory therapies in sepsis. *Intensive Care Med* **26** Suppl 1: S124–S128.
66. Pollak N, Sterns T, Echtenacher B, Mannel DN. (2005) Improved resistance to bacterial superinfection in mice by treatment with macrophage migration inhibitory factor. *Infect Immun* **73**(10): 6488–6492.
67. Arndt U, Wennemuth G, Barth P, *et al.* (2002) Release of macrophage migration inhibitory factor and CXCL8/interleukin-8 from lung epithelial cells rendered necrotic by influenza A virus infection. *J Virol* **76**(18): 9298–9306.

68. Kimura K, Nagaki M, Nishihira J, Satake S, Kuwata K, Moriwaki H. (2006) Role of macrophage migration inhibitory factor in hepatitis B virus-specific cytotoxic-T-lymphocyte-induced liver injury. *Clin Vaccine Immunol* **13**(3): 415–419.
69. de Mendonca-Filho HT, Gomes GS, Nogueira PM, *et al.* (2005) Macrophage migration inhibitory factor is associated with positive cultures in patients with sepsis after cardiac surgery. *Shock* **24**(4): 313–317.
70. Beishuizen A, Thijs LG, Haanen C, Vermes I. (2001) Macrophage migration inhibitory factor and hypothalamo-pituitary-adrenal function during critical illness. *J Clin Endocrinol Metab* **86**(6): 2811–2816.
71. Bozza FA, Gomes RN, Japiassu AM, *et al.* (2004) Macrophage migration inhibitory factor levels correlate with fatal outcome in sepsis. *Shock* **22**(4): 309–313.
72. Abraham E. Why immunomodulatory therapies have not worked in sepsis? (1999) *Intensive Care Med* **25**: 556–566.
73. Oddo M, Calandra T, Bucala R, Meylan PR. (2005) Macrophage migration inhibitory factor reduces the growth of virulent Mycobacterium tuberculosis in human macrophages. *Infect Immun* **73**(6): 3783–3786.
74. Yamada G, Shijubo N, Takagi-Takahashi Y, *et al.* (2002) Elevated levels of serum macrophage migration inhibitory factor in patients with pulmonary tuberculosis. *Clin Immunol* **104**(2): 123–127.
75. Chisari FV, Ferrari C. (1995) Hepatitis B virus immunopathogenesis. *Annu Rev Immunol* **13**: 29–60.
76. Zhang HY, Nanji AA, Luk JM, *et al.* (2005) Macrophage migration inhibitory factor expression correlates with inflammatory changes in human chronic hepatitis B infection. *Liver Int* **25**(3): 571–579.
77. Chen LC, Lei HY, Liu CC, *et al.* (2006) Correlation of serum levels of macrophage migration inhibitory factor with disease severity and clinical outcome in dengue patients. *Am J Trop Med Hyg* **74**(1): 142–147.
78. Meyer-Siegler KL, Iczkowski KA, Vera PL. (2006) Macrophage migration inhibitory factor is increased in the urine of patients with urinary tract infection: macrophage migration inhibitory factor-protein complexes in human urine. *J Urol* **175**(4): 1523–1528.
79. Chaiworapongsa T, Romero R, Espinoza J, *et al.* (2005) Macrophage migration inhibitory factor in patients with preterm parturition and microbial invasion of the amniotic cavity. *J Matern Fetal Neonatal Med* **18**(6): 405–416.
80. Kariya S, Okano M, Aoji K, *et al.* (2003) Role of macrophage migration inhibitory factor in otitis media with effusion in adults. *Clin Diagn Lab Immunol* **10**(3): 417–422.
81. Riedemann NC, Guo RF, Gao H, *et al.* (2004) Regulatory role of C5a on macrophage migration inhibitory factor release from neutrophils. *J Immunol* **173**(2): 1355–1359.

82. Morelli AE, Zahorchak AF, Larregina AT, *et al.* (2001) Cytokine production by mouse myeloid dendritic cells in relation to differentiation and terminal maturation induced by lipopolysaccharide or CD40 ligation. *Blood* **98**(5): 1512–1523.
83. Rossi AG, Haslett C, Hirani N, *et al.* (1999) Human circulating eosinophils secrete macrophage migration inhibitory factor (MIF). Potential role in asthma. *J Clin Invest* **101**(12): 2869–2874.
84. Wiener Z, Falus A, Toth S. (2004) IL-9 increases the expression of several cytokines in activated mast cells, while the IL-9-induced IL-9 production is inhibited in mast cells of histamine-free transgenic mice. *Cytokine* **26**(3): 122–130.
85. Zicari A, Ticconi C, Ietta F, *et al.* (2006) Macrophage migration inhibitory factor-nitric oxide interaction in human fetal membranes at term pregnancy. *J Soc Gynecol Investig* **13**(4): 263–270.

8

MIF in Parasitic and Helminthic Infections

Heidi M. Snider, Claudio Lezama-Davila*,*
Miriam Rodríguez-Sosa†, Luis I. Terrazas†,
*and Abhay R. Satoskar**

INTRODUCTION

Macrophage migration-inhibitory factor (MIF) is a pleiotropic cytokine that is produced by the pituitary gland and multiple different cell types, including macrophages and T-cells, during inflammatory responses. MIF is either taken up by cells via a cellular receptor (CD74) or by endocytosis. MIF phosphorylates the extracellular signal-regulated kinase (ERK) 1 and 2 proteins. It also promotes cell growth and activates transcription factors of the ETS family known to be crucial for the expression of the toll-like receptor 4 (TLR4) gene, which encodes the signal-transducing element of the lipopolysaccharide (LPS) receptor complex. It also binds to JUN-activation domain-binding protein 1 (JAB-1), preventing the JAB-1-induced activation of JUN, and the JAB-1-induced degradation of the cell-proliferation inhibitor KIP1, thereby leading to cell cycle arrest

*Department of Microbiology, Ohio State University, 484 West 12th Avenue, Columbus, OH 43210, USA.
†Unidad de Biomedicina, FEZ-Iztacala, U.N.A.M., Tlalnepantla, Edo. De Mexico 54090, Mexico.
Correspondence: Abhay R. Satoskar.

and apoptosis.[1] MIF inhibits the anti-inflammatory effects of corticosteroids, regulates inflammation of innate immune cells, and plays a critical role in the pathogenesis of sepsis. Additionally, MIF has a catalytically active N-terminal proline that can act as an enzyme to catalyze the tautomerization of several substrates.[1]

It has recently been demonstrated that MIF is primarily released by monocytes and macrophages, and once released it modulates the expression of several proinflammatory molecules, including cytokines (such as TNF (tumor necrosis factor)-α, IFN (interferon)-γ, IL (interleukin)-1β, IL-2, IL-6, IL-8, and macrophage inflammatory protein 2), nitric oxide, cyclooxygenase 2 (COX2), products of the arachidonic acid pathway, and several matrix metalloproteinases and their inhibitors. By generating oxidoreductase activity and COX2, MIF prevents activation-induced apoptosis mediated by the oxidative burst and by p53, which sustains the macrophage proinflammatory function. MIF is also MIF is also associated with the delayed-type hypersensitivity response, regulation of innate immunity, and promotion of protective immunity or as an aid in the development of adverse pathology.[1] These important discoveries, concerning the role of MIF in immune responses, have prompted investigators to study its role during protozoan and helminthic infections.

ROLE OF MACROPHAGE (MIF) IN PROTOZOAN INFECTIONS

Malaria

Malaria is a disease caused by intracellular parasitic protozoa of the genus *Plasmodium* and is transmitted by infected female *Anopheles* mosquitoes during their blood meals. Malaria is a major cause of death and severe illness in most parts of the world, with 300–500 million new infections per year, resulting in approximately 1–2 million deaths, mostly in children under the age of 5.[2] Complications of cerebral malaria and severe anemia are the major causes of morbidity and mortality. Of the four strains that infect humans, *Plasmodium falciparum* is the most prevalent one and accounts for most of the malaria-related deaths. The pathogenesis of

P. falciparum malarial anemia is complex and multifactorial, and remains poorly understood, despite being a major cause of death in regions of high endemicity.[2]

The association of malaria pathogenesis with MIF production has been studied.[3] MIF has been shown to inhibit erythropoiesis *in vitro* in the presence of erythropoietin. Furthermore, Martiney *et al.*[4] demonstrated that MIF production is increased in BALB/c mice during active *P. chabaudi* infection, and that serum MIF levels correlate with disease severity and anemia.[4] Clinical studies have also shown that MIF levels are highly elevated in the peripheral tissues in children and placenta of pregnant women suffering from severe *P. falciparum* infection.[5] Together, the above studies support the notion that MIF is involved in the pathogenesis of malaria by promoting promoting immunopathology. However, a recent study found that MIF levels are decreased in plasma and PBMCs from children with acute children with acute non-cerebral *falciparum* malaria, which suggests that reduced MIF production may also enhance disease severity, probably by impairing host resistance.[6]

Leishmaniasis

Leishmaniasis is caused by intracellular protozoan parasites of the genus *Leishmania*, which is transmitted by the female sand fly. This disease is endemic in 88 countries and an estimated 12 million people are presently infected worldwide.[7] More than 90% of cutaneous leishmaniasis (e.g. *L. major, L. mexicana*) cases occur in Iran, Afghanistan, Syria, Saudi Arabia, Brazil, and Peru, whereas most cases of visceral leishmaniasis (e.g. *L. donovani, L. chagasi*) occur in Bangladesh, Brazil, India and Sudan.[7]

Macrophages and other monocytic lineage cells are primary target cells for *Leishmania*, but they are also the principal effector cells that kill intracellular parasites.[8] Activated macrophages also function as APCs and secrete cytokines which play critical roles in regulation of innate and acquired immune responses during infection. Several *in vitro* and *in vivo* studies have shown that MIF plays a critical role in mediating host resistance to *Leishmania*.[9–11] For example, Xu *et al.*[11] found that oral administration of MIF alone, or together with TNF-α, via attenuated *Salmonella*, enhanced the resistance of BALB/c mice to *L. major*.[11] Furthermore, others demonstrated that MIF in conjunction with endogenous TNF-α also enhanced macrophage NO production and induced the *in vitro* killing of *L. major* by macrophages.[10] Interestingly, the same study also found that both susceptible BALB/c and resistant C57BL/6 mice express high levels of MIF mRNA in their draining lymph nodes following *L. major* infection.[10] Satoskar *et al.*[12] examined the cutaneous growth of *L. major* in genetically resistant C57BL/6 × 129/Sv mice lacking the MIF gene and demonstrated that MIF deficiency increases susceptibility to *L. major* because of impaired macrophage leishmanicidal activity rather than dysregulation of Th1 and Th2 responses.[12] Although little is known about the role of MIF in human leishmaniasis, a recent clinical study reported that $CD2^+$ T-cells isolated from *L. donovani*-infected patients after anti-leishmanial treatment produce high levels of MIF.[13] Collectively, these observations indicate that MIF Collectively, these observations indicate that MIF plays a critical role in mediating host resistance to *Leishmania* by enhancing macrophage leishmanicidal function.

Trypanosomiasis: Sleeping Sickness and Chagas Disease

Protozoan parasites of the genus *Trypanosoma* cause sleeping sickness in Africa and Chagas disease in Latin America. Sleeping sickness has been reported to be responsible for 20,000 deaths every year and presents an annual prevalence of 300,000 infected persons in Africa.[14] Chagas disease is the leading cause of cardiac death among young adults in parts of South America, with an incidence of 800,000 new cases per year, and a prevalence of 18 million infected individuals, and is responsible for 45,000 deaths every year.[14]

Little is known about the role of MIF in the regulation of the host immune response and resistance in African or American trypanosomiasis. Using the MIF inhibition test, anti-parasite immunity and tissue destruction was found to be associated with MIF production.[15] However, in a more recent study, we have found that MIF is critical for controlling acute *T. cruzi* infection. In the absence of MIF, C57BL/6 mice develop high tissue parasitemia, inflammatory cardiac pathology, and rapidly succumb to infection (Fig. 1).[16] Furthermore, increased susceptibility of $MIF^{-/-}$ mice to *T. cruzi* is associated with a significant reduction in titers of *T. cruzi*-specific antigen specific IgG2a (Th1-associated antibodies), and

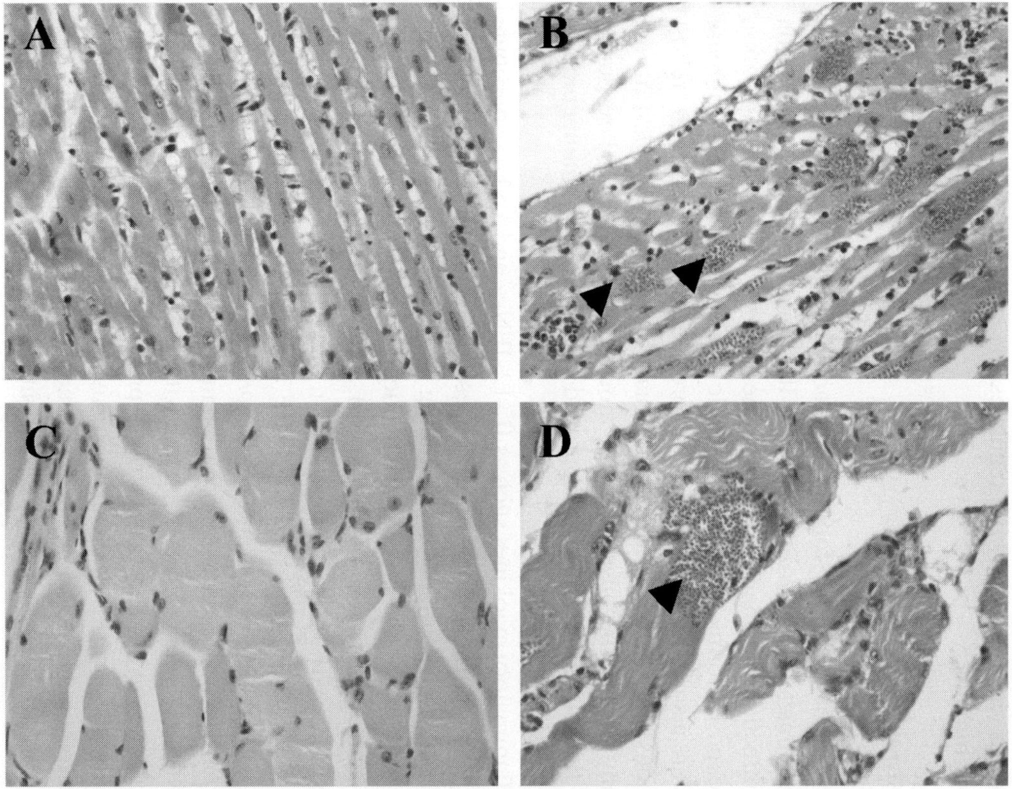

Fig. 1 MIF$^{-/-}$ mice are highly susceptible to *T. cruzi* infection. Histopathology of skeletal muscle (**A** and **B**) and heart (**C** and **D**) from *T. cruzi*-infected MIF+/+ and MIF$^{-/-}$ mice on day 22 after infection. Tissue sections from the heart and skeletal muscle were examined by routine hematoxyline and eosin staining. Black arrows show parasite-filled cells. Magnification $\times$ 40.

impaired production of pro-inflammatory cytokines such as TNF-α, IL-12, IL-18, IFN-γ, and IL-1β.[20] The role of MIF during the resolution or aggravation of African trypanosomiasis remains unclear and will require further investigation. Additionally, it will be of significant interest to know whether MIF regulates the outcome MIF regulates the outcome of trypanosome infection in humans.

Toxoplasmosis

Toxoplasmosis is mainly a food-borne parasitic infection caused by the protozoan parasite *Toxoplasma gondii.*[17] It is one of the most common parasitic infections of man and nearly one-third of the world population has been exposed to this parasite. Serologic surveys indicate that *T. gondii* infections are also common in wild carnivores, including pigs, bears, felids, fox, raccoons and skunks.[17] Most individuals exposed to *T. gondii* do not develop any disease because they can mount a protective Th1-response. However, this infection can be serious and even fatal in immunocompromised patients.

Pro-inflammatory cytokines such as IL-12, IFN-γ and TNF-α contribute to host resistance during toxoplasmosis, whereas anti-inflammatory cytokines such as IL-10 are required for controlling immunopathology.[18] A previous study found that the spleens from mice infected with *T. gondii* displayed enhanced MIF activity in the 3rd and 4th weeks post-infection, suggesting that MIF may be involved in mediating host resistance to *T. gondii.*[19] Recently, we found that MIF$^{-/-}$ BALB/c mice succumb to infection with the ME49 strain of *T. gondii*, which is non-pathogenic to WT BALB/c mice. Furthermore, enhanced susceptibility of MIF$^{-/-}$ mice to ME49 *T. gondii* is associated with the impaired production of pro-inflammatory cytokines (manuscript in preparation). These findings indicate that MIF plays a critical role in host resistance against *T. gondii.* Further studies are required to determine are required to determine whether the protective role of MIF in toxoplasmosis is parasite- or mouse strain-dependent.

Amoebiasis

Amoebiasis is caused by the protozoan parasite *Entamoeba histolytica* and is usually transmitted via a fecal-oral route, although it may also be transmitted by sexual contact in homosexual men. This parasitic disease is the second leading cause of death from parasitic diseases worldwide, with an estimated 40,000–100,000 people dying each year.[20] Although *E. histolytica* infection is primarily confined to the large intestine, the parasites can also disseminate to the systemic organs, such as the liver and form abscesses.

Little is known about the role of MIF in regulating host immune responses after *Entamoeba histolytica* infection. Experimental immunization of hamsters with *E. histolitica* antigens induced the release of MIF by spleen cells pulsed *in vitro* with amoebic antigens.[21] A human study revealed that patients with liver abscesses released higher levels of MIF than patients with amoebic colitis and no liver abscesses. Furthermore, patients with amoebic colitis released more MIF than patients with non-amoebic disease, who served as non-exposed controls.[22] It seems quite likely that future experimental therapeutic interventions in amoebiasis may consider targeting MIF production, which may be involved in the modulation of other cytokines that may not only mediate protection but also induce immunopathology during this disease.

ROLE OF MIF IN HELMINTHIC INFECTIONS

Helminthic infections are of major global health concern; an estimated 25% of the global population is infected with a helminth at some point in life, and an estimated 10% of the 1998 total world population of 5.7 billion people had a helminthic infection at that time. In 1997, an estimated 10,000 people died from food-borne helminthic infection. Most helminthic infections take place in developing nations, and in these countries 170 out of 1000 children die of helminthiases before they reach the age of 5.[23] In addition to mortality in children, helminths cause substantial morbidity. Helminthic infections in children can lead to malnutrition, growth stunting, and iron deficiency anemia, all of which have been correlated with problems in intellectual functioning.[23] Since helminthic infections represent such a ubiquitous global healthcare concern, immunological research has taken place to try to understand helminthic interactions with hosts, including research on the cytokine MIF, which appears to play an important role in various helminthic infections.

Schistosomiasis

Schistosomes are commonly referred to as blood flukes and have a complex life cycle that involves both a snail and a human host.[24] Two million people worldwide are infected with schistosomiasis, with 600 million people at risk. Most cases and at-risk individuals are in Africa.[25] Individuals infected with schistosomiasis often experience myalgia, anemia, dysentery, fever, eosinophilia, leukocytosis, cachexia and anorexia.[24,26] Morbidity is the result of worm fecundity,[26] and in *Schistosoma mansoni* and *S. japonicum*, parasite eggs become embolized primarily in the liver and intestines of their hosts. Disease then results from the inflammatory immune response against these eggs, with a delayed-type hypersensitivity response leading to granuloma formation.[27,28] Most pathology results from the host's own immune response, and this finding has actually led to the suggestion of creating a therapeutic vaccine to be used in place of, or in conjunction with, an antihelminthic vaccine.[27] One important difference to note between human disease and murine models is that fibrosis takes place in mice surrounding the deposited eggs, whereas in humans, fibrosis takes place in the portal tracts. Therefore, nonhuman primates have been used to study schistosomiasis in an experimental model that more closely resembles human disease.[26,27]

Researchers have found that baboons infected chronically with *S. mansoni* have lowered serum levels of the hypothalamic-pituitary-adrenocortical (HPA) axis hormones corticotrophin-releasing hormone (CRH), adrenocorticotrophic hormone (ACTH), cortisol, and dehydroepiandrosterone sulfate (DHEA-S).[26] It has been hypothesized that this is due to changes in cytokine mRNA expression in HPA-axis tissues, since mRNA levels of IL-1β were increased in the hypothalamus, TNF-α mRNA levels were increased in the hypothalamus and anterior pituitary, and IL-6 and MIF mRNA expression levels were increased in the anterior pituitary and adrenal cortex. Expression of these cytokines was not seen in non-HPA tissues, which indicates that they affect the HPA axis in a specific manner during schistosomiasis.[26] In addition, the HPA tissues of infected baboons showed no histological differences from uninfected controls, which indicates that these increases in cytokine mRNA expression may be due to a direct activation of these cells as a result of *S. mansoni* infection, and not as a result of changes or damage to HPA tissues.[26] Since

MIF counteracts the anti-inflammatory effects of glucocorticoids that are produced in the adrenal cortex,[1] the fact that MIF mRNA expression is upregulated not only in the adrenal cortex, but also in the anterior pituitary suggests that that *S. mansoni*-infected baboons may be able to override the immunosuppressive effects of glucocorticoids systemically. Clinically, these findings indicate that before glucocorticoid replacement or cytokine treatments are considered for patients suffering from schistosomiasis, the role of the HPA axis in schistosomiasis needs to be studied in more depth.[26] In addition, DHEA and cortisol are also capable of reducing both fecundity and viability of *S. mansoni*.[27] Since chronic infections lower the serum levels of these hormones, this may be a mechanism of immune evasion by this helminth.

In another study, mRNA cytokine expression was studied in the livers of *S. mansoni*-infected baboons to determine the cytokine players in hepatic granuloma formation. At six weeks post-infection, the mRNA expression of MIF was increased above the constitutive levels expressed in the control, and IL-1β, IL-6, and TNF-α were all expressed in only the livers of infected baboons. By week 12, the levels of those cytokines were either decreased or no longer expressed, and the mRNA expression showed an increase in Th1-associated IL-2, IL-12, and IFN-γ and Th2-associated IL-4, IL-5, and IL-10. This mixed Th1/Th2 response mimics human disease.[27] MIF could be involved in granuloma formation, since pro-inflammatory cytokines such as TNF and IFN-γ stimulate its release from macrophages, and because it is capable of inducing the production of several pro-inflammatory effectors, such as TNF, IFN-γ, such as TNF, IFN-γ, IL-1β, IL-2, IL-6, IL-8 and macrophage inflammatory protein 2.[1,27] It would be of interest to determine the levels of cytokine expression at earlier time points in *S. mansoni*-infected baboons. This would allow for a better understanding of the chronological sequence of cytokine expression and perhaps a better understanding of the influence of cytokines expressed early in infection, such as MIF, on later immune and pathological responses.

Stavitsky *et al.*[28] have studied the role of MIF in mice infected with *S. japonicum*. These researchers found that mice treated with rabbit anti-MIF IgG beginning at 4.5–6.5 weeks post-infection had 1.7–3 times more adult worms and half the number of ova per worm pair in their livers than that in isotype controls 6.5–7 weeks post-infection. It was found that the mice treated with anti-MIF antibodies had reduced TNF-α and IL-10 expression in granuloma cells.[28] Previous research have shown that severe combined immunodeficient (SCID) mice infected with *S. mansoni* had reduced worm fecundity that could be re-established upon treatment with TNF-α.[29] Hence, Stavitsky *et al.* hypothesize that the decrease in worm fecundity in their *S. japonicum*-infected C57BL/6 mice was due to the lack of TNF-α production that resulted from anti-MIF antibody anti-MIF antibody treatment.[1,28] Since MIF is capable of regulating innate immune responses by the upregulation of TLR4 expression on macrophages in response to LPS,[1] and schistosomal lyso-phosphatidylserine has been found to stimulate TLR2 on dendritic cells,[30] Stavitsky *et al.*[28] hypothesize that the increase in worm burden in anti-MIF antibody treated mice could aslo be due to decreased worm killing by macrophages activated by TLRs 4.5–6 weeks post-infection. In addition, MIF can also act on the adaptive immune response by reinforcing T-cell activation and proliferation, and IL-2 production,[1,28] which could also affect worm killing and account for the differences in adult worm account for the differences in adult worm burden between anti-MIF treated and isotype control mice.[28] This study supports the notion that schistosomal granuloma formation in both mice and humans is a very complicated immune process involving the interactions of several cytokines, and that simple anti-cytokine treatments are not viable treatment options for schistosomiasis.

Cysticercosis

Cestodes are more commonly known as tapeworms, and can infect a wide variety of vertebrate hosts.[31] Research concerning cestodes and MIF has focused on the helminth *Taenia crassiceps*, which has a life cycle in rodents and is therefore ideal for studying in laboratory settings. Experimental *T. crassiceps* infection in mouse models is used as an animal model to study *T. solium,* which infects humans and pigs, with pigs serving as an intermediate reservoir. Both *T. crassiceps* and *T. solium* can cause cysticercosis,[32] which is caused by eating pork contaminated with cysticerci (larvae) or by fecal-oral contamination from eggs deposited by carriers.[33] Cysticerci can infect several tissues, including subcutaneous connective tissue, muscle, heart, eye and brain tissue. Outside of the nervous system and the eye, cysticercosis is relatively asymptomatic, except for the presence of mobile, painless nodules in subcutaneous infections. Ophthalmic cysticercosis can result in vision loss, depending on the magnitude of damage to the retina. Neurocysticercosis is the most serious form of the disease which is widespread in Latin America, the majority of Asia, sub-Saharan Africa and portions of Oceania.[33] This disease can result in epileptic seizures, increased intracranial pressure, hydrocephalus, aseptic meningitis and alterations in mental status.[33,34]

Although STAT-6 dependent Th2-like responses are involved in protective responses against most helminthic infections, these responses mediate susceptibility to *T. crassiceps*.[35] On the other hand, Th1-like response associated with the production of IL-12 and IFN-γ mediates protective immunity to *T. crassiceps*.[35] Rodríguez-Sosa *et al.*[32] studied the role of MIF in the regulation of host resistance to *T. crassiceps*, using genetically engineered mice lacking the MIF gene. MIF$^{-/-}$ mice were very susceptible to *T. crassiceps* and exhibited higher parasite loads as compared to infected MIF$^{+/+}$ mice.[32] In addition, spleen cells harvested from infected mice showed interesting cytokine profiles when stimulated with soluble *T. crassiceps* antigen (TcAg), in which IL-4 levels were similar in both levels were similar in both MIF$^{+/+}$ and MIF$^{-/-}$ mice, but the latter produced significantly more Th2-associated IL-13 and Th1-associated IFN-γ.[32] These data indicate that MIF$^{-/-}$ mice are able to mount a mixed Th1/Th2-type response and that the susceptibility of MIF$^{-/-}$ is not due to a deficit in IFN-γ production. To determine whether the loss of MIF has an effect on macrophage cytokine and NO production, Rodríguez-Sosa *et al.*[32] analyzed production of IL-12, IL-6, TNF-α, and NO by peritoneal macrophages isolated from both MIF$^{-/-}$ and MIF$^{+/+}$ mice isolated from both MIF$^{-/-}$ and MIF$^{+/+}$ mice following *T. crassiceps* infection.[32] MIF$^{-/-}$ peritoneal macrophages produced significantly less IL-12, TNF-α, and nitric oxide, but more IL-6 than MIF$^{+/+}$ macrophages following *in vitro* LPS+IFN-γ stimulation.[32] These results and the observation that IL-6 can inhibit macrophage NO production[36] indicate that MIF may production[36] indicate that MIF may contribute to host resistance by upregulating IL-12, TNF-α, and NO production and downregulating IL-6.[32] *Taenia crassiceps* infected MIF$^{-/-}$ mice were also observed to produce low levels of both total IgG2a and TcAg-specific IgG2a, which is Th-1-associated. Taken together, these data indicate that murine $MIF^{-/-}$ macrophages do not respond to IFN-γ, which prevents them from becoming activated and producing the cytokines and NO that would enable them to effectively combat a *T. crassiceps* infection. Rodríguez-Sosa *et al.*[32] hypothesize that this inability to respond to IFN-γ could be due to changes in the IFN-γ R or STAT-1 signaling pathway, or as a result of inactivation by the high levels of IL-13 in $MIF^{-/-}$ high levels of IL-13 in $MIF^{-/-}$ mice.[32] These data also suggest that impairment in macrophage function may be responsible for the increased susceptibility to *T. crassiceps*, and not the lack of a Th1-response.[32] Clinically, these findings suggest that MIF may be a very important immune target in strategies to treat human *T. solium* infection.

Another research study investigating the role of MIF in *T. crassiceps* infection explored the observation that in early infections with *T. crassiceps*, female mice are more susceptible than their male counterparts. However, once infections have become chronic, both male and female mice exhibit similar parasite burdens and males undergo a striking feminization process. $MIF^{-/-}$ mice of both sexes infected with *T. crassiceps* had similar worm burdens, and no changes in sex-hormone levels, whereas female wild-type (WT) controls had a two-fold increase in worm burden, and WT male mice exhibited an 80% decrease in both testosterone and dihydrotestosterone, and a 100-fold increase in estradiol.[37] These results not only support the finding that female sex hormones can increase parasitic proliferation in both sexes,[38] but also implicate MIF as a major player in the feminization process of male mice, which leads to susceptibility in chronic *T. crassiceps* infection.[37] This study further supports the notion that the neuroendocrine pathways need pathways need further investigation so that the role of MIF in host–helminth interactions can be more fully understood.

Trichenellosis

Nematodes are some of the most ubiquitous creatures on earth. Free-living forms are found in marine and fresh water, soil, and environments as extreme as hot springs and Antarctic ice. Parasitic forms infect both plants and animals, and almost all insects carry nematode parasites.[31] Interestingly, several filarial and non-filarial nematodes secrete homologues of MIF. Most non-filarial nematode research studying MIF has focused on the parasite *Trichinella spiralis*, which causes trichinellosis in humans. Humans are infected with trichinellosis by eating undercooked pork or wild game contaminated with larval cysts. Human disease manifestation can range from asymptomatic cases to lethality. Symptoms can include myalgia, vomiting, diarrhea, constipation, fever, edema, headache, conjunctivitis, rashes, pruritus, cough and eosinophilia, as the larvae travel to and encyst in muscle tissue.[39] Once inside a muscle tissue, *T. spiralis* transforms myocytes into nurse cells that produce collagen capsules and stimulate angiogenesis.[40] Cases that result in death are often the result of heart and neurological complications.[39] The non-filarial nematodes *T. spiralis, T. pseudospiralis* (a human parasite), *Trichuris muris* (a murine nematode), and *Trichuris trichiura* (a human whipworm), all produce homologues of mammalian MIF.[40–42] The homologues of *T. spiralis* and *T. muris* have both been found to have L-dopachrome-methyl ester tautomerase activity similar to mammalian MIF.[41] Recombinant *T. spiralis* MIF (rTsMIF) is also similar to human MIF in its ability to inhibit the migration of human peripheral blood mononuclear cells *in vitro*. However, mononuclear cells *in vitro*. However, rTsMIF had no effect on the proliferation of anti-CD3 stimulated murine splenocytes or CD4+ cells, nor did it significantly change the amounts of IL-2, IL-4 or IFN-γ in the cell supernatants,[40] indicating that TsMIF may not be capable of directly altering T-cell dependent responses *in vivo*. However, other studies have implicated a role for rTsMIF in the modulation of macrophages. Histopathological analysis of muscle tissue from *T. spiralis*-infected mice revealed that although many mononuclear cells infiltrated to the site of infected myocytes, very few of these cells were actually macrophages. To determine if this reduction was due to TsMIF, Sephadex beads, which induce granulomatous reactions, were coated with rTsMIF (control mice received uncoated beads) and transplanted subcutaneously into BALB/c mice. Two days post-infection, fewer numbers of macrophages were found to accumulate around the beads treated with rTsMIF when compared with those in untreated beads, indicating that TsMIF may be capable of inhibiting macrophage migration to the sites of *T. spiralis* infection.[42] Taken together, these data, in addition to the observation that TsMIF is secreted extracellularly in levels higher than circulating human MIF, support the notion that TsMIF may be acting as an immunoevasive molecule in trichinellosis. Clinically, this could lead to the development of therapies that target the MIF homologues of parasites.[40]

Filariasis

Filarial nematodes infect the lymphatics and subcutaneous tissues of their hosts. The most important of these parasites from a human morbidity standpoint are *Brugia malayi* and *Wuchereria bancrofti*, which cause lymphatic filariasis, and *Onchocerca volvulus*, which causes river blindness (onchocerciasis). Geographically, *O. volvulus* is primarily found in Africa, although there are endemic areas in Latin America and the Middle East.[43] River blindness afflicts more than 17 million people globally and results in pruritus, disfiguring dermatitis, severe ocular lesions and blindness.[44] *Wuchereria bancrofti* is endemic in tropical areas worldwide, whereas *B. malayi* occurs only in Asia.[43] Lymphatic filariasis infects about 120 million people worldwide and is transmitted by arthropod vectors. Although most patients remain asymptomatic, lymphatic filariasis is commonly known as elephantiasis because of the profound enlargement of limbs, breasts and genitalia that can occur as a result of lymphoedema.[43,44] Most research involving the role of MIF in filarial infections has focused on the parasite *B. malayi*.

MIF homologues have been characterized in *B. malayi, B. pahangi, W. bancrofti* and *O. volvulus*, and *B. malayi* and *B. pahangi* MIFs have both been found to have tautomerase activity similar to mammalian MIF.[41,45] *Brugia malayi* MIF (BmMIF) has been isolated in both somatic extracts and secretory products of the worm and has been found to have chemoattractant properties similar to that of human MIF for human monocytes and macrophages.[46] In addition, researchers have found that BmMIF can activate human monocytes to generate IL-8, TNF-α and endogenous MIF.[45] In individuals infected with lymphatic filariasis, asymptomatic infections are characterized by Th2-type immune responses that include eosinophilia and high levels of IgE and IgG4, whereas chronically infected individuals develop Th1-type immune responses.[46] Researchers have found that *B. malayi* implanted into the peritoneum of mice induces the migration of eosinophils and alternatively activated macrophages — which are involved in down-regulatory responses — to the site of infection. In addition, BmMIF was found to increase the transcription of a recently discovered eosinophil chemotactic factor (ECF-L). This has led researchers to hypothesize that Bm-MIF may be involved in eosinophil recruitment and the skewing of host responses towards a Th2-phenotype.[46] Since hosts mounting a Th2-responses are asymptomatic, secretion of Bm-MIF would confer *B. malayi* a competitive advantage in its host.[47] It has also been speculated that at high levels of secretion, BmMIF may actually be able to block pro-inflammatory gene expression in the host.[45] These data indicate that MIF and its role in T-cell responses may serve as potential clinical targets of future treatments for filariasis.

CONCLUDING REMARKS

As discussed earlier, MIF plays a critical role in mediating host resistance against several protozoa such as *Leishmania, Trypanosoma* and *Toxoplasma.* On the other hand, experimental evidence suggests that MIF is also involved in the pathogenesis of malaria-associated anemia. Although inducing or blocking MIF during protozoan infections may be potential strategies to treat these infections or disease-associated pathology, caution needs be taken when considering approaches to manipulate MIF levels *in vivo*, since inconsistencies appear in the literature. For example, although MIF is involved in inducing anemia in mice infected with *P. chabaudi*,[4] blood MIF levels showed no correlation with either parasitemia or hemoglobin levels in children suffering from *P. falciparum* malaria.[6]

Interestingly, while host MIF appears to play an important role in immune responses against helminthic infections, helminthic MIF homologues appear to be playing important roles in immune evasion. It is important to note that certain species of free-living (non-helminthic) worms, such as *Caenorhabditis elegans*, do express homologues of MIF. Research has indicated that *C. elegans* MIF may play a role in cellular homeostasis during periods of environmental stress.[48] Taking this into consideration, perhaps the secretion of MIF by helminthes was initially a response designed simply to maintain cellular integrity under the attack of the host's immune system. This protective response may have later evolved into the evasive response that is seen now.

Further research exploring the impact of both host MIF and parasite MIF on the immune response needs to be performed in order to successfully develop efficacious clinical treatments that target MIF.

REFERENCES

1. Calandra T, Roger T. (2003) Macrophage migration inhibitory factor: a regulator of innate immunity. *Nat Rev Immunol* **3**: 791–800.
2. WHO. (2002) Severe falciparum malaria. World Health Organization, Communicable Diseases Cluster. *Trans R Soc Trop Med Hyg* **94**(Suppl 1): S1–S90.
3. Coleman RM, Bruce A, Rencricca NJ. (1976) Malaria: macrophage migration inhibition factor (MIF). *J Parasitol* **62**: 137–138.
4. Martiney JA, Sherry B, Metz CN, *et al.* (2000) Macrophage migration inhibitory factor release by macrophages after ingestion of *Plasmodium chabaudi*-infected erythrocytes: possible role in the pathogenesis of malarial anemia. *Infect Immun* **68**: 2259–2267.
5. Chaiyaroj SC, Rutta AS, Muenthaisong K, *et al.* (2004) Reduced levels of transforming growth factor-β1, interleukin-12 and increased migration inhibitory factor are associated with severe malaria. *Acta Tropica* **89**: 319–327.
6. Awandare GA, Hittner JB, Kremsner PG, *et al.* (2006) Decreased circulating macrophage migration inhibitory factor (MIF) protein and blood mononuclear cell MIF transcripts in children with *Plasmodium falciparum* malaria. *Clin Immunol* (In press).
7. WHO. (1996) The World Health Report. Geneva, p. 24.
8. Alexander J, Satoskar AR, Russell DG. (1999) Leishmania species: models of intracellular parasitism. *J Cell Sci* **18**: 2993–3002.
9. Weiser WY, Pozzi LM, David JR. (1991) Human recombinant migration inhibitory factor activates human macrophages to kill *Leishmania donovani*. *J Immunol* **147**: 2006–2011.
10. Jüttner S, Bernhagen J, Metz CN, *et al.* (1998) Migration inhibitory factor induces killing of *Leishmania major* by macrophages: dependence on reactive nitrogen intermediates and endogenous TNF-α. *J Immunol* **161**: 2383–2390.
11. Xu D, McSorley SJ, Tetley L, *et al.* (1998) Protective effect on *Leishmania major* infection of migration inhibitory factor TNF-α and IFN-γ administered orally via attenuated *Salmonella typhimurium*. *J Immunol* **160**: 1285–1289.
12. Satoskar AR, Bozza M, Rodríguez-Sosa M, *et al.* (2001) Migration-inhibitory factor gene-deficient mice are susceptible to cutaneous *Leishmania major* infection. *Infect Immun* **69**: 906–911.
13. Bimal S, Singh SK, Das VN, *et al.* (2005) *Leishmania donovani*: effect of therapy on expression of CD2 antigen and secretion of macrophage migration inhibitory factor by T cells in patients with visceral leishmaniasis. *Exp Parasitol* **111**: 130–132.
14. WHO. (1996) The World Health Report. Geneva, p. 53.
15. Ribeiro dos Santos, Hudson L. (1981) Denervation of the immune response in mice infected with *Trypanosoma cruzi*. *Clin Exp Immunol* **44**: 349–354.

16. Reyes JL, Terrazas LI, Espinoza B, *et al.* (2006) Macrophage migration inhibitory factor (MIF) contributes in host defense against acute *Trypanosoma cruzi* infection. *Infect Immun* (In press).
17. Montoya JG, Liesenfeld O. (2004) Toxoplasmosis. *The Lancet* **363**: 1965–1976.
18. Liesenfeld O. (1999) Immune response to *Toxoplasma gondii* in the gut. *Immunobiology* **201**: 229–239.
19. Igarashi I, Taguchi M, Suzuki N. (1979) Fundamental studies on macrophage migration inhibitory factor(s) in the supernatant from spleen cells in mice infected with *Toxoplasma gondii. Zentralbl Bakteriol* **244**: 374–382.
20. Stanley SL Jr. (2003) Amoebiasis. *Lancet* **361**: 1025–1034.
21. Haq A, Sharma A, Ahmad S. (1984) Increased macrophage migration inhibition factor production in hamsters sensitized with amoebic antigen and glucan. *Parasite Immunol* **6**: 391–396.
22. Vinayak VK, Jain P, Gupta B, *et al.* (1980) Cellular and humoral responses in amoebic patients. *Trop Geogr Med* **32**: 298–302.
23. Chowdhury N, Tada I. (2001) *Perspectives on Helminthology*. Science Publishers, Inc., Enfield, NH.
24. CDC Fact Sheet: http://www.dpd.cdc.gov/dpdx/HTML/Schistosomiasis.htm
25. Chitsulo L, Engels D, Montresor A, Savioli L. (2000) The global status of schistosomiasis and its control. *Acta Trop* **77**: 41–51.
26. Morales-Montor J, Mohamed F, Baghdadi A, *et al.* (2003) Expression of mRNA for interleukin-1 beta, interleukin-6, tumor necrosis factor-alpha and macrophage migration inhibitory factor in HPA-axis tissue in *Schistosoma mansoni*-infected baboons (*Papio cynocephalus*). *Int J Parasitol* **33**: 1515–1524.
27. Morales-Montor J, Mohamed F, Drake L, *et al.* (2004) Dynamics of the cytokine messenger RNA expression pattern in the liver of baboons infected with *Schistosoma mansoni. J Parasitol* **90**: 547–556.
28. Stavitsky AB, Metz C, Liu S, *et al.* (2003) Blockade of macrophage migration inhibitory factor (MIF) in *Schistosoma japonicum*-infected mice results in an increased adult worm burden and reduced fecundity. *Parasite Immunol* **25**: 369–374.
29. Amiri P, Locksley RM, Parslow TG, *et al.* (1992) Tumour necrosis factor alpha restores granulomas and induces parasite egg-laying in schistosome-infected SCID mice. *Nature* **356**: 604–607.
30. van der Kleij D, Latz E, Brouwers JF, *et al.* (2002) A novel host-parasite lipid cross-talk. Schistosomal lyso-phosphatidylserine activates toll-like receptor 2 and affects immune polarization. *J Biol Chem* **277**: 48122–48129.
31. Matthews BE. (1998*) An Introduction to Parasitology*. Cambridge University Press, Cambridge.

32. Rodríguez-Sosa M, Rosas LE, David JR, *et al.* (2002) Macrophage migration inhibitory factor plays a critical role in mediating protection against the helminth parasite *Taenia crassiceps. Infect Immun* **71**: 1247–1254.
33. García HH, Gonzalez AE, Evans CAW, Gilman RH. (2003) *Taenia solium* cysticercosis. *Lancet* **361**: 547–556.
34. Sciutto E, Frangoso G, Fleury A, *et al.* (2000) *Taenia solium* diseases in humans and pigs: an ancient parasitosis disease rooted in developing countries and emerging as a major health problem of global dimensions. *Microbes Infect* **2**: 1875–1890.
35. Rodríguez-Sosa M, David JR, Bojalil R, *et al.* (2002) Susceptibility to the larval stage of the helmith parasite *Taenia crassiceps* is mediated by Th2 response induced via STAT6 signaling. *J Immunol* **168**: 3135–3139.
36. Hatzigeorgiou DE, He S, Sobel J, *et al.* (1993) IL-6 down-modulates the cytokine-enhanced antileishmanial activity in human macrophages. *J Immunol* **151**: 3682–3692.
37. Morales-Montor J, Baig S, Kabbani A, Damian RT. (2002) Do interleukin-6 and macrophage-migration inhibitory factor play a role during sex-associated susceptibility in murine cysticercosis? *Parasitol Res* **88**: 901–904.
38. Terrazas LI, Bojalil R, Govezensky T, Larralde C. (1994) A role for 17β-estradiol in immunoendocrine regulation of cysticercosis (*Taenia crassiceps*). *J Parasitol* **80**: 563–568.
39. CDC. (2003) Trichinellosis Surveillance — United States, 1997–2001. In: *CDC Surveillance Summaries (25 July). MMWR* **52**(SS-6): 1–8.
40. Tan THP, Edgerton SAV, Kumari R, *et al.* (2001) Macrophage migration inhibitory factor of the parasitic nematode *Trichinella spiralis. Biochem J* **357**: 373–383.
41. Pennock JL, Behnke JM, Bickle QD, *et al.* (1998) Rapid purification and characterization of L-dopachrome-methyl ester tautomerase (macrophage-migration-inhibitory factor) from *Trichinella spiralis, Trichuris muris* and *Brugia pahangi. Biochem J* **335**: 495–498.
42. Wu Z, Boonmars T, Nagano I, *et al.* (2003) Molecular expression and characterization of a homologue of host cytokine macrophage migration inhibitory factor from *Trichinella* spp. *J Parasitol* **89**: 507–515.
43. CDC Fact Sheet: http://www.dpd.cdc.gov/dpdx/HTML/Filariasis.htm
44. Bhebehani K. (1999) Candidate parasitic diseases. *MMWR Suppl* **48**(SU01): 80–85.
45. Zang X, Taylor P, Wang JM, *et al.* (2002) Homologues of human macrophage migration inhibitory factor from a parasitic nematode. *J Biol Chem* **277**: 44261–44267.
46. Pastrana DV, Raghavan N, Fitzgerald P, *et al.* (1998) Filarial nematode parasites secrete a homologue of the human cytokine macrophage migration inhibitory factor. *Infect Immun* **66**: 5955–5963.

47. Falcone FH, Loke P, Zang X, *et al.* (2001) A *Brugia malayi* homolog of macrophage migration inhibitory factor reveals an important link between macrophages and eosinophil recruitment during nematode infection. *J Immunol* **167**: 5348–5354.
48. Marson AL, Tarr DEK, Scott AL. (2001) Macrophage migration inhibitory factor (*mif*) transcription is significantly elevated in *Caenorhabditis elegans* dauer larvae. *Gene* **278**: 53–62.

9

MIF in the Pathogenesis of Kidney Disease

*Gregory H. Tesch,**,† *Robert C. Atkins**,† *and*
*David J. Nikolic-Paterson**,†

INTRODUCTION

Renal inflammation is a hallmark of most kidney diseases[1] and is associated with a persistent infiltrate of macrophages and T-lymphocytes. The density of the interstitial leukocytic infiltrate in human glomerulonephritis correlates with the degree of renal function at the time of biopsy and can predict outcome.[1] The pathogenic role of leukocytes in kidney diseases has been demonstrated in animal models of experimental glomerulonephritis, where leukocyte depletion[2] and adoptive transfer studies[3,4] have established roles for macrophages and T-cells in causing renal injury. Since MIF can promote both macrophage accumulation and the activation of macrophages and T-cells,[5] it has been postulated that MIF promotes leukocyte-mediated renal injury.[6]

The importance of MIF bioactivity in the pathogenesis of renal disease was first suggested when T lymphocytes from patients with anti-glomerular basement membrane (GBM) disease were shown to release a protein with

*Department of Nephrology, Monash Medical Centre, Clayton, Victoria, Australia.
†Monash University Department of Medicine, Monash Medical Centre, Clayton, Victoria, Australia.
Correspondence to: Dr. Greg Tesch. E-mail: gtesch@hotmail.com.

MIF bioactivity upon exposure to a GBM antigen.[7] More recent studies have demonstrated up-regulation of MIF mRNA and protein expression within the kidney in human and experimental glomerular and tubulointerstitial renal diseases (Tables 1 and 2), in association with leukocyte accumulation. Furthermore, studies blocking MIF activity in an experimental model of crescentic glomerulonephritis have now provided direct evidence that MIF plays a pathological role in leukocyte-mediated renal injury.[8,9]

MIF EXPRESSION IN HUMAN KIDNEY DISEASE

An initial study of normal renal tissue from a carcinoma nephrectomy specimen revealed expression of MIF mRNA by RT-PCR and the presence of MIF protein in tubules and parietal epithelial cells by

Table 1 MIF in Human Renal Disease

Condition	Location of MIF Expression	References
Normal Human Kidney	weak expression in glomerular podocytes, parietal epithelial cells and tubular epithelial cells	10, 11
Non-Proliferative Renal Disease	**Site of increased MIF expression**	**References**
Minimal change disease	same as normal kidney	11
Membranous GN	podocytes	11
IgA nephropathy	tubular epithelial cells	11
Proliferative Renal Disease	**Site of increased MIF expression**	**References**
IgA and non-IgA mesangioproliferative GN	tubular epithelial cells, proliferative mesangial lesions	11
Mesangiocapillary proliferative GN, Focal segmental glomerulosclerosis, Lupus nephritis, Crescentic GN	glomerular endothelial cells, glomerular parietal epithelial cells, podocytes, tubular epithelial cells, macrophages, T-cells, crescents	11
Myeloperoxidase anti-neutrophil cytoplasmic antibody associated glomerulonephritis	mesangial proliferative lesions, cellular crescents	12
Other Renal Diseases	**Site of increased MIF expression**	**References**
Renal allograft rejection	vascular endothelial and smooth muscle cells, tubular epithelial cells, macrophages, T cells	31
Congenital nephrotic syndrome of the Finnish type	tubular epithelial cells, interstitial cells	14

GN = glomerulonephritis.

Table 2 MIF in Experimental Renal Disease

Experimental Model	Site of increased MIF Expression	References
Rat anti-GBM nephritis	glomerular podocytes and parietal epithelial cells, tubular epithelial cells, macrophages, endothelial cells	8, 9, 20
Rat tubulointerstitial nephritis induced by Tamms-Horsfall Protein	glomerular podocytes and parietal epithelial cells, tubular epithelial cells, macrophages	27
Exogenous Hypercholesterolemic (ExHC) rat	podocytes, macrophages, foam cells	24
Rat anti-Thy-1 nephritis	mesangial cells, podocytes	22
Rat renal allograft rejection	podocytes, tubular epithelial cells, macrophages, T-cells	33
Uric acid nephropathy	tubular epithelial cells, macrophages	28
Rat hypertensive nephropathy (subtotal nephrectomy)	tubular epithelial cells, some glomerular and interstitial cells	25
Unilateral ureteric obstruction	cortical tubular epithelial cells	29
db/db mouse diabetic nephropathy	tubular epithelial cells, podocytes	26

immunohistochemistry.[10] A comprehensive study of MIF expression in human glomerulonephritis (GN)[11] confirmed that MIF mRNA and protein was restricted to glomerular and tubular epithelial cells in normal kidney. The study also revealed that in non-proliferative forms of GN such as minimal change disease and membranous GN, there was little change in the pattern of MIF expression from normal. However, in proliferative forms of GN, such as mesangioproliferative GN, lupus nephritis, mesangiocapillary GN and crescentic GN, there was a marked increase in tubular and glomerular epithelial cell MIF expression. In addition, there was de novo expression in glomerular capillary endothelial cells and mesangial cells as well as MIF expression by infiltrating macrophages and T cells. The macrophage and T cell infiltrate was largely restricted to areas of marked up-regulation of MIF in both the glomerulus and the interstitium. In addition, the loss of renal function and the severity of glomerular lesions (crescent formation and sclerosis) correlated with glomerular and tubular MIF expression in human GN.[11] This suggests that MIF may be an important mediator of renal injury in proliferative forms of GN through its ability to cause the accumulation and activation of leukocytes.

Other researchers examining human kidney diseases have reported similar findings. Biopsy specimens obtained from patients with myeloperoxidase anti-neutrophil cytoplasmic antibody (MPO-ANCA)-associated GN show increases in MIF protein expression, macrophage accrual and cellular crescents in proportion to the severity of disease.[12] Monocytes and T cells isolated from patients with IgA nephropathy have been found to produce more MIF protein *in vitro* than those from normal healthy individuals and this production correlated with disease severity and was reduced by steroid therapy.[13] A study of congenital nephrotic syndrome of the Finnish type (NPHS1), a genetic disease characterized by a mutation in the kidney protein nephrin, found that MIF protein levels increased 4-fold in the nephrectomized kidneys of infants in association with severe tubulointerstitial lesions, macrophage accumulation and massive proteinuria.[14] These findings emphasize that renal MIF expression is common to kidney diseases in both children and adults, and appears whenever macrophage infiltration is a feature of the injury.

Serum MIF levels vary widely in normal healthy individuals and are elevated in a number of immune-mediated diseases, but not in human GN.[15,16] However, serum MIF levels are increased in patients with type 2 diabetes,[17] which can lead to renal injury. The mechanism of MIF transportation in circulation, its metabolic turnover and its access to the kidney are not fully understood and therefore it is difficult to determine the relative importance of serum-derived MIF compared to kidney-derived MIF in promoting renal injury. Current data suggests that measurement of serum MIF is unlikely to help in the diagnosis of renal disease.

Urine MIF levels in humans are unchanged in non-proliferative GN but are elevated in proliferative GN,[24] especially in crescentic disease.[15] The concentration of MIF in the urine correlates with MIF protein expression within the kidney, but not with the concentration of MIF in the serum, indicating that urinary MIF is derived from production within the kidney. The urine MIF concentration also correlates with the degree of renal dysfunction, histological damage and leukocytic infiltration.[15,18] Therefore, monitoring urine MIF levels may be a useful non-invasive tool for monitoring the disease activity and/or the response to treatment in patients with crescentic GN.

Recent genetic studies of patients have identified polymorphisms in the promoter of the MIF gene which may influence disease susceptibility (see Chapters 13–15). Having a C allele at position −173 of the MIF gene is known to increase MIF expression through the creation of an activator protein 4 response element in the MIF promoter. The MIF-173 G/C polymorphism and C allele frequency have recently been investigated in pediatric idiopathic nephrotic syndrome (INS).[19] Compared to healthy individuals, patients with INS have a significant increase in the GC phenotype and in the C allele frequency. Furthermore, a 20-fold higher frequency of the CC phenotype was found in INS patients who are steroid resistant. Increases in C allele frequency were also detected in patients with focal segmental glomerulosclerosis and those who have permanent renal function failure. Those with a CC-genotype were found to be at an increased risk of progressing to end-stage renal failure and developing an earlier onset of proteinuria compared to patients with GC or GG phenotype. These results show that the MIF-173 C allele confers an increased risk of susceptibility to INS and plays an important role in determining the glucocorticoid responsiveness.

MIF EXPRESSION IN EXPERIMENTAL KIDNEY DISEASE

MIF mRNA and protein expression is up-regulated in intrinsic renal cells and inflammatory cells in animal models of GN induced by immunological insults, including anti-GBM nephritis and anti-Thy-1 mesangioproliferative nephritis[20–22] (Table 2). Antibody-based blocking studies in animal models have provided evidence that MIF plays a pathological role in GN. The administration of neutralizing anti-MIF antibodies during the induction phase (day 0 to 14) of rat crescentic anti-GBM nephritis significantly inhibited leukocyte infiltration, histological damage, proteinuria and prevented the loss of renal function. These effects were attributed to the ability of MIF blockade to: 1) prevent up-regulation of interleukin (IL)-1β and inducible nitric oxide synthase (iNOS); 2) suppress the induction of the leukocyte adhesion molecules intercellular adhesion molecule (ICAM)-1 and vascular adhesion molecule (VCAM)-1; and 3) inhibit the delayed-type hypersensitivity (DTH) response.[8] However, MIF blockade had no effect on the humoral immune response in this model.[8] To determine whether MIF blockade is feasible in the clinical setting, a study was also performed using MIF antibody treatment as an intervention therapy in established rat crescentic anti-GBM nephritis. Treatment with neutralizing MIF antibodies between day 7 and 21 partially reversed the disease by restoring normal renal function and reducing the inflammation and histological damage compared with untreated animals at day 7.[9] In this study, MIF antibody treatment caused a significant increase in endogenous serum corticosterone levels, which correlated with the reversal of disease parameters.[9] In addition, neutralizing MIF antibody has also been shown to reduce glomerular TGF-β1 expression in a model of IgA nephropathy in BALB/c mice.[23] Thus, blocking MIF activity using neutralizing antibodies or small molecule inhibitors, may be an attractive therapeutic target in the treatment of human proliferative GN.

MIF expression is also elevated in experimental models of renal injury induced by non-immune insults. These include the exogenous hypercholesterolemic (ExHC) rat model of nephrosclerosis, hypertensive nephropathy in subtotal nephrectomized rats, type 2 diabetic nephropathy in db/db mice, and tubulointerstitial nephritis induced by either Tamm-Horsfall proteinuria, hyperuricemia or unilateral ureteric obstruction,[24–29] (Table 2). Similar to proliferative GN, macrophage accumulation is prominent in these models and co-localizes to areas of MIF expression, which occurs mainly at sites of injured renal epithelia.

Recent studies in MIF transgenic and MIF knockout mice have provided additional information about the role of MIF in renal injury. Transgenic mice overexpressing MIF in glomerular podocytes have been generated using murine MIF cDNA driven by a hybrid promoter consisting of a cytomegalovirus enhancer and a β-actin/β-globin promoter.[30] These MIF transgenic mice showed progressive glomerular matrix accumulation accompanied by collagen IV deposition, but had no increase in glomerular cellularity. Proteinuria and declining renal function developed in association with alterations in podocyte morphology (cell flattening, contracted foot processes, villous transformation) and a decrease in synaptopodin expression. Glomerular leukocytes were unaltered. These findings suggest that podocyte MIF production may injure podocytes themselves and either directly or indirectly promote glomerular matrix production.

Mice which are genetically deficient in MIF develop normally and have morphologically normal kidneys and normal renal function.[29] Studies in these MIF knockout mice have demonstrated that although MIF expression is increased in obstructed kidneys, it is not required for the development of renal inflammation or renal fibrosis following unilateral ureteric obstruction (UUO).[29] The redundancy of MIF in this model indicates that strategies targeting MIF are unlikely to benefit patients with nephropathy resulting from ureteric obstruction. This is in contrast to the role of MIF in rat crescentic anti-GBM nephritis, where it appears to play a pivotal role in the development and progression of renal injury following an immunological insult.[8]

MIF IN RENAL ALLOGRAFT TRANSPLANTATION

Renal allograft rejection is a T cell-dependent process in which the graft is infiltrated with T cells and macrophages. T cells cause injury directly through cytotoxic mechanisms and indirectly via the recruitment and activation of macrophages.[31] In humans, MIF mRNA and protein expression is essentially unchanged in non-rejecting transplant kidneys. However, in acute and chronic renal allograft rejection, MIF mRNA and protein expression is up-regulated in tubular epithelial cells, vascular endothelial cells and podocytes as well as being expressed by smooth muscle cells and infiltrating T lymphocytes and macrophages. As in GN, inflammatory cells are largely limited to areas of marked MIF up-regulation, and tubular MIF up-regulation correlates with interstitial macrophage and T cell infiltrate, severity of graft rejection and loss of renal function.[32] In contrast, kidneys damaged by cyclosporine toxicity show a pattern of MIF protein expression similar to that seen in normal human kidneys.[33]

Serum MIF levels are not elevated in renal transplant patients as compared with normal healthy volunteers, regardless of whether they are stable renal transplant recipients, patients with acute rejection or patients with cyclosporine nephrotoxicity.[33] In contrast, immediately post-transplantation, urinary MIF levels increase, and fall in parallel with serum creatinine.[33] The mean urine MIF levels in a group of stable transplant patients are not different to a group of transplant patients with acute rejection or cyclosporine toxicity. However, measurements of serial urine MIF concentration in an individual may be useful in monitoring acute rejection and as a discriminator from cyclosporine nephrotoxicity.[33] This is due to the fact that a patient with an acute rejection episode may have an elevation in their urinary MIF level prior to the rejection episode or the urine MIF level may fail to fall post-transplantation. This pattern is not seen in cyclosporine nephrotoxicity.

Studies in a rat model of acute renal allograft rejection found similar patterns of increased MIF expression to those observed in rejected human transplant kidneys. Furthermore, the MIF expression seen in the rat renal allografts did not appear in rat renal isografts, confirming it is the rejection process which up-regulates MIF and not stress from the surgical procedure.[34] This finding suggested that MIF may be an important mediator in allograft rejection.

Recent studies with MIF gene knockout mice have examined the role of MIF in acute renal allograft rejection. Transplantation of $MIF^{+/+}$ or $MIF^{-/-}$ C57BL/6 kidneys into BALB/c recipients demonstrated similar leukocytic infiltrate, graft dysfunction and histological damage, suggesting that the acute rejection was not dependent on local MIF production.[35] BALB/c kidneys transplanted into $MIF^{+/+}$ or $MIF^{-/-}$ mice showed similar parameters of rejection, indicating that systemic MIF was also not required for acute rejection.[35] Therefore, neither local nor systemic MIF are necessary for the rejection of fully mismatched renal allografts.

MIF PRODUCTION AND ACTIVITY IN CULTURED RENAL CELLS: IMPLICATIONS FOR HUMAN KIDNEY DISEASE

In renal diseases, infiltrating macrophages and T cells express MIF but the major source of MIF is intrinsic renal cells. The cytokines interferon (IFN)-γ and tumor necrosis factor (TNF)-α, as well as endotoxin, are strong inducers of MIF synthesis and secretion by macrophages.[36] Similarly, TNF-α increases MIF mRNA and protein in tubular epithelial cells *in vivo*,[21] while *in vitro*, IFN-γ increases MIF mRNA and protein in mesangial cells.[22] A recent study has also shown that IFN-γ causes the rapid secretion of MIF from renal tubular epithelial cells,[37] which may be an important mechanism for inhibiting interstitial macrophage migration. It is feasible that these cytokines could also be involved in MIF production in human kidney diseases. Levels of TNF-α and IFN-γ are increased systemically in patients with diseases that exhibit elevated kidney MIF expression in association with renal injury including diabetes, SLE and IgA immune complex disease.[38–40] In addition, polymorphisms in IFN-γ and TNF-α are associated with increased susceptibility to IgA nephropathy.[41,42] Elevated expression of these cytokines is also detected in the kidneys of animals with experimental GN.[43,44] Consequently, TNF-α and IFN-γ may be important regulatory mechanisms of MIF expression and MIF-mediated renal injury in human kidney disease.

Other biologically active substances such as transferrin, a component of proteinuria, as well as polymeric and monomeric IgA have been shown to directly up-regulate MIF expression in cultured proximal tubular epithelial cells and mesangial cells, respectively.[45,46] In addition, a recent study has demonstrated that IgA-stimulated human mesangial cells induce tubular MIF synthesis and secretion through the release of TNF-α,[47] indicating that IgA can also indirectly promote tubular MIF production. These *in vitro* studies together provide potential pathogenic mechanisms by which proteinuria and IgA deposits could lead to MIF-mediated renal injury in proteinuric renal diseases and IgA nephropathy.

Recently, it has been shown that angiotensin II, an important molecule in progressive renal injury, also induces MIF mRNA production and protein secretion in cultured tubular epithelial cells.[25] This finding suggests that angiotensin II may promote leukocyte accumulation and activation via induction of MIF synthesis and secretion. Furthermore, it demonstrates that reduction in MIF expression represents a novel mechanism which, in part, may account for the beneficial effects of blockade of the renin-angiotensin system in reducing renal injury.

In vitro studies have also shown that exogenous MIF can stimulate responses in cells found in the kidney. Recombinant MIF induces mesangial cell secretion of TGF-β1 in a time and dose-dependent fashion.[23] MIF also stimulates fibroblast proliferation in association with activation of the p44/p42 extracellular signal-regulated kinase (ERK) signaling pathway.[48] Therefore, MIF may play a role in the development of fibrosis and sclerosis in kidney disease. However, these potential mechanisms of MIF-induced renal fibrosis are not evident in the obstructed kidney where TGF-β1-dependent renal fibrosis is unaltered by the absence of MIF.[29]

MIF-RECEPTOR SIGNALING: IMPLICATIONS FOR MIF ACTIVITY IN KIDNEY DISEASE

Recently, it has been reported that the extracellular domain of CD74, the MHC class II invariant chain, is a high affinity MIF receptor which facilitates MIF-induced activation of the ERK signaling pathway, cell proliferation and PGE2 production.[49] In normal kidneys, MHC class II is expressed on a small subpopulation of glomerular and interstitial cells and on the vascular endothelium.[50] In patients with primary and systemic renal diseases, expression of MHC class II is markedly increased on glomerular and interstitial cells and is also detected on many tubular cells.[50,51] Consequently, there are many renal cells in diseased kidneys that could respond to MIF via CD74 signaling. Furthermore, a recent study has shown that tubular MHC class II expression correlates strongly with interstitial inflammation and declining renal function in progressive kidney diseases, suggesting a pathological role for MHC class II and CD74 in renal injury.[51]

There is also growing evidence that MIF signaling requires interaction between CD74 and a second transmembrane protein, CD44.[52] Examination of renal biopsies by immunostaining has shown that patients with glomerular and interstitial kidney diseases exhibit increased expression of CD44 on glomerular, tubular and interstitial cells which correlates strongly with MHC class II expression.[51] *In vitro* studies have identified CD44 and MHC class II on macrophages,[53] mesangial cells[54,55] and tubular epithelial cells,[56,57] and each of these cell types can both produce MIF and respond to it. Consequently, MIF may act in an autocrine or paracrine manner in the diseased kidney. Furthermore, since IFN-γ is a potent inducer of MIF, CD44 and MHC class II in renal cells,[22,37,54–57] MIF-mediated renal injury may be highly dependent on kidney expression of IFN-γ.

FUTURE DIRECTIONS OF MIF RESEARCH IN KIDNEY DISEASE

There is still much to be learned about the role of MIF in kidney disease. Although, a number of *in vitro* studies have examined the factors which induce MIF production by renal cells, very few have identified the effects of MIF on intrinsic renal cells. In particular, renal glomerular and tubular epithelial cells are major sources of MIF in the injured kidney and it is unknown whether this MIF production has an autocrine effect that either promotes damage or could protect these cells from damage. MIF may induce proinflammatory, proliferative or profibrotic events in resident renal cells by regulating cell signaling events. A number of renal cell functions (cytokine production, proliferation, epithelial to mesenchymal transdifferentiation) are known to be dependent on ERK signaling[58–60] and, since MIF can induce ERK activity in other cell types (macrophages, B-cells, embryonic fibroblasts cells),[48,49] it is possible that MIF may also promote ERK-mediated effects in renal cells.

Current evidence implicates MIF involvement in the pathogenesis of lupus nephritis. TNF and IFN-γ are known to be critical mediators of disease in animal models of lupus nephritis and these cytokines are the most potent inducers of MIF production by infiltrating and resident renal cells.[22,37] In addition, immunostaining studies have shown that MIF expression is markedly increased in the kidneys of lupus patients.[11] Therefore, it will be important for future studies to examine the effects of MIF blockade or MIF gene deletion in experimental models of lupus nephritis and determine the functional importance of MIF in this disease.

Diabetic nephropathy is now the single major cause of end-stage renal failure in the USA and Europe and it will be important to establish whether MIF has a significant role in this disease. Recently, it has been shown that renal macrophage accumulation contributes to the pathogenesis of diabetic nephropathy and is associated with elevated kidney MIF expression.[61,26] In diabetic kidneys, macrophages have increased iNOS expression and interstitial fibroblast proliferation is detected.[26,62] Both of these events are consistent with the known activities of MIF.[48,63] In addition, *in vitro* studies have shown that advanced glycation end products (AGEs) stimulate transdifferentiation of tubular epithelial cells to myofibroblasts via an ERK-dependent mechanism which promotes interstitial fibrosis.[60] It is possible that tubular MIF activity may also help mediate this event. Indeed, there is little known about the mechanisms of MIF production in diabetic kidneys and this is an area that warrants further investigation. Future studies examining MIF blockade or deficiency in animal models of diabetes will help to determine the functional role of MIF in diabetic nephropathy.

Current evidence from human and animal studies suggests that MIF plays an important role in the development and progression of proliferative forms of GN. The potential involvement of MIF in non-proliferative forms of kidney disease is less clear and will require further investigation. Future studies aimed at developing a better understanding of the mechanisms of MIF-mediated renal injury will need to examine the relative importance of MIF-induced CD74/CD44 signaling, MIF enzymatic activity, and MIF production by intrinsic versus infiltrating kidney cells in the development of kidney disease. These studies will help to identify novel therapeutic strategies for preventing MIF-mediated injury in human kidney disease.

REFERENCES

1. Nikolic-Paterson DJ, Lan HY, Atkins RC. (1997) Macrophages in immune renal injury. In Neilson EG, Couser WG (eds). *Immunologic Renal Disease*, Raven Press, pp. 575–592.
2. Schreiner GF, Cotran RS, Pardo V, Unanue ER. (1978) A mononuclear cell component in experimental immunological glomerulonephritis. *J Exp Med* **147**: 369–384.
3. Ikezumi Y, Hurst LA, Masaki T, *et al.* (2003) Adoptive transfer studies demonstrate that macrophages can induce proteinuria and mesangial cell proliferation. *Kidney Int* **63**: 83–95.
4. Wu J, Hicks J, Borillo J, *et al.* (2002) CD4(+) T cells specific to a glomerular basement membrane antigen mediate glomerulonephritis. *J Clin Invest* **109**: 517–524.
5. Bernhagen J, Calandra T, Bucala R. (1998) Regulation of the immune response by macrophage migration inhibitory factor: biological and structural features. *J Mol Med* **76**: 151–161.
6. Atkins RC, Nikolic-Paterson DJ, Bucala R, Lan HY. (1998) Macrophage migration inhibitory factor (MIF)-potential perspectives for immune-intervention in renal disease. *Nephrol Dial Transplant* **13**: 2719–2720.
7. Rocklin RE, Lewis EJ, David JR. (1970) *In vitro* evidence for cellular hypersensitivity to glomerular-basement-membrane antigens in human glomerulonephritis. *N Engl J Med* **283**: 497–501.
8. Lan HY, Bacher M, Yang N, *et al.* (1997) The pathogenic role of macrophage migration inhibitory factor in immunologically induced kidney disease in the rat. *J Exp Med* **185**: 1455–1465.
9. Yang N, Nikolic-Paterson DJ, Ng YY, *et al.* (1998) Reversal of established rat crescentic glomerulonephritis by blockade of macrophage migration

inhibitory factor (MIF): potential role of MIF in regulating glucocorticoid production. *Mol Med* **4**: 413–424.
10. Imamura K, Nishihira J, Suzuki M, *et al.* (1996) Identification and immunohistochemical localization of macrophage migration inhibitory factor in human kidney. *Biochem Mol Biol Int* **40**: 1233–1242.
11. Lan HY, Yang N, Nikolic-Paterson DJ, *et al.* (2000) Expression of macrophage migration inhibitory factor in human glomerulonephritis. *Kidney Int* **57**: 499–509.
12. Taniguchi Y, Yorioka N, Yamashita K, *et al.* (2000) Expression of macrophage migration inhibitory factor in patients with myeloperoxidase antineutrophil cytoplasmic antibody-associated glomerulonephritis. *Clin Nephrol* **54**: 147–150.
13. Matsumoto K, Kanmatsuse K. (2001) Increased production of macrophage migration inhibitory factor by T cells in patients with IgA nephropathy. *Am J Nephrol* **21**: 455–464.
14. Kuusniemi AM, Lapatto R, Holmberg C, *et al.* (2005) Kidneys with heavy proteinuria show fibrosis, inflammation and oxidative stress, but no tubular phenotypic change. *Kidney Int* **68**: 121–132.
15. Brown FG, Nikolic-Paterson DJ, Hill PA, *et al.* (2002) Urine macrophage migration inhibitory factor reflects the severity of renal injury in human glomerulonephritis. *J Am Soc Nephrol* **13**: S7–S13.
16. Honda K, Nishihira J, Nitta K, *et al.* (2000) Serum levels of macrophage migration inhibitory factor in various types of glomerulonephritis. *Nephron* **86**: 91–92.
17. Herder C, Kolb H, Koenig W, *et al.* (2006) Association of systemic concentrations of macrophage migration inhibitory factor with impaired glucose tolerance and type 2 diabetes: results from the Cooperative Health Research in the Region of Augsburg, Survey 4 (KORA S4). *Diabetes Care* **29**: 368–371.
18. Matsumoto K, Kanmatsuse K. (2002) Urinary levels of macrophage migration inhibitory factor in patients with IgA nephropathy. *Nephron* **92**: 309–315.
19. Berdeli A, Mir S, Ozkayin N, *et al.* (2005) Association of macrophage migration inhibitory factor −173C allele polymorphism with steroid resistance in children with nephrotic syndrome. *Pediatr Nephrol* **20**: 1566–1571.
20. Lan HY, Mu W, Yang N, *et al.* (1996) De Novo renal expression of macrophage migration inhibitory factor during the development of rat crescentic glomerulonephritis. *Am J Pathol* **149**: 1119–1127.
21. Lan HY, Yang N, Metz C, *et al.* (1997) TNF-alpha up-regulates renal MIF expression in rat crescentic glomerulonephritis. *Mol Med* **3**: 136–144.
22. Tesch GH, Nikolic-Paterson DJ, Metz CN, *et al.* (1998) Rat mesangial cells express macrophage migration inhibitory factor *in vitro* and *in vivo*. *J Am Soc Nephrol* **9**: 417–424.

23. Leung JC, Chan LY, Tsang AW, *et al.* (2004) Anti-macrophage migration inhibitory factor reduces transforming growth factor-β1 expression in experimental IgA nephropathy. *Nephrol Dial Transplant* **19**: 1976–1985.
24. Miyazaki K, Isbel NM, Lan HY, *et al.* (1997) Up-regulation of macrophage colony-stimulating factor (M-CSF) and migration inhibitory factor (MIF) expression and monocyte recruitment during lipid-induced glomerular injury in the exogenous hypercholesterolaemic (ExHC) rat. *Clin Exp Immunol* **108**: 318–323.
25. Rice EK, Tesch GH, Cao Z, *et al.* (2003) Induction of MIF synthesis and secretion by tubular epithelial cells: a novel action of angiotensin II. *Kidney Int* **63**: 1265–1275.
26. Chow F, Ozols E, Nikolic-Paterson DJ, *et al.* (2004) Macrophages in mouse type 2 diabetic nephropathy: correlation with diabetic state and progressive renal injury. *Kidney Int* **65**: 116–128.
27. Song Q, Nikolic-Paterson DJ, Atkins RC, *et al.* (1996) Delayed-type hypersensitivity mediates Bowman's capsule rupture in Tamm-Horsfall protein-induced tubulointerstitial nephritis in the rat. *Nephrology* **2**: 417–427.
28. Kim YG, Huang XR, Suga S, *et al.* (2000) Involvement of macrophage migration inhibitory factor (MIF) in experimental uric acid nephropathy. *Mol Med* **6**: 837–48.
29. Rice EK, Nikolic-Paterson DJ, David JR, *et al.* (2004) Macrophage accumulation and renal fibrosis are independent of MIF in mouse obstructive nephropathy. *Nephrology* **9**: 278–287.
30. Sasaki S, Nishihira J, Ishibashi T, *et al.* (2004) Transgene of MIF induces podocyte injury and progressive mesangial sclerosis in the mouse kidney. *Kidney Int* **65**: 469–481.
31. Wyburn KR, Jose MD, Wu H, *et al.* (2005) The role of macrophages in allograft rejection. *Transplantation* **80**: 1641–1647.
32. Lan HY, Yang N, Brown FG, *et al.* (1998) Macrophage migration inhibitory factor expression in human renal allograft rejection. *Transplantation* **66**: 1465–1471.
33. Brown FG, Nikolic-Paterson DJ, Chadban SJ, *et al.* (2001) Urine macrophage migration inhibitory factor concentrations as a diagnostic tool in human renal allograft rejection. *Transplantation* **71**: 1777–1783.
34. Brown FG, Nikolic-Paterson DJ, Metz C, *et al.* (1999) Up-regulation of macrophage migration inhibitory factor in acute renal allograft rejection in the rat. *Clin Exp Immunol* **118**: 329–336.
35. Jose MD, David JR, Atkins RC, Chadban SJ. (2003) Blockade of macrophage migration inhibitory factor does not prevent acute renal allograft rejection. *Am J Transplant* **3**: 1099–1106.
36. Calandra T, Bernhagen J, Mitchell RA, Bucala R. (1994) The macrophage is an important and previously unrecognized source of macrophage migration inhibitory factor. *J Exp Med* **179**: 1895–1902.

37. Rice EK, Nikolic-Paterson DJ, Hill PA, *et al.* (2003) IFN-g induces MIF Synthesis and Secretion by Renal Parenchymal Cells. *Nephrology* **8**: 156–161.
38. Ozer G, Teker Z, Cetiner S, *et al.* (2003) Serum IL-1, IL-2, TNFalpha, INFgamma levels of patients with type 1 diabetes mellitus and their siblings. *J Pediatr Endocrinol Metab* **16**: 203–210.
39. Robak E, Smolewski P, Wozniacka A, *et al.* (2004) Relationships between peripheral blood dendritic cells and cytokines involved in the pathogenesis of systemic lupus eryhematosus. *Eur Cytokine Netw* **15**: 222–230.
40. Yano N, Endoh M, Nomoto Y, *et al.* (1997) Phenotypic characterization of cytokine expression in patients with IgA nephropathy. *J Clin Immunol* **17**: 396–403.
41. Masutani K, Miyake K, Nakashima H, *et al.* (2003) Impact of interferon-gamma and interleukin-4 gene polymorphisms on the development and progression of IgA nephropathy in Japanese patients. *Am J Kidney Dis* **41**: 371–379.
42. Tuglular S, Berthoux P, Berthoux F. (2003) Polymorphisms of tumor necrosis factor alpha gene at position −308 and TNFd microsatellite in primary IgA nephropathy. *Nephrol Dial Transplant* **18**: 724–731.
43. Fan X, Wuthrich RP. (1997) Up regulation of lymphoid and renal interferon-gamma mRNA in autoimmune MRL-Fas(lpr) mice with lupus nephritis. *Inflammation* **21**: 105–112.
44. Kim JH, Ha IS, Hwang CI, *et al.* (2004) Gene expression profiling of anti-GBM glomerulonephritits model: the role of NF-kappaB in immune complex kidney disease. *Kidney Int* **66**: 1826–1837.
45. Tang S, Leung JC, Tsang AW, *et al.* (2002) Transferrin up-regulates chemokine synthesis by human proximal tubular epithelial cells: implication on mechanism of tubuloglomerular communication in glomerulopathic proteinuria. *Kidney Int* **61**: 1655–1665.
46. Leung JC, Tang SC, Chan LY, *et al.* (2003) Polymeric IgA increases the synthesis of macrophage migration inhibitory factor by human mesangial cells in IgA nephropathy. *Nephrol Dial Transplant* **18**: 36–45.
47. Chan LY, Leung JC, Tsang AW, *et al.* (2005) Activation of tubular epithelial cells by mesangial-derived TNF-alpha: glomerulotubular communication in IgA nephropathy. *Kidney Int* **67**: 602–612.
48. Mitchell RA, Metz CN, Peng T, Bucala R. (1999) Sustained mitogen-activated protein kinase (MAPK) and cytoplasmic phospholiase A2 activation by macrophage migration inhibitory factor (MIF). Regulatory role in cell proliferation and glucocorticoid action. *J Biol Chem* **274**: 18100–18106.
49. Leng L, Metz CN, Fang Y, *et al.* (2003) MIF signal transduction initiated by binding to CD74. *J Exp Med* **197**: 1467–1476.
50. Muller CA, Markovic-Lipkovski J, Risler T, *et al.* (1989) Expression of HLA-DQ, -DR, and -DP antigens in normal kidney and glomerulonephritis. *Kidney Int* **35**: 116–124.

51. Kirimca F, Sarioglu S, Camsari T, Kavucku S. (2001) Expression of CD44 and major histocompatibility complex class II antigens correlate with renal scarring in primary and systemic renal diseases. *Scand J Urol Nephrol* **35**: 509–514.
52. Leng L, Bucala R. (2006) Insight into the biology of macrophage migration inhibitory factor (MIF) revealed by cloning of its cell surface receptor. *Cell Res* **16**: 162–168.
53. Natale VA, McCullough KC. (1998) Macrophage culture: influence of species-specific incubation temperature. *J Immunol Methods* **214**: 165–174.
54. Nikolic-Paterson DJ, Jun Z, Tesch GH, *et al.* (1996) De novo CD44 expression by proliferating mesangial cells in rat anti-Thy-1 nephritis. *J Am Soc Nephrol* **7**: 1006–1014.
55. Nikolic-Paterson DJ, Tesch GH, Lan HY, *et al.* (1995) Deoxyspergualin inhibits mesangial cell proliferation and major histocompatibility complex class II expression. *J Am Soc Nephrol* **5**: 1895–1902.
56. Takazoe K, Foti R, Tesch GH, *et al.* (2000) Up-regulation of the tumor-associated marker CD44V6 in experimental kidney disease. *Clin Exp Immunol* **121**: 523–532.
57. Jevnikar AM, Wuthrich RP, Takei F, *et al.* (1990) Differing regulation and function of ICAM-1 and class II antigens on renal tubular cells. *Kidney Int* **38**: 417–425.
58. Leonard M, Ryan MP, Watson AJ, *et al.* (1999) Role of MAP kinase pathways in mediating IL-6 production in human primary mesangial and proximal tubular cells. *Kidney Int* **56**: 1366–1377.
59. Bokemeyer D, Panek D, Kramer HJ, *et al.* (2002) *In vivo* identification of the mitogen-activated protein kinase cascade as a central pathogenic pathway in experimental mesangioproliferative glomerulonephritis. *J Am Soc Nephrol* **13**: 1473–1480.
60. Li JH, Wang W, Huang XR, *et al.* (2004) Advanced glycation end products induce tubular epithelial-myofibroblast transition through RAGE-ERK1/2 MAP kinase signaling pathway. *Am J Pathol* **164**: 1389–1397.
61. Chow FY, Nikolic-Paterson DJ, Ozols E, *et al.* (2006) Monocyte chemoattractant protein-1 promotes diabetic renal injury in streptozotocin-treated mice. *Kidney Int* **69**: 73–80.
62. Chow FY, Nikolic-Paterson DJ, Atkins RC, Tesch GH. (2004) Macrophages in streptozotocin-induced diabetic nephropathy: potential role in renal fibrosis. *Nephrol Dial Transplant* **19**: 2987–2996.
63. Juttner S, Bernhagen J, Metz CN, *et al.* (1998) Migration inhibitory factor induces killing of Leishmania major by macrophages: dependence on reactive nitrogen intermediates and endogenous TNF-alpha. *J Immunol* **161**: 2383–2390.

10

MIF and CNS Autoimmune Disease

Claudia M. Denkinger and Thomas G. Forsthuber†*

Part I. AUTOIMMUNE CNS DISEASE

MIF has been established as a key molecule in the defence against infectious organisms. However, its potency and pleiotropic effects on the immune system also frequently place it at the center of undesired inflammatory conditions that lead to immune-mediated tissue damage and disease in humans.[1] Importantly, MIF is produced by a number of immune and non-immune cells implicated in CNS autoimmune diseases. Its broad proinflammatory actions and ability to counter-regulate immunosuppressive glucocorticoid effects therefore make it a prime suspect for the pathogenesis of CNS autoimmunity and a potential therapeutic target. In the following, the authors will outline the salient immunologic mechanisms believed to propagate CNS autoimmune diseases and discuss the evidence supporting a role for MIF in the disease process.

The central nervous system (CNS) is of critical importance for members of all mammalian species. For this reason, the CNS is protected by multiple mechanisms, including physical and chemical barriers, such as the blood brain barrier (BBB). While the CNS needs to be protected from

*Department of Medicine, Beth Israel Deaconess Medical Center, 330 Brookline Avenue, Boston, MA 02215.
†Department of Biology, University of Texas at San Antonio, 6900 North Loop 1604 West, San Antonio, TX 78249.

microbial infections, the extent of damage caused by immune activation in this tissue must be restricted as much as possible since the potential for "self regeneration" of the CNS is limited. Therefore, a number of regulatory mechanisms are in place in the CNS to control the activation and function of immune cells. For example, high expression levels of FasL induces apoptosis of activated T cells via the Fas-mediated death pathway and limits the expansion and function of these cells in the CNS.[2] Furthermore, expression of inhibitory molecules in the CNS, such as PD-L1, down-modulate immune cell activation and thereby also contribute to the prevention of excessive tissue damage.[3]

Finally, T lymphocytes are "educated" during their development in the thymus to be tolerant to self antigens, including CNS proteins. T cell tolerance to self antigens is furthermore enforced by peripheral tolerance mechanisms, and, in addition, the access of immune cells to the CNS is restricted. However, despite the strict control of immune activation in the CNS, the immune system sometimes mounts an erroneous attack on the CNS in some individuals and causes autoimmune disease pathology.[4,5]

Disease presentation primarily depends on the nature and site of the immune-mediated damage, resulting in a wide range of symptoms. Lesions in CNS specific autoimmune diseases can be confined to certain areas of the brain, such as the glia of the optical nerve in some patients with multiple sclerosis (MS), especially in the early stage of the disease. Later during the disease, lesions can also be distributed randomly throughout the brain and spinal cord. While Schilder's, Marburg's and Balo's diseases are considered separate disease entities from MS, they share a common CNS lesion with MS, namely the pathognomonic, sharp-edged plaque (Fig. 1).[6] ADEM and Devic's disease are also distinct from MS. They are monophasic demyelinating diseases that present themselves in childhood and appear to be mostly reversible. A pathophysiological continuum between ADEM and MS is often suggested, despite different pathological characteristics, and a clearer understanding of the monophasic immuno-dysregulative response in ADEM would be important to facilitate treatment of MS.[6]

Some systemic autoimmune diseases, such as systemic lupus erythematosus (SLE), or mixed connective tissue disorder, can also affect the CNS. In these diseases, CNS lesions are similar to lesions of MS, and in

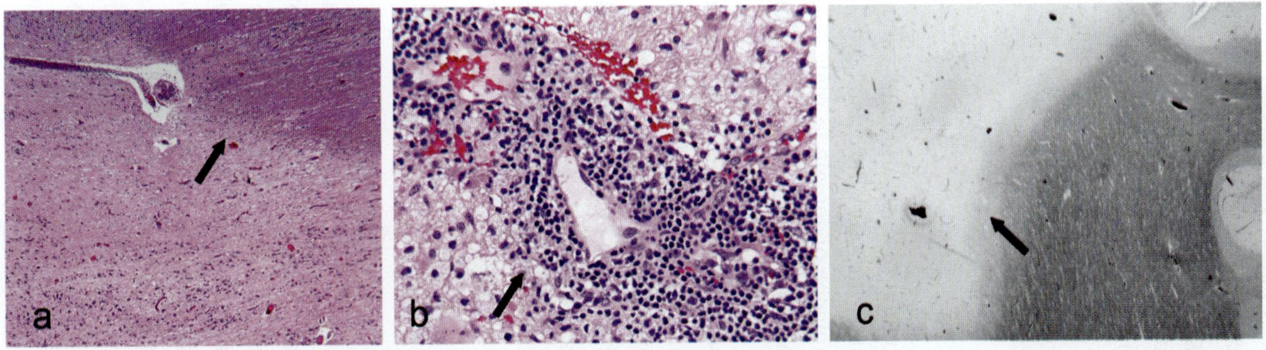

Fig. 1 The pathological characteristic of MS is the demyelinized lesion, which is characterized by a well defined hypocellular area with preserved, but demyelinized axons and scaring of astrocytes (**a**, **c**, *arrows*). Several medium-sized vessels can often be found in the close surrounding. In the peripheral area of the lesion, depending on the stage of the disease, various inflammatory cells, usually lymphocytes and macrophages, can be found (**a**, **b**, *arrows*). Regions predisposed to lesions are the optic nerve, periventricular white matter, the brain stem, cerebellum, and the white matter of the spinal cord. **(a)** MS lesion, H&E, low magnification; **(b)** MS lesion with inflammatory cells, H&E, higher magnification; **(c)** MS lesion with demyelination, myelin staining (pictures courtesy of Mark L. Cohen, M.D., Case Western Reserve University, Cleveland).

fact, it is clinically often unclear whether the patients do have SLE and concomitant MS, or whether the disease symptoms manifest an "overlap syndrome."

MS is the most frequent autoimmune disease affecting the CNS. This chapter will argue for the potential role of MIF in the pathogenesis of MS and as a therapeutic target.

MULTIPLE SCLEROSIS

MS is the most common neurological disease affecting young and middle aged people in North America and Europe. It affects about 1 out of 1000 people in the US, with a predominance of female to male patients of 2:1.[4] The disease is characterized by a chronic, relapsing — remitting course in 80% of patients.

The clinical picture of the disease is very heterogeneous and definitive diagnosis was difficult until the introduction of MRI scanning into clinical practice. Due to the heterogeneity of the disease, a number of clinicians and pathologists favor the view that several pathogenic mechanisms lead to a common disease picture that is lumped under the term MS.[6] This view must be taken into account and be addressed therapeutically in order to successfully treat the disease.[4]

The Pathogenesis of MS

MS is a chronic inflammatory and demyelinating disease of the central nervous system (CNS). The cause of MS is unknown. However, substantial evidence suggests that, for still unknown reasons, the immune system launches an erroneous attack on the CNS, which in turn leads to demyelination and axonal degeneration. The salient pathologic features are demyelination (Fig. 1), axonal loss, gliosis and varying degrees of inflammation consisting of CD4+ and CD8+ T cells, macrophages, B cells and plasma cells.[4,5]

The mechanism underlying autoimmune-mediated damage in the CNS

What triggers the autoimmune attack on the CNS in humans remains unresolved. What has become clear, from the work of many investigators, is that susceptibility to MS and other autoimmune diseases is linked to the genetic background of an individual and to environmental factors that remain undefined.[4]

Among the environmental factors implicated in the pathogenesis of MS are infectious diseases. An elegant hypothesis, termed molecular mimicry,[7] proposes that some infectious organisms could possess certain structural and/or sequence homologies with self antigens. Upon infection with these microorganisms, the ensuing immune response designed at clearing the pathogen could show (undesired) cross-reactivity to self antigens and lead to tissue damage. Subsequently, the autoimmune response becomes chronic and autoimmune disease develops. Viral and bacterial epitopes which activate myelin basic protein (MBP) reactive T cells have been identified.[8] Among the viruses, the herpesvirus group, specifically herpes virus 6 (HHV6), has attracted significant attention.[9] Among the bacteria, *Chlamydia pneumoniae* was found to be more prevalent in MS patients in several studies.[10] Examples of molecular mimicry have been established in animal models, such as induction of EAE subsequent to infection with Theiler's murine encephalomyelitis virus.[11] Most likely, molecular mimicry is but one of several mechanisms that lead to the induction of autoimmune disease. Failure of tolerance to self antigens may be another mechanism, as evidenced by autoimmune pathology in humans with polyglandular autoimmune syndrome due to mutations in an autoimmune regulator gene (AIRE).[12] Furthermore, a number of protective mechanisms involved in preventing damage, for example, by regulating the activity of inflammatory cells or the effects of oxidative stress during inflammation, have been implicated in the pathogenesis of MS.[13,14] Failure of the latter may lead, via a self amplificatory loop, to enhanced release of CNS antigens and enhanced activation of autoreactive lymphocytes.[15]

Overall, it is clear that the destructive power of the immune system needs to be carefully harnessed in the CNS. In the following, the authors will discuss the major contributors believed to play a role in the failure to harness this destructive power.

The role of immune cells in MS

Patients with MS, as well as their healthy counterparts, frequently show T and B cell immune-reactivity directed against different myelin and non-myelin antigens. Specifically, a number of investigators have shown that T cells isolated from MS patients proliferate and produce cytokines *in vitro* upon stimulation with antigens specifically found in the CNS. Among these antigens are myelin basic protein (MBP), myelin-associated glycoprotein (MAG), myelin oligodendrocyte glycoprotein (MOG), proteolipid protein (PLP), $\alpha\beta$-crystallin and S-100-protein.[16] Importantly, injection of these myelin antigens into rodents and primates induces specific T cell immunity and results in a neurological disease, termed experimental autoimmune encephalomyelitis (EAE),[17] which has a number of similarities to MS.[17] EAE can be transferred between animals by transferring neuroantigen-specific T cells, thus establishing the link between T cells and EAE and possibly MS.[17]

Historically, $CD4^+$ helper T cells have been suspected of playing an important role in EAE/MS, and they have been at the center of research for the last decades. Recently, however, the role of $CD8^+$ T cells in EAE and MS has also been appreciated. Sun and Goverman and colleagues independently showed the importance of CD8+ T-cells in the pathogenesis of EAE in two different models using C57BL/6 and CH3shiv mice.[18,19]

Other cells involved could be B cells producing anti-myelin antibodies.[5] and natural killer T cells (NKT cells), which have been shown to contribute to the pathogenesis of MS. NKT cells spontaneously produce cytokines, such as IFN-γ and IL-4, and biasing these cells towards production of IL-4 over IFN-γ results in the suppression of EAE.[20] Similarly, mast cells.[21] and the complement system, may well have a role in MS/EAE.[22]

What are the mechanisms that promote undesirable autoimmune activation in the CNS? Clearly, migration of immune cells to the CNS is important, as may be their effector functions at this site, such as cytokine production, cytotoxicity and their cell survival. Furthermore, the mechanisms that regulate the activation of inflammatory cells in the CNS are important.

It is interesting that autoreactive T cells found in the peripheral blood of MS patients occur in greater cell numbers, are easier to activate, and are less dependent on costimulation via CD80/CD86-molecules.[23] as compared with their counterparts in healthy individuals. The significance of the latter observation is that costimulatory molecules, such as CD80/86, CD40 or Ox40, on antigen presenting cells (APCs), are critically required for the activation of myelin reactive T cells.[22] A decreased dependency on these interactions may allow these cells to avoid a critical regulatory checkpoint on the way to MS.[24]

Recently, additional molecules have been identified on T cells, APCs and non-immune cells that could play an important role in fine-balancing the activation and inhibition of T cells during immune function. For example, the programmed death (PD)-1 family of molecules exerts negative regulatory functions on T lymphocytes and has been implicated in the pathogenesis of EAE. Specifically, *in vivo* blockade of PD-1 resulted in increased antigen-specific T cell expansion, activation, cytokine production and enhanced EAE.[3]

"Epitope spreading" is another pathogenetic mechanism in MS and EAE.[15,25] Epitope spreading (also called determinant spreading) primarily plays a role in the chronic and relapsing-remitting phase of the disease.[15,25] Thus, during the initial stages of EAE the immune response is directed against the immunodominant epitope of myelin antigens, while the subsequent T cell autoimmune response "snow balls" and the T cells become sensitized against multiple epitopes of myelin antigens. As a consequence, a wide variety of T cells with differing T cell receptor (TCR) repertoires are engaged, which renders specific targeting of individual pathogenic T cell clones not feasible. Epitope spreading may be due to self antigens released from the tissues during the initial destructive inflammatory reaction, which then continues to supply autoantigens and to promote immune activation.[26,27]

Finally, autoantibodies produced by B cells may contribute to the pathogenesis of MS by binding to and destroying the myelin sheath with the help of the complement system. Autoantibodies against MBP and MOG can be found in lesions in the CNS and the cerebrospinal fluid of MS patients. Antibodies against MOG were shown to mediate hyperacute demyelinating disease in marmoset monkeys following T helper type 2 (Th2) immune deviation.[28] Several publications suggest that anti-MOG and anti-MBP antibodies could be used as a prognostic marker for patients who will progress from an isolated clinical syndrome to clinically definite MS.[29]

Migration of immune cells into the CNS

Adhesion molecules

Perivascular infiltration with inflammatory cells is one of the hallmarks of MS lesions in the brain. Immune cell infiltration requires the adhesion, rolling and transmigration of the cells to and through the endothelial cell layer of the blood-brain barrier.[30] ICAM-1 on endothelial cells, which binds to the integrin receptor LFA-1 on leukocytes, and VCAM-1, which interacts with VLA-4, play an important role in the endothelium–leukocyte interaction and leukocyte extravasation, particularly into the CNS.[31] Qing and colleagues have recently suggested that PECAM-1 (CD31) also might play a role in regulating antigen-specific T cells trafficking in CNS inflammatory diseases.[32] The increased expression of ICAM-1 and VCAM-1 on endothelial cells and its ligands on inflammatory cells and microglial cells were found both in acute as well as in chronic MS lesions.[33] Kuchroo and colleagues,[34] and later Theien and colleagues,[35] were able to show that treatment with anti-VLA-4 mAb before disease onset prevents EAE. These findings support a critical role for the VCAM-1–VLA-4 axis in the pathogenesis of EAE. In concordance with this finding, and underlining the importance of the VLA-4/VCAM-1 interaction in the pathogenesis of MS, alpha4-integrin antagonists were recently certified as a new therapy for relapsing-remitting MS.[36] Unfortunately, several cases of progressive multifocal leukoencephalopathy were described among patients treated with alpha4-integrin-antagonists, and the treatment had to be suspended until further studies can identify the relation between the leukoencephalopathy, caused by the JC virus, and treatment with the antagonists.[36]

Chemokines

Chemokines facilitate the transmigration of activated leukocytes across the BBB into the CNS and therefore play an important role in the pathogenesis of MS.[37] Chemokines support CNS migration of T cells and macrophages directly via chemoattraction, and indirectly, through activation of leukocyte integrins, which bind adhesion receptors on the surface of the endothelial cells. Various α- and β-chemokines were found to be increased in the cerebrospinal fluid (CSF) in MS patients, as well as in the CNS in EAE models.[38]

Matrix metalloproteinases

Matrix metalloproteinases (MMP) are a family of tightly regulated proteolytic enzymes, which are secreted into the extracellular matrix. MMP are produced by inflammatory cells and microglial cells and are found to be increased in patients with MS. They disintegrate the basal membrane of the BBB and thereby facilitate the extravasation of inflammatory cells. They also degrade the extracellular matrix and facilitate deeper infiltration of inflammatory cells in the neutrophil. Furthermore, they cleave and activate membrane-bound cytokines such as TNF, and thereby damage the myelin sheath directly.[39] Inflammatory cytokines are able to induce the expression of MMP, while corticosteroids decrease the MMP concentration. IFN-β1b inhibits T cell secretion of MMP and thereby decreases the progression of disease in MS.[40]

Cytokines

The effector mechanisms by which immune cells mediate damage in autoimmune diseases have generated substantial interest and research efforts as potentially important contributors to MS/EAE. Cytokines, for example those produced by autoimmune T cells, are of particular interest. In MS patients, T helper 1 (Th1) cell-associated cytokines, such as TNF, IFN-γ and IL-12, are increased during the acute phase of MS.[5] While the number of T cells which secrete Th2-cytokines is comparable in MS patients and their healthy counterparts,[41] some evidence has suggested that Th2 cytokines may also contribute to the disease pathogenesis.[42]

A recent shift in paradigm suggests that cytokines other than the well established Th1 cytokines have to be considered in the pathogenesis of EAE/MS. As such, IL-17 is upregulated in CNS lesions *in situ*[43] and an essential role has been demonstrated for IL-23 in the pathogenesis of EAE.[44] While IL-12 was considered to play a crucial role in determining the Th1/Th2 balance and disease outcome.[45] recent studies showed induction of EAE in the absence of IL-12.[46] Subsequently, the crucial role of IL-23 in autoimmune CNS pathology was established.[44] In particular, genetic disruption of IL-23 (which shares the p40 subunit with IL-12) renders mice resistant to EAE.[44] Furthermore, the potential of T cells to induce EAE segregates with the production of IL-17, not IFN-γ.[44] These observations are tied together by the finding that IL-23 induces the production of IL-17 by T cells.[47] Thus, while IL-12 seems to be critical for the priming of Th1 cells, IL-23 induces a subset of T cells with the potential to produce IL-17.[48]

The exact mechanisms leading to MS are still unclear, but it is likely that the disease is of multifactorial origin, with the possibility that the above-mentioned factors contribute at one point or another to the pathogenesis of MS.

Recent work by several groups has suggested an important role for MIF in EAE/MS,[49,50] as will be discussed below. Importantly, MIF has broad actions and can affect the function, migration, survival and glucocorticoid regulation of immune and non-immune cells in EAE and MS, and therefore could play an important role in the pathogenesis of these diseases and in CNS autoimmune disease in general.

Part II. MIF IN THE PATHOGENESIS OF EAE AND MS

The Role of MIF in EAE/MS

MIF is produced by a number of the immune and non-immune cells associated with the pathogenesis of CNS autoimmune diseases.[51,52] What sets MIF apart from other cytokines implicated in the disease process is its broad range of proinflammatory actions, and its ability to counter-regulate immunosuppressive glucocorticoid effects. MIF affects the production of cytokines, the expression of adhesion and costimulatory molecules, the activity of immune cells, and the survival of these cells, and additionally, it regulates the effects of glucocorticoids in these systems.

The first line of evidence supporting a role for MIF in MS stems from the observation that MIF was found to be markedly increased in the cerebral fluid of MS patients as compared with healthy individuals.[53] Based on this observation, and the fact that MIF plays such a pivotal role in innate and adaptive immunity, a number of laboratories, including the authors' started to investigate the role of MIF in EAE, an animal model with many similarities with MS.[49,50,54] Importantly, the results have shown that MIF plays an important role in mediating pathology in EAE, possibly by counter-regulating the suppressive effects of glucocorticoids in this disease model.[49,50]

Using EAE induced by immunization with proteolipid peptide amino acid 139-151 (PLP139-151) in SJL mice as the model, Denkinger and colleagues showed that neutralization of MIF *in vivo* with anti-MIF monoclonal antibodies (mAb) substantially ameliorates EAE severity and decreases the disease incidence, duration and lethality.[49] Anti-MIF mAb ameliorates EAE when the treatment is started before the onset of EAE, and, of clinical relevance, also when the anti-MIF antibody is given after the onset of disease.[49] The results show that MIF plays an important role in sustaining the autoimmune T cell population. Specifically, the number of autoreactive IFN-γ producing Th1 cells in EAE decreases in anti-MIF mAb treated mice and in MIF-KO mice.[49,50] The decreased number of effector and memory T cells could be due to a reduced activation of naive myelin-reactive T cells, since MIF is important in the priming and the activation of T cells.[52] However, studies by Denkinger and colleagues argue against an effect of MIF in the priming of naïve T cells during the autoimmune T cell responses,[49] and instead suggest a requirement for MIF during the effector (disease inducing) phase of EAE.[49] An important observation in these studies was that the expression of VCAM-1, but not CD31 or LFA-1, was substantially decreased in anti-MIF antibody treated mice with EAE, but not in the control antibody treated mice. As outlined above, VCAM-1 is critically required for the recruitment and migration of T lymphocytes into the CNS. Thus, the effect of neutralization of MIF on the disease inducing phase of EAE could be due to a decreased expression of VCAM-1, and an impaired ability of autoreactive T cells to migrate to the CNS. Consistent with this hypothesis, inflammatory infiltrates were dramatically reduced in anti-MIF mAb treated mice, as were the number of T cells, and (to a lesser extend) macrophages.

Several lines of evidence suggested an additional role for MIF in promoting CNS of EAE. First, not only was the absolute number of T cells decreased in the anti-MIF mAb treated mice, but also the number of myelin-specific T cells as well. Moreover, the myelin-specific T cells that could be recovered from the CNS of these mice needed higher concentrations of myelin antigen to become activated *in vitro*, suggesting a more rapid decay of functionally highly avid neuroantigen-specific T cells (very low doses of antigen required for their activation).[55] Specifically, the production of IFN-γ by CD4+ T cells recovered from the CNS of anti-MIF antibody treated mice required 10- to 100-fold higher concentrations of PLP139-151 peptide when compared with T cells from the control antibody-injected mice. Furthermore, adoptive transfer of PLP-specific T cells from anti-MIF antibody treated mice to naive recipients demonstrated that these cells were substantially less pathogenic as compared with T cells from the control antibody treated mice (CMD and TGF, unpublished observations).

Second, MIF also seemed to be important for the survival of autoimmune T lymphocytes during EAE. Along these lines, the authors have observed increased apoptosis of T cells in the CNS of anti-MIF treated mice by TUNEL staining (CMD and TGF, unpublished observations). Furthermore, as noted earlier, VCAM-1 was downregulated in anti-MIF mAb treated mice (see below). Interestingly, VLA-4, the ligand for VCAM-1 on T cells, was shown to function as a co-stimulatory molecule and to provide survival signals to T cells.[56,57] Hence, anti-MIF mAb treatment could block VCAM-1–VLA-4 signals required for the survival of the autoimmune T cells by downregulating the expression of VCAM-1.

Along these lines, MIF regulates the transcriptional activity of the pro-apoptotic molecule p53,[14] which could also play a role in promoting the survival of immune cells during EAE. Furthermore, stimulation with recombinant MIF is associated with the sustained activation of the p42/p44 extracellular-signal-regulated (ERK) MAP kinase pathway,[58] which was recently shown to be mediated by reduction of MAPK phosphatase 1 (MKP-1), a phosphatase that inactivates MAPK activity in response to proinflammatory stimuli.[59] In turn, it is well established that activation of the MAPK/ERK pathway has anti-apoptotic function.[60]

Ultimately, MIF-mediated survival signals may prolong the lifespan and/or effector function of autoimmune T lymphocytes and lead to enhanced autoimmune pathology. This view is consistent with the observation that the frequencies of high-avidity PLP139-151 peptide-reactive T cells are substantially decreased in anti-MIF mAb treated mice.[49] High-avidity T cells, upon migration to the CNS, are more likely to be activated by TCR recognition of myelin antigens. Subsequently, it is these activated cells that need additional survival signals, such as those mediated by MIF, to counter-balance apoptosis and to protect them from activation-induced cell death (AICD).[61]

While it is likely that MIF affects T cell effector functions by promoting their survival, there are additional mechanisms that may be involved. T cells are not only dependent on the TCR–MHC:peptide-mediated "signal 1" for activation, but also on a co-stimulatory (e.g. CD80/86) mediated "signal 2." Thus, if MIF participates in the regulation of costimulatory molecules, such as CD80/86, then treatment with anti-MIF mAb may prevent the critical signal 2 provided by APCs, and the T cells may become functionally unresponsive (anergic).[62] Consistent with this view, it has been shown that MIF affects the expression of CD40-CD40L and CD28-CD80/86 in a concentration dependent manner.[63,64] In addition, it was shown that MIF regulates the expression of MHC II.[63] Hence, treatment with anti-MIF mAb leads to decreased expression of MHC-II on the surface of the immune cells.[63] and subsequently decreased signalling via the TCR, resulting in decreased activation of T cells. Lastly, MIF could promote other APC functions, such as antigen processing, which remain to be investigated.

MIF and the Migration of Immune Cells into the CNS

In order to induce EAE and MS, immune cells, including autoreactive T cells, must migrate into the CNS. The migration of autoreactive T cells into areas of the inflammation is critically dependent on interaction with adhesion molecules and chemokines. As noted earlier, the migration of

T cells into the CNS requires interaction of VLA-4 on T cells with VCAM-1 expressed on endothelial cells.[65,66] Importantly, the expression of VCAM-1 on CNS vessels was substantially decreased by anti-MIF mAb treatment in mice with PLP139-151-induced EAE.[49] Therefore, the observed reduction in EAE severity upon anti-MIF mAb treatment could be due to the decreased expression of VCAM-1 and concomitant impaired T cell migration into the brain.[49,67] In support of this view, decreased VCAM-1 expression in areas of inflammation has also been described in a glomerulonephritis model in rats upon anti-MIF mAb treatment.[68] Analogous to these findings, and underlining the importance of the VLA-4/VCAM-1-interaction in the pathogenesis of MS, Natalizumab, an alpha4-integrin antagonist of VLA-4, has shown great efficacy for the treatment of MS.[36,69] Similarly, treatment with antibodies directed against VLA-4 prevented EAE if given before the onset of the disease symptoms.[35] However, the picture may be more complicated than this. Treatment of EAE with anti-VLA-4 antibodies after the onset of EAE worsens the disease and results in an increased accumulation of T cells in the brain.[35] The authors of these studies interpreted this finding to be due to the increased activation of Th1 cells and an increased production of IFN-γ and IL-2.[35]

Importantly, and pointing towards an additional effect of MIF beyond the regulation of VCAM-1-mediated CNS migration, is the observation that treatment with anti-MIF mAb is efficacious when applied during or after the onset of EAE.[49] In this scenario, cell populations other than T lymphocytes may be modified by MIF. Among the cell populations in question are microglial cells, oligodendrocytes, and APCs such as CNS dendritic cells.

MIF-blockade in these cells could, for instance, modify the expression of chemokines such as MIP-1α or MCP-1, which are thought to regulate T cell recruitment to inflammatory loci. The important role of these chemokines in the pathogenesis of EAE was demonstrated by Karpus and colleagues.[38,70] The function of MIF in the regulation of chemokines or chemokine receptor expression in the CNS is not well understood.

Overall, regulation of migration to the CNS during autoimmune inflammation is a likely contribution of MIF to the pathogenesis of EAE/MS. However, regulation of migration is probably not the only role of MIF in promoting EAE and MS.

MIF and Cytokines

A widely recognized feature of MIF is its ability to promote the production and release of proinflammatory cytokines by a number of immune and non-immune cells.

Hence, it is prudent to ask whether MIF affects the balance of proinflammatory to anti-inflammatory cytokines in EAE and MS. For example, does the blockade of MIF cause a shift of the proinflammatory autoimmune Th1 or "Th-17" cell response towards a protective Th2-cytokine pattern, as has been described after treatment with monoclonal antibodies against CD80?[71] Currently, this issue is not resolved. *In vivo* neutralization of MIF with anti-MIF mAb leads to a decrease in the production of myelin antigen-induced Th1 cytokines, without an increase in or shift towards Th2 cytokines.[49] This observation is supported by studies of Kitaichi and colleagues, who showed that anti-MIF mAb treatment of different T cell subpopulations *in vitro* specifically inhibits IFN-γ producing T cells.[72] The inhibition of IFN-γ production in this system appears to be mediated via blockade of TCR-mediated signaling.

Anti-MIF mAb, or knockout of the MIF gene, was also shown to decrease TNF production,[50,73,74] an effect which may be explained by modulation of Toll-like receptor 4 (TLR4) signaling.[75] A role for MIF on TNF production in the CNS has not been established. However, TNF is known to play a pivotal role in a number autoimmune diseases, including MS and rheumatoid arthritis (RA). In RA, therapy with Infliximab, an anti-TNF antibody, has proven to be highly efficacious. In contrast, clinical trials aimed at TNF inhibition in MS patients were not successful and had to be abandoned prematurely due to increased disease severity and side effects. And, even more disturbing is the appearance of case reports of MS diagnosed in patients after treatment with anti-TNF antibody for other conditions.[76]

The different outcomes of anti-TNF treatment in RA and MS trials emphasize the complicated and frequently contradictory roles of cytokines and other effector molecules in specific cell types and tissues: TNF is important for the induction of pathology in MS/EAE via TNFR1. However, TNF is required for the subsequent remyelination via TNFR2 signaling.[77] Hence, it must be cautioned that MIF may have disease promoting effects in the induction phase of EAE/MS, but later on it may play a role in the recovery from disease. The latter possibility has not been formally tested. However, MIF regulates TNF, which as pointed out above, is required for the remyelination of oligodendrocytes.[77]

MIF AND GLUCOCORTICOIDS

The most unique feature of MIF, in terms of the regulation of autoimmune pathology, is its ability to counter-regulate the immunosuppressive effects of glucocorticoids. As such, MIF is the only known proinflammatory cytokine to be induced by glucocorticoids.[78]

Small quantities of the protein are constitutively produced in many cells of the CNS,[51,79,80] and may therefore be of specific relevance to glucocorticoid regulation in general, and glucocorticoid regulation in the CNS in particular. The immunosuppressive effects of glucocorticoids are well documented.[81] Glucocorticoids inhibit T cell activation and proliferation, and the production of a number of cytokines including IL-2, GM-CSF and TNF.[81] They do so via binding to the glucocorticoid receptor, translocation to the nucleus, and gene regulation by interaction with glucocorticoid response elements via the zinc fingers of its DNA binding domain, or indirectly by protein-protein interactions with transcription factors such as AP-1 and NFkB.[82] Another well established property of glucocorticoids is the induction of apoptosis in T cells, particularly in immature thymocytes.[81,82] and in a significant population of mature splenic T cells.[81] Moreover, glucocorticoids seem to be critical for superantigen-mediated deletion of Vβ8+ T cells.

Importantly, glucocorticoids facilitate the secretion of MIF, while MIF counteracts the glucocorticoid-induced inhibition of inflammatory cytokines in macrophages, synovial fibroblasts and T cells, subsequently leading to an increase in the expression of TNF, IL-1β, IL-6, IL-8, IFN-γ and IL-2.[52,73,83]

The molecular pathways underlying this mutual regulation are under investigation. One important pathway involves the extracellular-signal-regulated-kinases-(ERK)-subfamily of the mitogen-activated protein kinases (MAPK) that result in activation of a cascade of cytoplasmatic proteins, like phopholipase A2 (cPLA2). cPLA2 is a target molecule in the action of glucocorticoids,[58] while MIF is a known regulator of the p44/p42-ERK family of the MAPK.[58] Recently, Roger and colleagues found that MIF alters the ERK/MAPK pathway by inhibiting MAPK phosphatase-1 (MKP-1), a phosphatase that inactivates MAPK activity in response to proinflammatory stimuli. MKP-1 appears to be a critical target of the MIF-glucocorticoid crosstalk.[59]

Another mechanism by which MIF affects inflammation, and could counteract glucocorticoid effects, is via regulation of NF-κB. NF-κB is an important modulator of inflammatory cytokine gene expression.[84] and the expression of surface molecules on immune and non-immune cells.[85] Glucocorticoids decrease NF-κB activity by inhibiting NF-κB-release from the NF-κB/IκB-complex,[50,78,82] while MIF promotes NF-κB activity.[86] As a consequence, MIF induced by glucocorticoids subsequently increases the activity of NF-κB and enhances the expression of NF-κB induced cytokines, such as TNF, IL-1β or IL-8. Roger and colleagues[75] showed that the effects of MIF, both on the NF-κB-activity and on the production of TNF, are mediated by the TLR4 via an increased activity of the transcription factor PU.1, which is necessary for the optimal expression of the TLR4-gene in myeloid cells.

As pointed out earlier, MMP are involved in the pathogenesis of MS[87] and EAE.[39] Interestingly, MIF induces the expression of MMP,[88] while glucocorticoids decrease their expression. Therefore, the therapy with anti-MIF mAb may also counteract glucocorticoids in this pathway of the pathogenesis of MS.

Finally, direct evidence for the counter-regulation of glucocorticoids by MIF in EAE, and possibly in MS, stems from data by Papenfuss, Whitacre and colleagues. In their elegant study, MIF KO mice were resistant to EAE relapses, as compared with wild type mice. Interestingly, the levels of glucocorticoids are elevated in MIF KO mice, and a shift towards Th2 cytokine production is observed.[50]

To summarize, it is highly likely that MIF has a unique role in EAE/MS by counter-regulating the immunosuppressive effects of glucocorticoids on the autoimmune response. Combination therapy with anti-MIF mAb and glucocorticoids may prove to be of substantial benefit in the management of autoimmune CNS diseases.

MIF AS A TARGET OF THERAPY

MIF immuno-neutralization is clearly beneficial for the treatment of EAE, independently of the neuroantigen and the mouse strain used.[49] An advantage of MIF over more selective, antigen-specific therapies, such as altered peptide ligands, is that no prior knowledge of the autoantigens targeted by the autoimmune response is necessary. From our studies, it is not evident that MIF blockade leads to a Th2 shift in cytokine patterns, which could be undesirable in some circumstances (e.g. induction of allergic responses). Nevertheless, anti-MIF mAb treatment appears to selectively inactivate high avidity autoimmune T cells, the long sought after "holy grail" of autoimmune therapy.

Glucocorticoids are still the mainstay of autoimmune therapy, but their usefulness is limited by dose-dependent side effects and steroid resistance. MIF is likely a contributing factor to the limited long-term efficacy of glucocorticoids. Resistance to glucocorticoid therapy of autoimmune diseases in humans is associated with increased AP-1 and NF-κB-activity[89] MIF also acts on the AP-1 and NF-κB signalling pathways.[63,75,86,90] Neutralization of MIF could therefore enhance the efficacy of steroids, and simultaneously decrease the dosage of glucocorticoids necessary and thereby limit glucocorticoid side effects. MIF blockade may also be a means of overcoming steroid resistance.[90]

In conclusion, MIF seems to play a unique role in the pathogenesis of CNS autoimmune diseases. Its pleiotrophic effects on (auto)immune cells and non-immune cells, and its unique ability to counter-regulate immunosuppressive glucocorticoid effects make it a highly relevant molecule for the investigation of the mechanisms underlying autoimmune diseases. For the same reasons, it should be a molecule of great promise for the therapy of CNS autoimmune diseases.

REFERENCES

1. Denkinger CM, Metz C, Fingerle-Rowson G, *et al.* (2004) Macrophage migration inhibitory factor and its role in autoimmune diseases. *Arch Immunol Ther Exp (Warsz)* **52**: 389–400.

2. D'Souza SD, Bonetti B, Balasingam V, *et al.* (1996) Multiple sclerosis: Fas signaling in oligodendrocyte cell death. *J Exp Med* **184**: 2361–2370.
3. Salama AD, Chitnis T, Imitola J, *et al.* (2003) Critical role of the programmed death-1 (PD-1) pathway in regulation of experimental autoimmune encephalomyelitis. *J Exp Med* **198**: 71–78.
4. Pender MP. (2000) Neurology 4: multiple sclerosis. *Med J Aust* **172**: 556–562.
5. Sospedra M, Martin R. (2005) Immunology of multiple sclerosis. *Annu Rev Immunol* **23**: 683–747.
6. Poser CM, Brinar VV. (2004) The nature of multiple sclerosis. *Clin Neurol Neurosurg* **106**: 159–171.
7. Fujinami RS, Oldstone MB. (1985) Amino acid homology between the encephalitogenic site of myelin basic protein and virus: mechanism for autoimmunity. *Science* **230**: 1043–1045.
8. Wucherpfennig KW, Strominger JL. (1995) Molecular mimicry in T cell-mediated autoimmunity: viral peptides activate human T cell clones specific for myelin basic protein. *Cell* **80**: 695–705.
9. Soldan SS, Berti R, Salem N, *et al.* (1997) Association of human herpes virus 6 (HHV-6) with multiple sclerosis: increased IgM response to HHV-6 early antigen and detection of serum HHV-6 DNA. *Nat Med* **3**: 1394–1397.
10. Sriram S, Stratton CW, Yao S, *et al.* (1999) Chlamydia pneumoniae infection of the central nervous system in multiple sclerosis. *Ann Neurol* **46**: 6–14.
11. Dal Canto MC, Melvold RW, Kim BS and Miller SD. (1995) Two models of multiple sclerosis: experimental allergic encephalomyelitis (EAE) and Theiler's murine encephalomyelitis virus (TMEV) infection. A pathological and immunological comparison. *Microsc Res Tech* **32**: 215–229.
12. Peterson P, Peltonen L. (2005) Autoimmune polyendocrinopathy syndrome type 1 (APS1) and AIRE gene: new views on molecular basis of autoimmunity. *J Autoimmun* **25**(Suppl): 49–55.
13. Nguyen MT, Lue H, Kleemann R, Thiele M, *et al.* (2003) The cytokine macrophage migration inhibitory factor reduces pro-oxidative stress-induced apoptosis. *J Immunol* **170**: 3337–3347.
14. Mitchell RA, Liao H, Chesney J, *et al.* (2002) Macrophage migration inhibitory factor (MIF) sustains macrophage proinflammatory function by inhibiting p53: regulatory role in the innate immune response. *Proc Nat Acad Sci USA* **99**: 345–350.
15. Tuohy VK, Yu M, Yin L, *et al.* (1999) Spontaneous regression of primary autoreactivity during chronic progression of experimental autoimmune encephalomyelitis and multiple sclerosis. *J Exp Med* **189**: 1033–1042.
16. Steinman L. (1996) Multiple sclerosis: a coordinated immunological attack against myelin in the central nervous system. *Cell* **85**: 299–302.

17. Mokhtarian F, McFarlin DE, Raine CS. (1984) Adoptive transfer of myelin basic protein-sensitized T cells produces chronic relapsing demyelinating disease in mice. *Nature* **309**: 356–358.
18. Sun D, Whitaker JN, Huang Z, *et al.* (2001) Myelin antigen-specific CD8+ T cells are encephalitogenic and produce severe disease in C57BL/6 mice. *J Immunol* **166**: 7579–7587.
19. Goverman J, Woods A, Larson L, *et al.* (1993) Transgenic mice that express a myelin basic protein-specific T cell receptor develop spontaneous autoimmunity. *Cell* **72:** 551–60.
20. Miyamoto K, Miyake S, Yamamura T. (2001) A synthetic glycolipid prevents autoimmune encephalomyelitis by inducing TH2 bias of natural killer T cells. *Nature* **413**: 531–534.
21. Secor VH, Secor WE, Gutekunst CA, Brown MA. (2000) Mast cells are essential for early onset and severe disease in a murine model of multiple sclerosis. *J Exp Med* **191**: 813–822.
22. Liu J, Miwa T, Hilliard B, Chen Y, *et al.* (2005) The complement inhibitory protein DAF (CD55) suppresses T cell immunity *in vivo*. *J Exp Med* **201**: 567–577.
23. Lovett-Racke AE, Trotter JL, Lauber J, *et al.* (1998) Decreased dependence of myelin basic protein-reactive T cells on CD28-mediated costimulation in multiple sclerosis patients. A marker of activated/memory T cells. *J Clin Invest* **101**: 725–730.
24. Karandikar NJ, Vanderlugt CL, Bluestone JA, Miller SD. (1998) Targeting the B7/CD28:CTLA-4 costimulatory system in CNS autoimmune disease. *J Neuroimmunol* **89**: 10–18.
25. McRae BL, Vanderlugt CL, Dal Canto MC, Miller SD. (1995) Functional evidence for epitope spreading in the relapsing pathology of experimental autoimmune encephalomyelitis. *J Exp Med* **182**: 75–85.
26. Lehmann PV, Forsthuber T, Miller A, Sercarz EE. (1992) Spreading of T-cell autoimmunity to cryptic determinants of an autoantigen. *Nature* **358**: 155–157.
27. Yu M, Johnson JM, Tuohy VK. (1996) A predictable sequential determinant spreading cascade invariably accompanies progression of experimental autoimmune encephalomyelitis: a basis for peptide-specific therapy after onset of clinical disease. *J Exp Med* **183**: 1777–1788.
28. Genain CP, Abel K, Belmar N, *et al.* (1996) Late complications of immune deviation therapy in a nonhuman primate. *Science* **274**: 2054–2057.
29. Berger T, Rubner P, Schautzer F, *et al.* (2003) Antimyelin antibodies as a predictor of clinically definite multiple sclerosis after a first demyelinating event. *N Engl J Med* **349**: 139–145.
30. Engelhardt B, Conley FK, Butcher EC. (1994) Cell adhesion molecules on vessels during inflammation in the mouse central nervous system. *J Neuroimmunol* **51**: 199–208.

31. Lee SJ, Benveniste EN. (1999) Adhesion molecule expression and regulation on cells of the central nervous system. *J Neuroimmunol* **98**: 77–88.
32. Qing Z, Sandor M, Radvany Z, *et al.* (2001) Inhibition of antigen-specific T cell trafficking into the central nervous system via blocking PECAM1/CD31 molecule. *J Neuropathol Exp Neurol* **60**: 798–807.
33. Cannella B, Raine CS. (1995) The adhesion molecule and cytokine profile of multiple sclerosis lesions. *Ann Neurol* **37**: 424–435.
34. Kuchroo VK, Martin CA, Greer JM, *et al.* (1993) Cytokines and adhesion molecules contribute to the ability of myelin proteolipid protein-specific T cell clones to mediate experimental allergic encephalomyelitis. *J Immunol* **151**: 4371–4382.
35. Theien BE, Vanderlugt CL, Eagar TN, *et al.* (2001) Discordant effects of anti-VLA-4 treatment before and after onset of relapsing experimental autoimmune encephalomyelitis. *J Clin Invest* **107**: 995–1006.
36. Steinman L. (2005) Blocking adhesion molecules as therapy for multiple sclerosis: natalizumab. *Nat Rev Drug Discov* **4**: 510–518.
37. Mahad D, Callahan MK, Williams KA, *et al.* (2006) Modulating CCR2 and CCL2 at the blood-brain barrier: relevance for multiple sclerosis pathogenesis. *Brain* **129**: 212–223.
38. Karpus WJ, Ransohoff RM. (1998) Chemokine regulation of experimental autoimmune encephalomyelitis: temporal and spatial expression patterns govern disease pathogenesis. *J Immunol* **161**: 2667–2671.
39. Kieseier BC, Clements JM, Gearing AJ, Hartung HP. (1999) Matrix metalloproteinases exhibit different expression patterns in inflammatory demyelinating diseases of the central and peripheral nervous system. *Ann NY Acad Sci* **878**: 567–570.
40. Leppert D, Waubant E, Burk MR, *et al.* (1996) Interferon beta-1b inhibits gelatinase secretion and in vitro migration of human T cells: a possible mechanism for treatment efficacy in multiple sclerosis. *Ann Neurol* **40**: 846–852.
41. Leonard JP, Waldburger KE, Schaub RG, *et al.* (1997) Regulation of the inflammatory response in animal models of multiple sclerosis by interleukin-12. *Crit Rev Immunol* **17**: 545–553.
42. Raine CS, Cannella B, Hauser SL, Genain CP. (1999) Demyelination in primate autoimmune encephalomyelitis and acute multiple sclerosis lesions: a case for antigen-specific antibody mediation. *Ann Neurol* **46**: 144–160.
43. Lock C, Hermans G, Pedotti R, *et al.* (2002) Gene-microarray analysis of multiple sclerosis lesions yields new targets validated in autoimmune encephalomyelitis. *Nat Med* **8**: 500–508.
44. Langrish CL, Chen Y, Blumenschein WM, *et al.* (2005) IL-23 drives a pathogenic T cell population that induces autoimmune inflammation. *J Exp Med* **201**: 233–240.

45. Segal BM, Dwyer BK, Shevach EM. (1998) An interleukin (IL)-10/IL-12 immunoregulatory circuit controls susceptibility to autoimmune disease. *J Exp Med* **187**: 537–546.
46. Becher B, Durell BG, Noelle RJ. (2002) Experimental autoimmune encephalitis and inflammation in the absence of interleukin-12. *J Clin Invest* **110**: 493–497.
47. Aggarwal S, Ghilardi N, Xie MH, *et al.* (2003) Interleukin-23 promotes a distinct CD4 T cell activation state characterized by the production of interleukin-17. *J Biol Chem* **278**: 1910–1914
48. Park H, Li Z, Yang XO, Chang SH, *et al.* (2005) A distinct lineage of CD4 T cells regulates tissue inflammation by producing interleukin 17. *Nat Immunol* **6**: 1133–1141.
49. Denkinger CM, Denkinger M, Kort JJ, *et al.* (2003) *In vivo* blockade of macrophage migration inhibitory factor ameliorates acute experimental autoimmune encephalomyelitis by impairing the homing of encephalitogenic T cells to the central nervous system. *J Immunol* **170**: 1274–1282.
50. Powell ND, Papenfuss TL, McClain MA, *et al.* (2005) Cutting edge: macrophage migration inhibitory factor is necessary for progression of experimental autoimmune encephalomyelitis. *J Immunol* **175**: 5611–5614.
51. Ogata A, Nishihira J, Suzuki T, *et al.* (1998) Identification of macrophage migration inhibitory factor mRNA expression in neural cells of the rat brain by *in situ* hybridization. *Neurosci Lett* **246**: 173–177.
52. Bacher M, Metz CN, Calandra T, *et al.* (1996) An essential regulatory role for macrophage migration inhibitory factor in T-cell activation. *Proc Nat Acad Sci USA* **93**: 7849–7854.
53. Niino M, Ogata A, Kikuchi S, Tashiro K, Nishihira J. (2000) Macrophage migration inhibitory factor in the cerebrospinal fluid of patients with conventional and optic-spinal forms of multiple sclerosis and neuro-Behcet's disease. *J Neurol Sci* **179**: 127–131.
54. Baker D, O'Neill JK, Turk JL. (1991) Cytokines in the central nervous system of mice during chronic relapsing experimental allergic encephalomyelitis. *Cell Immunol* **134**: 505–510.
55. Targoni OS, Lehmann PV. (1998) Endogenous myelin basic protein inactivates the high avidity T cell repertoire. *J Exp Med* **187**: 2055–2063.
56. Zaitseva MB, Mojcik CF, Salomon DR, *et al.* (1998) Co-ligation of alpha4beta1 integrin and TCR rescues human thymocytes from steroid-induced apoptosis. *International Immunology* **10**: 1551–1561.
57. Damle NK Aruffo A. (1991) Vascular cell adhesion molecule 1 induces T-cell antigen receptor-dependent activation of CD4+T lymphocytes. *Proc Nati Acad Sci USA* **88**: 6403–6407.
58. Mitchell RA, Metz CN, Peng T, Bucala R. (1999) Sustained mitogen-activated protein kinase (MAPK) and cytoplasmic phospholipase A2 activation by macrophage migration inhibitory factor (MIF). Regulatory

role in cell proliferation and glucocorticoid action. *J Biol Chem* **274**: 18100–18106.
59. Roger T, Chanson AL, Knaup-Reymond M, Calandra T. (2005) Macrophage migration inhibitory factor promotes innate immune responses by suppressing glucocorticoid-induced expression of mitogen-activated protein kinase phosphatase-1. *Eur J Immunol* **35**: 3405–3413.
60. Zhang B, Hirahashi J, Cullere X, Mayadas TN. (2003) Elucidation of molecular events leading to neutrophil apoptosis following phagocytosis: cross-talk between caspase 8, reactive oxygen species, and MAPK/ERK activation. *J Biol Chem* **278**: 28443–28454.
61. Alderson MR, Tough TW, Davis-Smith T, *et al.* (1995) Fas ligand mediates activation-induced cell death in human T lymphocytes. *J Exp Med* **181**: 71–77.
62. Jenkins MK, Mueller D, Schwartz RH, *et al.* (1991) Induction and maintenance of anergy in mature T cells. *Adv Exp Med Biol* **292**: 167–176.
63. Stavitsky AB, Xianli J. (2002) *In vitro* and *in vivo* regulation by macrophage migration inhibitory factor (MIF) of expression of MHC-II, costimulatory, adhesion, receptor, and cytokine molecules. *Cellular Immunology* **217**: 95–104.
64. Van Kooten C, Banchereau J. (1996) CD40-CD40 ligand: a multifunctional receptor-ligand pair. *Adv Immunol* **61**: 1–77.
65. Brocke S, Piercy C, Steinman L, *et al.* (1999) Antibodies to CD44 and integrin alpha4, but not L-selectin, prevent central nervous system inflammation and experimental encephalomyelitis by blocking secondary leukocyte recruitment. *Proc Nat Acad Sci USA* **96**: 6896–6901.
66. Irani DN, Griffin DE. (2001) Regulation of T cell responses during central nervous system viral infection. *Adv Virus Res* **56**: 175–197.
67. Amin MA, Haas CS, Zhu K, *et al.* (2005) Migration inhibitory factor upregulates vascular cell adhesion molecule-1 and intercellular adhesion molecule-1 via Src, PI3 kinase, and NF{kappa}B. *Blood*
68. Lan HY, Bacher M, Yang N, *et al.* (1997) The pathogenic role of macrophage migration inhibitory factor in immunologically induced kidney disease in the rat. *J Exp Med* **185**: 1455–1465.
69. O'Connor PW, Goodman A, Willmer-Hulme AJ, *et al.* (2004) Randomized multicenter trial of natalizumab in acute MS relapses: clinical and MRI effects. *Neurology* **62**: 2038–2043.
70. Karpus WJ, Lukacs NW, McRae BL, *et al.* (1995) An important role for the chemokine macrophage inflammatory protein-1 alpha in the pathogenesis of the T cell-mediated autoimmune disease, experimental autoimmune encephalomyelitis. *Journal of Immunology* **155**: 5003–5010.
71. Kuchroo VK, Das MP, Brown JA, *et al.* (1995) B7-1 and B7-2 costimulatory molecules activate differentially the Th1/Th2 developmental pathways: application to autoimmune disease therapy. *Cell* **80**: 707–718.

72. Kitaichi N, Ogasawara K, Iwabuchi K, *et al.* (2000) Different influence of macrophage migration inhibitory factor (MIF) in signal transduction pathway of various T cell subsets. *Immunobiology* **201**: 356–367.
73. Calandra T, Echtenacher B, Roy DL, *et al.* (2000) Protection from septic shock by neutralization of macrophage migration inhibitory factor. *Nature Medicine* **6**: 164–170.
74. Nishihira J, Mizue Y, Sakamoto W. (1999) Induction of T-kininogen and tumor necrosis factor-alpha by macrophage migration inhibitory factor *in vivo*. *Semin Thromb Hemost* **25**: 557–562.
75. Roger T, David J, Glauser MP, Calandra T. (2001) MIF regulates innate immune responses through modulation of Toll-like receptor 4. *Nature* **414**: 920–924.
76. Sicotte NL, Voskuhl RR. (2001) Onset of multiple sclerosis associated with anti-TNF therapy. *Neurology* **57**: 1885–1888.
77. Arnett HA, Mason J, Marino M, *et al.* (2001) TNF alpha promotes proliferation of oligodendrocyte progenitors and remyelination. *Nat Neurosci* **4**: 1116–1122.
78. Calandra T, Bernhagen J, Metz CN, *et al.* (1995) MIF as a glucocorticoid-induced modulator of cytokine production. *Nature* **377**: 68–71.
79. Bacher M, Meinhardt A, Lan HY, *et al.* (1998) MIF expression in the rat brain: implications for neuronal function. *Mol Med* **4**: 217–230.
80. Bernhagen J, Calandra T, Mitchell RA, *et al.* (1993) MIF is a pituitary-derived cytokine that potentiates lethal endotoxaemia. [erratum appears in *Nature* 1995 Nov 23; **378**(6555): 419]. *Nature* **365**: 756–759.
81. Franchimont D. (2004) Overview of the actions of glucocorticoids on the immune response: a good model to characterize new pathways of immunosuppression for new treatment strategies. *Ann N Y Acad Sci* **1024**: 124–137.
82. Auphan N. (1995) Immunosuppression by glucocorticoids: inhibition of NF-kappa B activity through induction of I kappa B synthesis. *Science* **270**: 286–290.
83. Leech M, Metz C, Bucala R, Morand EF. (2000) Regulation of macrophage migration inhibitory factor by endogenous glucocorticoids in rat adjuvant-induced arthritis. *Arthritis & Rheumatism* **43**: 827–833.
84. Barnes PJ, Karin M. (1997) Nuclear factor-kappaB: a pivotal transcription factor in chronic inflammatory diseases. *New Engl J Med* **336**: 1066–1071.
85. Tak PP, Firestein GS. (2001) NF-kappaB: a key role in inflammatory diseases. *J Clin Invest* **107**: 7–11.
86. Daun JM, Cannon JG. (2000) Macrophage migration inhibitory factor antagonizes hydrocortisone-induced increases in cytosolic IkappaBalpha. *Am J Physiol Regul Integr Comp Physiol* **279**: R1043–1049.

87. Gijbels K, Masure S, Carton H, Opdenakker G. (1992) Gelatinase in the cerebrospinal fluid of patients with multiple sclerosis and other inflammatory neurological disorders. *J Neuroimmunol* **41**: 29–34
88. Onodera S, Kaneda K, Mizue Y, *et al.* (2000) Macrophage migration inhibitory factor up-regulates expression of matrix metalloproteinases in synovial fibroblasts of rheumatoid arthritis. *J Biol Chem* **275**: 444–450.
89. Bantel H, Schmitz ML, Raible A, *et al.* (2002) Critical role of NF-kappaB and stress-activated protein kinases in steroid unresponsiveness. *FASEB Journal* **16**: 1832-1834.
90. Kleemann R, Hausser, A, Geiger G, *et al.* (2000) Intracellular action of the cytokine MIF to modulate AP-1 activity and the cell cycle through Jab1. *Nature* **408**: 211–216.

11

MIF: Wound Repair

Matthew J. Hardman and Gillian S. Ashcroft

INTRODUCTION

The proinflammatory cytokine macrophage migration inhibitory factor (MIF) was first identified 40 years ago as a T-cell-derived factor that inhibited random macrophage migration.[1,2] Many years passed before the cloning of human MIF cDNA[3] rekindled interest in the subject. Subsequent studies identified MIF as a pituitary-derived cytokine, released in a similar fashion to a hormone after exposure to endotoxin.[4] MIF has profound proinflammatory activity, inducing the expression of classical proinflammatory cytokines such as IL-1β, IL-6 and TNF-α, nitric oxide and matrix metalloproteinases.[5] It is, however, unique in its ability to counteract the antiinflammatory activity of glucocorticoids, leading to an increased monocyte production of IL-6 and IL-1β.[6] Moreover, MIF is itself induced by glucocorticoids. In fact, MIF and glucocorticoids are now thought to synergistically modulate the balance between innate and acquired immunity.

It has become apparent that many aspects of MIF's structure and function are unique. MIF does not belong to any known cytokine family, but

Faculty of Life Sciences, University of Manchester, Michael Smith Building, Oxford Road, Manchester, M13 9PT. E-mails: Gillian.s.ashcroft@manchester.ac.uk; Matthew.j.hardman@manchester.ac.uk is highly conserved between species.[7] It is also the only known cytokine to possess tautomerase and oxidoreductase enzymatic activities. All known cytokines act by binding cognate cell surface receptors and inducing downstream signaling pathways. Although a potential cell surface receptor for MIF (CD74) has recently been identified,[8] the mechanism of receptor-mediated MIF signaling remains controversial as CD74 lacks an intercellular signal transduction domain. Intriguingly, MIF has instead been shown to be endocytosed by target cells and to interact directly with the intercellular protein Jab1,[9] a co-activator of the AP1 transcription factor. This is again a mode of action that is unique to MIF. Unlike other cytokines, MIF is constitutively expressed by a range of cell types, including macrophages. Significant quantities of MIF are stored in preformed intracellular pools within macrophages, which can be rapidly secreted in response to low levels of LPS, in contrast to other cytokines which require *de novo* synthesis. Moreover, MIF upregulates the expression of TLR4, the LPS receptor responsible for the detection of bacterial endotoxin.[10] It appears that the constitutive expression of MIF by tissues in contact with the external environment allows cells of the host antimicrobial defence system to respond rapidly to infection.

In recent years, MIF has been implicated in the pathogenesis of numerous disease states such as cancer, arthritis, sepsis, asthma, gastrointestinal inflammation and atherosclerosis,[4,11–15] and is associated with non-cutaneous tissue injury.[16–20] However, the authors have recently started to appreciate the key role that MIF plays in the healing of cutaneous wounds, particularly in circumstances where wound healing is delayed. This chapter will attempt to outline our current understanding of the complex role that MIF plays in wound healing.

WOUND HEALING AND HORMONES

The skin is the largest organ in the human body and the first line of defence between an organism and the external environment. As such, it has an extremely specialized structure consisting of an outer epidermal layer adapted for barrier function and protection from dessication, microorganisms, percutaneous absorption and ultraviolet radiation, and an underlying

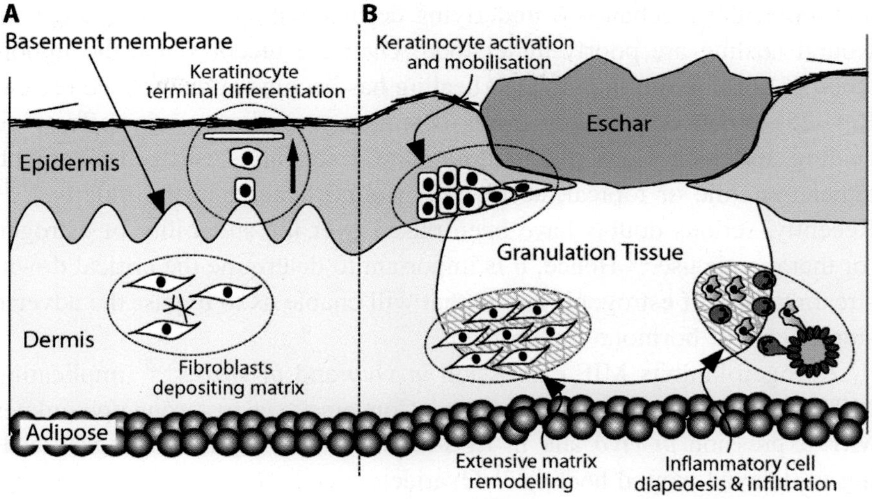

Fig. 1 Diagrammatic representation of normal skin and a healing wound. (**A**) The skin comprises an outer epidermal layer and an inner dermis. The epidermis, composed of terminally differentiating keratinocytes, provides barrier function, while the dermis cushions, supports and nourishes. (**B**) Wound healing is a complex multi-step process regulated by a plethora of growth factors and cytokines. Depicted here are the main stages of inflammation, re-epithelialization, and matrix deposition.

dermis dedicated to cushioning, support and nutrition (Fig. 1A). When the skin is compromised, a complex and highly evolved innate wound healing response is initiated. This response, involving numerous tissue and cell types, comprises several overlapping phases: inflammation, matrix deposition, re-epithelialization, angiogenesis and matrix remodeling. (Fig. 1B).[21] In young individuals, these events are tightly regulated by the carefully controlled levels of cytokines and growth factors[22]; wounds heal quickly; and the risk of infection is low. However, in the elderly this regulation is disrupted and healing is delayed.[23] In fact, in the developed world a rapid (and continuing) expansion of the elderly population has led to a substantial recent increase in age-impaired wound healing states, and a considerable financial burden on the health services.

Acute wounds from elderly subjects have an altered inflammatory response and reduced matrix deposition (Fig. 1B).[24] Age-impaired healing frequently results in wounds that fail to heal entirely, leading to various clinical conditions, such as chronic venous ulcers. The cellular and molecular mechanisms underlying detrimental age-related changes in wound healing are poorly understood. Over the recent years, the importance of hormones in age-related healing has become apparent (see review, Ref. 25). Reduced estrogen levels postmenopause have major effects on healing, inducing a proinflammatory state. Exogenous estrogen treatment, either systemic or topical, accelerates wound healing in the elderly.[26,27] Recently, serious doubts have been raised over the suitability of estrogen for therapeutic use.[28] Hence, it is important to determine the critical downstream targets of estrogenic action that will enable us to bypass the adverse side effects of hormone replacement.

estrogen inhibits MIF expression *in vivo* and *in vitro*,[25,29] implicating MIF as a downstream target candidate. Conversely, testosterone upregulates MIF expression *in vivo* and in LPS-activated murine macrophages, leading to delayed wound healing.[31] Ovariectomized (Ovx) mice, by virtue of the dramatically reduced systemic and local estrogen levels, provide an excellent model of human age-associated delayed healing.[26] Ovx mouse wounds display increased inflammation, reduced matrix deposition and substantially delayed healing. Clever use of the Ovx mouse model has provided substantial insight into the role of MIF in cutaneous healing[29,30] (see below).

MIF'S ROLE IN SKIN AND THE CUTANEOUS HEALING RESPONSE

MIF expression in the skin was first reported by Shimizu *et al.*[32] Using a polyclonal antibody against the human antigen, MIF was immunolocalized to the proliferative basal epidermal layer and endothelial cells. This observation was quickly followed by the demonstration that epidermal MIF expression is increased in several inflammatory dermatoses, including atopic dermatitis and psoriasis.[33,34] In accord with MIF's role as a proinflammatory cytokine, both skin conditions are characterized by chronic and persistent cutaneous inflammation. MIF is upregulated by both fibroblasts and keratinocytes in response to solar ultraviolet (UV) radiation.[35,36] UV radiation induces a proinflammatory response in the skin which leads to increased matrix metalloproteinase (MMP expression) and, with long-term exposure, reduced dermal collagen and altered dermal elastase. In cultured fibroblasts MIF directly upregulates MMP-1 expression, via PKC- and AP-1-dependent pathways, indicating a direct role in the photoaging process.[36] The first indication that MIF may be important in the cutaneous wound healing process came from Abe *et al.*,[37] who reported that local MIF mRNA levels in rats increased immediately after wounding. However, the authors monitored only the very early stages of healing, providing no information on the changes in MIF levels as the wound heals.

Our detailed studies demonstrate that in both rodent and humans, MIF is upregulated upon wounding, not only in the epidermis, but also in inflammatory and endothelial cells; later a second phase of MIF induction occurs that is confined to wound site fibroblasts[29,30] (Hardman and Ashcroft, unpublished observations; Fig. 2). For example, seven days post-wounding in human, we see strong inflammatory and endothelial cell expression, and reduced MIF in wound edge epidermis. In contrast, at one month post-wounding, inflammation has resolved (and hence the associated MIF expression disappeared) and fibroblasts within the wound granulation tissue express high levels of MIF. Interestingly, epidermal MIF expression is also increased (Fig. 3). Comparing MIF expression between normally healing acute mouse wounds and delayed healing Ovx mouse wounds clearly indicates that an increased and persistent MIF expression correlates with delayed healing and an increased inflammatory response.[29] The treatment of mouse wounds with recombinant human MIF impairs healing, while in the wounds treated with anti-MIF antibody, healing is accelerated, clearly demonstrating a functional role for MIF in the healing process.[29] Interestingly, the treatment with anti-MIF antibody has also been shown to be beneficial in an arthritis model[38]; various other disease models[14,39–41]; and a model of lung injury.[16] Moreover, MIF is increased in acute wounds in elderly human subjects and spectacularly upregulated in chronic non-healing venous ulcers, correlating with excessive inflammation.[30] In humans, estrogen downregulates MIF expression both systemically and locally in wound tissue.[30] In addition, ulcers that respond to four-layer compression bandage therapy and begin to heal downregulate MIF (Hardman and Ashcroft, unpublished observations). *In vitro*, estrogen inhibits MIF expression directly through an ER-mediated mechanism dependent on the proinflammatory transcription factor NFκB.[30]

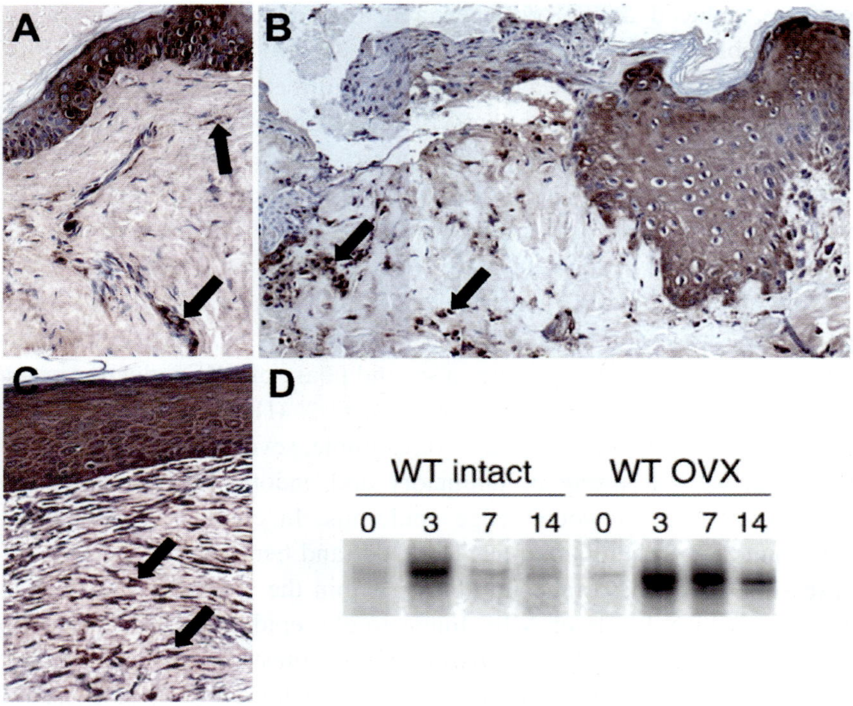

Fig. 2 MIF expression during wound healing. (**A–C**) MIF immunolocalization in human tissue. (**A**) In normal skin MIF localizes to the epidermis and endothelial cells (*arrows*). (**B**) At 7 days post-wounding, strong expression is detected in inflammatory cells (*arrows*) with reduced epidermal expression. (**C**) By 1 month post-wounding, fibroblasts strongly express MIF (*arrows*) and epidermal expression is also increased. (**D**) In normal mouse wounds, MIF mRNA expression peaks at day 3. In contrast, wounds from Ovx mice have stronger and temporally extended MIF upregulation.

This correlates mechanistically with the findings from other biological systems where, for example, TNF-α upregulates MIF directly via an NFκB dependent mechanism.[6,42,43]

MIF AS A CENTRAL REGULATOR OF WOUND HEALING

MIF null mice, simultaneously generated by three independent laboratories,[44–46] are viable and fertile, and were initially reported to have

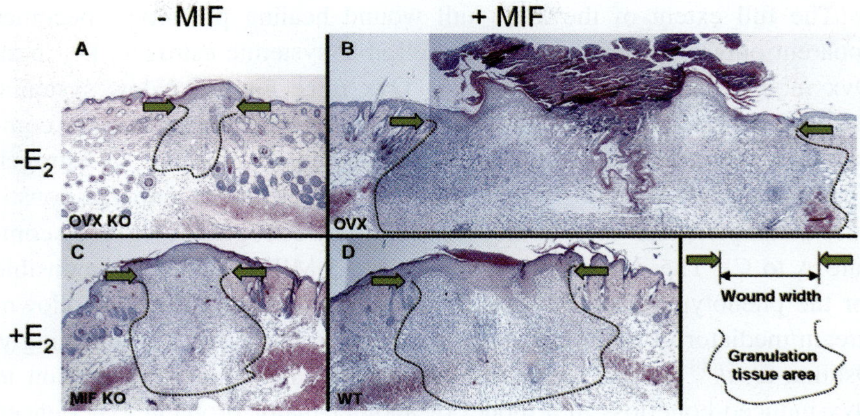

Fig. 3 Delayed healing due to reduced estrogen does not occur in MIF null mice. At 7 days post wounding, wounds from Ovx mice are larger with increased granulation tissue area and inflammatory cell number (**B**) compared with mice with intact ovaries (**D**). In contrast, healing in Ovx *MIF* null mice (**A**) is similar to that in *MIF* null mice with ovaries (**D**). Note that wounds from both *MIF* null groups have smaller wound width and area than those from wild-type mice.

a normal phenotype.[45] Subsequent studies have, however, demonstrated that mice lacking MIF are hyporesponsive in a variety of models, displaying a dampened inflammatory reponse and an increased survival rate, e.g., following severe endotoxic shock.[10] The presence of a wound healing phenotype in *MIF* null mice is somewhat controversial. Our own data indicate that healing is marginally (statistically non-significantly) accelerated in *MIF* null mice,[30] in line with MIF's role as a proinflammatory cytokine (i.e. detrimental to healing). In contrast, Zhao *et al.*[47] reported delayed healing in *MIF* null mice. This finding is somewhat counter-intuitive and at odds with our data and findings from numerous other model systems where the presence of MIF is associated with delayed repair, increased inflammation or accelerated tissue damage.[4,11–13,15,16,18–20,39] In addition, excessive inflammation is clearly implicated in delayed healing and the development of non-healing wounds.[24,27,29,48] The reason for this discrepancy is unclear, but it could in part be explained by the wound models employed — an incisional full-thickness wound model.[29,30] vs. an excisional model which is subject to increased variability in healing measurements[47]; or it could perhaps be due to differing mouse genetic backgrounds.

The full extent of the *MIF* null wound healing phenotype becomes apparent only on a background of negligible systemic estrogen, i.e. in an Ovx mouse. In contrast to wild-type Ovx mice, in which low systemic estrogen is linked to high MIF levels and delayed healing (Fig. 3 — compare B to D), *MIF* null mice heal normally in the absence of estrogen (Fig. 3 — compare A to B); i.e. the *MIF* null wound phenotype constitutes a complete lack of inhibition of healing by estrogen[29] (Fig. 3 — compare A to C). These findings suggest that: (1) MIF is directly responsible for the phenotype of delayed healing; and (2) MIF is a necessary downstream mediator of estrogenic modulation of wound healing. Interestingly, Oshima *et al.*[49] have recently shown that *MIF* null mice are resistant to Ovx-induced bone loss in a model of another major problem experienced by postmenopausal women—osteoporosis (bone loss). In this context, MIF is required for the effects of estrogen on osteoclast activation and an increased bone turnover.

The authors have used a microarray-based approach to dissect apart the relative contribution of MIF and estrogen to wound healing.[30] In light of the major contribution that estrogen makes to the outcome of healing,[26,27,29,30] the authors were intrigued to find that in a cutaneous wound, very few genes (6%) are regulated by estrogen independently of MIF, the vast majority of genes being regulated by estrogen through MIF interaction (94%). Thus, virtually all of estrogen's influence on wound healing is exerted through MIF, making MIF the major downstream mediator of age-associated delayed healing. Equally intriguing was the finding that a large proportion of the genes identified are regulated by MIF independently of estrogen (see below).

Gene clusters identified in our microarray study contained multiple statistically-significant over-represented biologically-related gene groupings (Fig. 4). The data highlight the involvement of MIF in every aspect of the wound healing process[30] and are highly relevant to other MIF-mediated disease processes. Many genes are up- or downregulated by MIF with estrogenic interaction (i.e. those important for the MIF-mediated age-associated delayed healing phenotype). Genes upregulated in delayed healing wounds are predominantly responsible for cell motility/muscle contraction, immune response or adhesion.

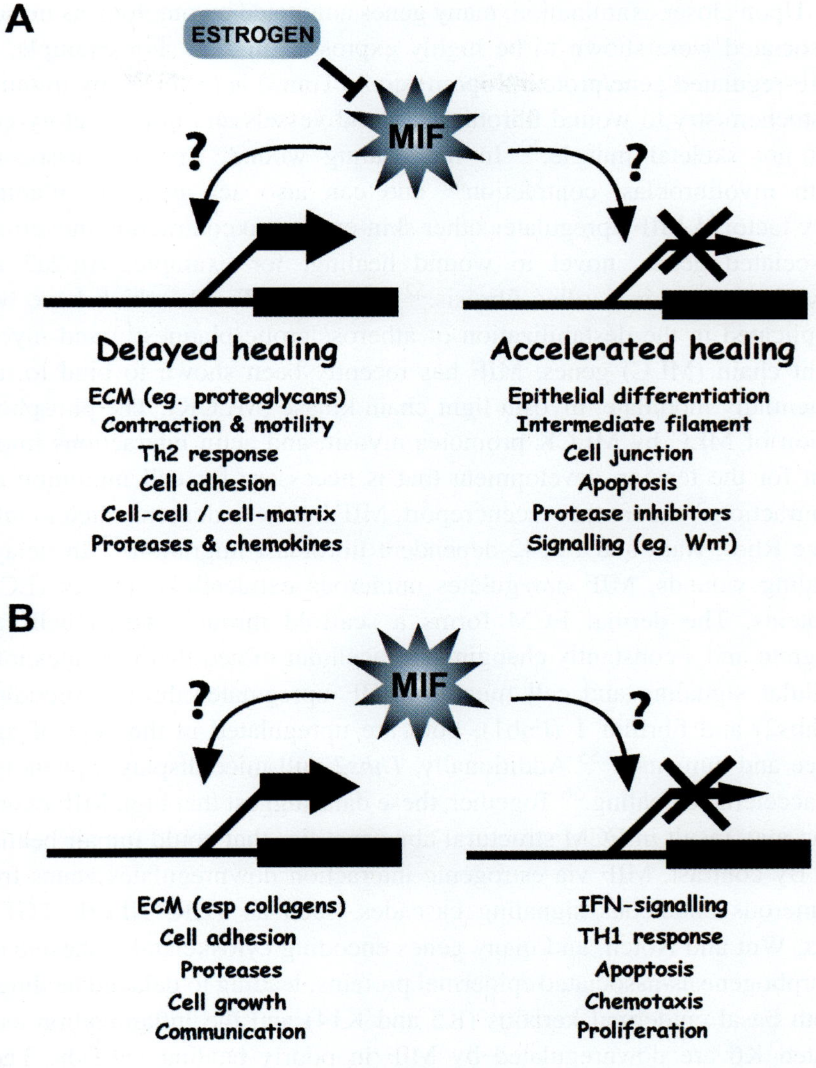

Fig. 4 A summary of the effects of MIF on wound gene expression. (**A**) estrogen-dependent MIF regulated gene ontology (GO) groups associated with accelerated healing are involved in epidermal migration, resolution of inflammation and inhibition of degradation. In contrast, in delayed healing wounds, MIF upregulates fibroblast- and Th2-associated genes. (**B**) Independently of estrogen MIF is involved in the balance between cell growth and apoptosis, Th1 and Th2 response, and proteolytic activity.

Upon closer examination, many genes annotated by ontology as muscle-associated were shown to be highly expressed in skin. For example, the MIF-regulated gene/protein tropomysin 3 (Tpm3) is localized by immunohistochemistry to wound fibroblasts, blood vessels and inflammatory cells but not skeletal muscle.[30] In the healing wound, Tpm3 is associated with myofibroblast contraction[50] and can also act as a proinflammatory factor.[51] MIF upregulates other skin-expressed contraction/movement-associated genes, novel to wound healing, for example, Atp2a2 and Myh7 involved in cardiac fibrosis;[52] Myh6 and Tnnt1, which have been implicated in the destabilization of atherosclerotic plaques;[13] and myosin light chain (MLC) genes. MIF has recently been shown to bind to, and potentially modulate, myosin light chain kinase (MLCK). The phosphorylation of MLC by MLCK promotes myosin and actin interactions important for the tension development that is necessary for cell migration and contraction.[53] In another recent report, MIF has been demonstrated to influence Rho-, Rac- and Cdc42-dependent fibroblast migration.[47] In delayed healing wounds, MIF upregulates numerous extracellular matrix (ECM) proteins. The dermal ECM forms a scaffold through which cells can migrate and a constantly changing extracellular milieu that regulates intercellular signaling and cell function. MIF upregulates thrombospondin 2 (Thbs2) and fibrillin 1 (Fnb1); both are upregulated in the skin of aged mice and humans.[54,55] Additionally, *Thbs2* null mice display a phenotype of accelerated healing.[56] Together, these data suggest that high MIF expression may result in ECM structural abnormalities that could impair healing.

By contrast, MIF via estrogenic interaction downregulates genes from numerous cutaneous signaling cascades, such as TLR, NFκB, TGF-β, Akt, Wnt and Notch, and many genes encoding cytoskeletal, adhesive and morphogenesis-associated epidermal proteins, leading to delayed healing.[30] Both basal epidermal keratins (K5 and K14) and the inflammation associated K6 are downregulated by MIF in poorly healing wounds. These basal keratins are associated with an increased cell motility, and are upregulated in the wound margins to promote epithelialization.[57,58] These findings are in agreement with the observed phenotype of chronic human wounds, in which the lack of epithelialization is a major problem. Recently Stojadinovic *et al.*[59] have reported that stabilized β-catenin, i.e. altered Wnt signaling, in human chronic ulcers can inhibit keratinocyte migration and delay wound healing.

Several Wnt family members are expressed in the skin, most notably Wnt5a, which has also recently been shown to be upregulated in human leukocytes via a TLR/NFκB-dependent mechanism in response to bacterial infection.[60] Using conditional knockouts, Cheon *et al.*[61] have shown that β-catenin (the major downstream effector of wnt signaling) regulates wound healing and mediates the effects of TGF-β. The exact role of β-catenin in the healing wound is complex as it inhibits keratinocyte migration, yet activates fibroblast proliferation. In this model, increased β-catenin activity was associated with delayed healing, probably through a direct effect on mysenchymal cells as opposed to keratinocytes. Notably, no difference was observed in wound inflammatory response. This was most likely because the retrovirus-mediated, conditional knockout system used was locally acting, so circulating leukocytes would still express β-catenin. β-catenin is certainly involved in the inflammatory response. Lobov *et al.*[62] describe a novel role for the canonical WNT pathway in macrophages, which activaties cell death in target cells, via a short range Wnt7b signal. In a detailed study Fathke *et al.*[63] suggest that β-catenin-dependent and -independent signaling may have differing roles in wound epithelium. These authors found than, rather that exhibiting delayed healing, lithium chloride-treated wounds develop abnormal epithelial structures, a phenotype previously reported in a selection of transgenic mice where β-catenin activity is upregulated in normal skin.[64] Furthermore, forced Wnt5a expression produced a regenerative phenotype.[63] These observations are suggestive of cell- and context-dependent effects. In other MIF-associated pathological conditions such as rheumatoid arthritis, increased Wnt signaling contributes directly to disease progression by inducing proinflammatory cytokines and promoting the expression of matrix degrading MMPs.[65]

MIF promotes tumorogenesis, and regulates the balance between proliferation and apoptosis, through multiple mechanisms,[66] one of which involves downregulation of the tumor suppressor oncogene p53.[11] In various cell lines MIF has been shown to be a potent inducer of cell proliferation.[67] In tumorogenesis MIF functions as a growth factor that stimulates uncontrolled growth and promotes neovascularization,[68] the latter being equally important in wound healing. In the wound, MIF's pro-proliferative/antiapoptotic influence on inflammatory cells is suggested to prolong the inflammatory response, while in fibroblasts MIF induces phosphorlyation of ERK/MAPK, thereby overriding p53-dependent cell cycle control and stimulating the proliferation of quiescent cells. The MIF-dependent downregulation of Notch during wound healing suggests a link between MIF, Notch and cutaneous carcinogenesis.[69] Other genes downregulated by MIF in delayed healing wounds include novel protease inhibitors, antimicrobial peptides and anti-inflammatory genes, such as Gpx2, previously described as the predominant hydroperoxide reducing enzyme in the epithelium of the gastrointestinal tract. The MIF-mediated repression of these genes in the skin is likely to contribute to the excessive inflammation observed in delayed healing wounds. Indeed, *Gpx2* null mice are highly susceptible to bacterial infection and inflammation of the gut.[70]

In addition to its role as a mediator of hormonally-modulated healing, MIF regulates a plethora of wound healing-associated genes independently of estrogen. These genes may be particularly important in other MIF-regulated processes such as systemic sepsis,[4] or degenerative diseases such as rheumatoid arthritis.[12] In this context too, MIF upregulates the expression of key ECM genes, including most skin-expressed members of the collagen family, small proteoglycans, basement membrane components and matrix-associated proteins. This is accompanied by the upregulation of proteolytic matrix remodelling enzymes and genes associated with leukocyte-endothelial adhesion and the acute phase response.[30] These fibroblast-associated ECM promoting activities correlate well with the observed increase in fibroblast-dependent MIF expression during the latter stages of human healing. In this context, MIF is likely to be important in a range of fibrotic diseases, including cutaneous disorders of fibrosis, such as the development of keloid scars. It is of note that the fibroblasts from patients with diffuse systemic sclerosis (a connective tissue disease characterized by excessive and inappropriate collagen deposition in numerous organs including the skin) display an increased MIF expression.[71] Conversely, MIF independently of estrogen downregulates many IFN-γ-regulated Th1 response genes indicating a potential cross-talk between MIF and INF-γ during wound healing.[30] In macrophages MIF is induced by numerous proinflammatory mediators, including IFN-γ.[42] Hagemann *et al.*[72] recently presented a fascinating paper which could perhaps provide substantial insight into the fundamental processes occurring in a healing wound. They use a macrophage/epithelial tumor cell co-culture model which may parallel the effects of macrophages on the "tumor-like," invasive, activated wound edge epithelium. In their model, macrophages induced tumor cell invasiveness via a TNF-α-, JNK- and NFκB-dependent mechanism that induced, and was dependent upon, MIF.

The effects of MIF on the process of wound healing, as in other systems, appear complex. The wound is a dynamic and heterogeneous tissue with an intricate extracellular milieu containing numerous interacting cytokines and growth factors. Combining this with the complexity of potential MIF signaling pathways may provide a possible explanation for the varied and, in some cases, contrasting reported effects of MIF on wound healing, and in other systems. This year, exactly 40 years after MIF was first described, Lue *et al.*[73] reported an entirely novel, rapidly and transiently activated, ERK and Jab1-dependent, MIF signaling pathway involving an upstream Src kinase. Intriguingly, the rapid and transient activation of MIF can have entirely opposite effects on cellular functions. For example, sustained ERK activation by high MIF concentrations can inhibit proliferation,[9] in contrast to MIF's generally accepted pro-proliferative influence, which may instead be mediated by transient ERK activation at low MIF concentrations.[73] Nonetheless, MIF's implication in leukocyte recruitment, activation, proliferation and production of proinflammatory cytokines; matrix deposition, remodeling, and degredation; epithelial activation/migration; contraction; and cell movement in the healing wound makes this most interesting of factors an excellent candidate for the therapeutic modulation of delayed healing.

MIF, INFLAMMATION AND REPAIR IN OTHER TISSUES

MIF, in addition to its major role in delayed wound healing, is a key cytokine in numerous other inflammatory-degenerative diseases, such as rheumatoid arthritis (RA).[74] Indeed, many of the key pathological features of arthritis are shared with wounds that fail to heal: excessive and inappropriate infiltration of leukocytes (neutrophils, monocytes, macrophages and Th1 lymphocytes) into the dermis/granulation tissue in wounds and the synovium in RA, which is mediated by similar cytokines and chemokines, such as TNF-α, IL-1 and IL-6; destruction of extracelluar matrix through the upregulation of MMPs and other proteases; and inhibition of resident cell function. In each system MIF has been independently implicated in these events.[29,30,38,75]

The use of anti-MIF antibodies decreases the severity and/or increases the survival rate in a number of disease models,[16] including both arthritis[76] and age-associated delayed healing.[29] The correlation between reduced MIF activity and decreased disease severity is sustained in *MIF* null mice, which are protected against joint destruction in an RA model[77] and against the development of delayed healing in a cutaneous wound model.[29] In addition to its pro-proliferative effects on dermal (wound) fibroblasts,[67] MIF also promotes the proliferation of fibroblast-like synoviocytes, cells which play a crucial role in the pathogenesis of RA.[78] In RA, MIF has been shown to regulate leukocyte recruitment into the joint.[79] An analogous direct effect on cell recruitment is likely to contribute to the excessive inflammation observed in wounds expressing high levels of MIF.[30] Although MIF promoter polymorphisms associated with high systemic MIF expression have been linked to RA severity and incidence[80,81] to date, no such association has been made with the quality of cutaneous healing.

MIF could potentially be implicated in simultaneous susceptibility to various disease states. For instance, it has long been known that RA patients have an increased susceptibility to atherosclerosis.[82] Circulating MIF concentrations are elevated in the elderly, in chronic ulcer patients and in patients with certain inflammatory skin conditions or other non-cutaneous inflammatory diseases, such as RA.[5,30] Under these conditions, the observed serum MIF concentrations are sufficiently high to induce the pivotal events that lead to atherosclerosis development, such as monocyte-endothelial adhesion and consequent localized inflammation.[13] Indeed, localized MIF expression has recently been reported in human atheroma lesions.[83] In an animal model of atherosclerosis, anti-MIF antibodies once again block an inappropriate inflammatory response and reduce lesion formation.[84] It is highly likely that elevated serum MIF levels, whether due to infection, disease or simply aging, predispose an individual to a range of inflammatory disorders, including delayed wound healing.

CONCLUSIONS

The multifunctional cytokine MIF is emerging as a key mediator of the delayed healing and excessive inflammatory response observed in wounds of the elderly. That MIF should act as a master regulator of estrogenic effects may be of critical importance in light of the recent finding that in many chronic ulcer patients, estrogen receptor (ERβ) function may be disrupted,[85] and estrogen therapy ineffective. It may also constitute a potential mechanism for estrogen-mediated action in other systems and numerous disease states. Manipulation of MIF, a potential modulator of ERβ function, or genes downstream of MIF should provide an effective therapeutic strategy to accelerate wound healing while avoiding the numerous unwanted side effects currently associated with estrogen treatment.

REFERENCES

1. Bloom BR, Bennett B. (1966) Mechanism of a reaction *in vitro* associated with delayed-type hypersensitivity *Science* **153**: 80–82.
2. David JR. (1996) Delayed hypersensitivity *in vitro*: its mediation by cell-free substances formed by lymphoid cell-antigen interaction. *Proc Natl Acad Sci USA* **56**: 72–77.
3. Weiser WY, Temple PA, Witek-Giannotti JS, *et al.* (1989) Molecular cloning of a cDNA encoding a human macrophage migration inhibitory factor. *Proc Natl Acad Sci USA* **86**: 7522–7526.
4. Bernhagen J, Calandra T, Mitchell RA, *et al.* MIF is a pituitary-derived cytokine that potentiates lethal endotoxaemia. *Nature* **365**: 756–759.
5. Calandra T, Roger T. (2003) Macrophage migration inhibitory factor: a regulator of innate immunity. *Nat Rev Immunol* **3**: 791–800.
6. Calandra T, Bernhagen J, Metz CN, *et al.* (1995) MIF as a glucocorticoid-induced modulator of cytokine production. *Nature* **377**: 68–71.
7. Du J, Yu Y, Tu H, *et al.* (2006) New insights on macrophage migration inhibitory factor: based on molecular and functional analysis of its homologue of Chinese amphioxus. *Mol Immunol* **43**: 2083–2088.
8. Leng L, Metz CN, Fang Y, *et al.* MIF signal transduction initiated by binding to CD74. *J Exp Med* **197**: 1467–1476.
9. Kleemann R, Hausser A, Geiger G, *et al.* (2000) Intracellular action of the cytokine MIF to modulate AP-1 activity and the cell cycle through Jab1. *Nature* **408**: 211–216.
10. Roger T, David J, Glauser MP, Calandra T. (2001) MIF regulates innate immune responses through modulation of Toll-like receptor 4. *Nature* **414**: 920–924.
11. Hudson JD, Shoaibi MA, Maestro R, *et al.* (1999) A pro-inflammatory cytokine inhibits p53 tumor suppressor activity. *J Exp Med* **190**: 1375–1382.

12. Morand EF, Leech M. (2005) Macrophage migration inhibitory factor in rheumatoid arthritis. *Front Biosci* **10**: 12–22.
13. Schober A, Bernhagen J, Thiele M, *et al.* (2004). Stabilization of atherosclerotic plaques by blockade of macrophage migration inhibitory factor after vascular injury in apolipoprotein E-deficient mice. *Circulation* **109**: 380–385.
14. Kobayashi M, Nasuhara Y, Kamachi A, *et al.* (2006) Role of macrophage migration inhibitory factor in ovalbumin-induced airway inflammation in rats. *Eur Respir J* **27**: 726–734.
15. Ohkawara T, Miyashita K, Nishihira J, *et al.* (2005) Transgenic overexpression of macrophage migration inhibitory factor renders mice markedly more susceptible to experimental colitis. *Clin Exp Immunol* **140**: 241–248.
16. Tanino Y, Makita H, Miyamoto K, *et al.* (2002) Role of macrophage migration inhibitory factor in bleomycin-induced lung injury and fibrosis in mice. *Am J Physiol Lung Cell Mol Physiol* **283**: L156–162.
17. Brown FG, Nikolic-Paterson DJ, Hill PA, *et al.* (2002) Urine macrophage migration inhibitory factor reflects the severity of renal injury in human glomerulonephritis. *J Am Soc Nephrol* **13** Suppl 1: S7–13.
18. Yu CM, Lai KW, Chen YX, *et al.* (2003) Expression of macrophage migration inhibitory factor in acute ischemic myocardial injury. *J Histochem Cytochem* **51**: 625–631.
19. Koda M, Nishio Y, Hashimoto M, *et al.* (2004) Upregulation of macrophage migration-inhibitory factor expression after compression-induced spinal cord injury in rats. *Acta Neuropathol (Berl)* **108**: 31–36.
20. Nakajima H, Takagi H, Horiguchi N, *et al.* (2006) Lack of macrophage migration inhibitory factor protects mice against concanavalin A-induced liver injury. *Liver Int* **26**: 346–351.
21. Martin P. (1997) Wound healing — aiming for perfect skin regeneration. *Science* **276**: 75–81.
22. Werner S, Grose R. (2003). Regulation of wound healing by growth factors and cytokines. *Physiol Rev* **83**: 835–870.
23. Ashcroft GS, Mills SJ, Ashworth JJ. (2002) Ageing and wound healing. *Biogerontology* **3**: 337–345.
24. Ashcroft GS, Horan MA, Ferguson MW. (1997) Aging is associated with reduced deposition of specific extracellular matrix components, an upregulation of angiogenesis, and an altered inflammatory response in a murine incisional wound healing model. *J Invest Dermatol* **108**: 430–443.
25. Hardman MJ, Ashcroft GS. (2005) Hormonal influences on wound healing: a review of current experimental data. *Wounds* **17**: 313–320.
26. Ashcroft GS, Dodsworth J, van Boxtel E, *et al.* (1997) Estrogen accelerates cutaneous wound healing associated with an increase in TGF-beta1 levels. *Nat Med* **3**: 1209–1215.

27. Ashcroft GS, Greenwell-Wild T, Horan MA, *et al.* (1999) Topical estrogen accelerates cutaneous wound healing in aged humans associated with an altered inflammatory response. *Am J Pathol* **155**: 1137–1146.
28. Rossouw JE, Anderson GL, Prentice RL, *et al.* (2002) Risks and benefits of estrogen plus progestin in healthy postmenopausal women: principal results from the women's health initiative randomized controlled trial. *JAMA* **288**: 321–333.
29. Ashcroft GS, Mills SJ, Lei K, *et al.* (2003) Estrogen modulates cutaneous wound healing by downregulating macrophage migration inhibitory factor. *J Clin Invest* **111**: 1309–1318.
30. Hardman MJ, Waite A, Zeef L, *et al.* (2005) Macrophage migration inhibitory factor: a central regulator of wound healing. *Am J Pathol* **167**: 1561–1574.
31. Ashcroft GS, Mills SJ, Flanders KC, *et al.* (2002) Role of Smad3 in the hormonal modulation of *in vivo* wound healing responses. *Wound Rep Regen* **11**: 468–473.
32. Shimizu T, Ohkawara A, Nishihira J, Sakamoto W. (1996) Identification of macrophage migration inhibitory factor (MIF) in human skin and its immunohistochemical localization. *FEBS Lett* **381**: 199–202.
33. Shimizu T, Abe R, Ohkawara A, *et al.* (1997) Macrophage migration inhibitory factor is an essential immunoregulatory cytokine in atopic dermatitis. *Biochem Biophy Res Commun* **240** 173–178.
34. Steinhoff M, Meinhardt A, Steinhoff A, *et al.* (1999) Evidence for a role of macrophage migration inhibitory factor in psoriatic skin disease. *Br J Dermatol* **141**: 1061–1066.
35. Shimizu T, Abe R, Nakamura H, *et al.* (1999) High expression of macrophage migration inhibitory factor in human melanoma cells and its role in tumor cell growth and angiogenesis. *Biochem Biophys Res Commun* **264**: 751–758.
36. Shimizu T, Abe R, Ohkawara A, Nishihira J. (1999) Ultraviolet B radiation upregulates the production of macrophage migration inhibitory factor (MIF) in human epidermal keratinocytes. *J Invest Dermatol* **112**: 210–215.
37. Abe R, Shimizu T, Ohkawara A, Nishihira J. (2000) Enhancement of macrophage migration inhibitory factor (MIF) expression in injured epidermis and cultured fibroblasts. *Biochim Biophys Acta* **1500**: 1–9.
38. Santos LL, Hall P, Metz CN, *et al.* (2001) Role of macrophage migration inhibitory factor (MIF) in murine antigen-induced arthritis: interaction with glucocorticoids. *Clin Exp Immunol* **123**: 309–314.
39. Kobayashi S, Nishihira J, Watanabe S, Todo S. (1999) Prevention of lethal acute hepatic failure by antimacrophage migration inhibitory factor antibody in mice treated with bacille Calmette-Guerin and lipopolysaccharide. *Hepatology* **29**: 1752–1759.

40. Ohkawara T, Nishihira J, Takeda H, *et al.* (2002) Amelioration of dextran sulfate sodium-induced colitis by anti-macrophage migration inhibitory factor antibody in mice. *Gastroenterology* **123**: 256–270.
41. Sakai Y, Masamune A, Satoh A, *et al.* (2003) Macrophage migration inhibitory factor is a critical mediator of severe acute pancreatitis. *Gastroenterology* **124**: 725–736.
42. Calandra T, Bernhagen J, Mitchell RA, Bucala R. (1994) The macrophage is an important and previously unrecognized source of macrophage migration inhibitory factor. *J Exp Med* **179**: 1895–1902.
43. Cao WG, Morin M, Sengers V, *et al.* (2006) Tumour necrosis factor-alpha upregulates macrophage migration inhibitory factor expression in endometrial stromal cells via the nuclear transcription factor NF-kappaB. *Hum Reprod* **21**: 421–428.
44. Bozza M, Satoskar AB, Lin G, *et al.* (1999) Targeted disruption of migration inhibitory factor gene reveals its critical role in sepsis. *J Exp Med* **189**: 341–346.
45. Honma N, Koseki H, Akasaka T, *et al.* (2000) Deficiency of the macrophage migration inhibitory factor gene has no significant effect on endotoxaemia. *Immunology* **100**: 84–90.
46. Fingerle-Rowson G, Petrenko O, Metz CN, *et al.* (2003) The p53-dependent effects of macrophage migration inhibitory factor revealed by gene targeting. *Proc Natl Acad Sci USA* **100**: 9354–9359.
47. Zhao Y, Shimizu T, Nishihira J, *et al.* (2005) Tissue regeneration using macrophage migration inhibitory factor-impregnated gelatin microbeads in cutaneous wounds. *Am J Pathol* **167**: 1519–1529.
48. Ashcroft GS, Horan MA, Ferguson MW. (1998) Aging alters the inflammatory and endothelial cell adhesion molecule profiles during human cutaneous wound healing. *Lab Invest* **78**: 47–58.
49. Oshima S, Onodera S, Amizuka N, *et al.* (2006) Macrophage migration inhibitory factor-deficient mice are resistant to ovariectomy-induced bone loss. *FEBS Lett* **580**: 1251–1256.
50. Tomasek JJ, Gabbiani G, Hinz B, *et al.* (2001) Myofibroblasts and mechano-regulation of connective tissue remodelling. *Nat Rev Mol Cell Biol* **3**: 349–363.
51. Takubo T, Tatsumi N. (1999) Microscopic co-distributions of myosin, actin, alpha-actinin and vinculin in human neutrophils during movement. *Haematologia* **29**: 285–294.
52. Richard P, Charron P, Carrier L, *et al.* (2003) Hypertrophic cardiomyopathy: distribution of disease genes, spectrum of mutations, and implications for a molecular diagnosis strategy. *Circulation* **107**: 2227–2232.
53. Wadgaonkar R, Dudek SM, Zaiman AL, *et al.* (2005) Intracellular interaction of myosin light chain kinase with macrophage migration inhibition factor (MIF) in endothelium. *J Cell Biochem* **95**: 849–858.

54. Ashcroft GS, Kielty CM, Horan MA, Ferguson MW. (1997) Age-related changes in the temporal and spatial distributions of fibrillin and elastin mRNAs and proteins in acute cutaneous wounds of healthy humans. *J Pathol* **183**: 80–89.
55. Agah A, Kyriakides TR, Letrondo N, *et al.* (2004) Thrombospondin 2 levels are increased in aged mice: consequences for cutaneous wound healing and angiogenesis. *Matrix Biol* **22**: 539–547.
56. Kyriakides TR, Tam JW, Bornstein P. (1999). Accelerated wound healing in mice with a disruption of the thrombospondin 2 gene. *J Invest Dermatol* **113**: 782–787.
57. Mazzalupo S, Wong P, Martin P, Coulombe PA. (2003) Role for keratins 6 and 17 during wound closure in embryonic mouse skin. *Dev Dyn* **226**: 356–365.
58. Patel GK, Wilson CH, Harding KG, *et al.* (2006) Numerous keratinocyte subtypes involved in wound re-epithelialization. *J Invest Dermatol* **126**: 497–502.
59. Stojadinovic O, Brem H, Vouthounis C, *et al.* (2005) Molecular pathogenesis of chronic wounds: the role of beta-catenin and, c-myc in the inhibition of epithelialization and wound healing. *Am J Pathol* **167**: 59–69.
60. Blumenthal A, Ehlers S, Lauber J (2006) The Wingless homologue Wnt5a and its receptor Frizzled-5 regulate inflammatory responses of human mononuclear cells induced by microbial stimulation. *Blood* Apr 6; (Epub ahead of print).
61. Cheon SS, Wei Q, Gurung A, *et al.* (2006) Beta-catenin regulates wound size and mediates the effect of TGF-beta in cutaneous healing. *FASEB J* **20**: 692–701.
62. Lobov IB, Rao S, Carroll TJ, *et al.* (2005) WNT7b mediates macrophage induced programmed cell death in patterning of the vasculature. *Nature* **437**: 417–421.
63. Fathke C, Wilson L, Shah K, *et al.* (2006) Wnt signaling induces epithelial differentiation during cutaneous wound healing. *BMC Cell Biol* **7**: 4.
64. Hardman MJ, Liu K, Avilion AA, *et al.* (2005) Desmosomal cadherin misexpression alters beta-catenin stability and epidermal differentiation. *Mol Cell Biol* **25**: 969–978.
65. Sen M (2005) Wnt signalling in rheumatoid arthritis. *Rheumatology* (Oxford) **44**: 708–713.
66. Mitchell RA. (2004) Mechanisms and effectors of MIF-dependent promotion of tumourigenesis. *Cell Signal* **16**: 13–19.
67. Mitchell RA, Metz CN, Peng T, Bucala R. (1999) Sustained mitogen-activated protein kinase (MAPK) and cytoplasmic phospholipase A2 activation by macrophage migration inhibitory factor (MIF). Regulatory role in cell proliferation and glucocorticoid action. *J Biol Chem* **274**: 18100–18106.

68. Watanabe H, Shimizu T, Nishihira J, *et al.* (2004) Ultraviolet A-induced production of matrix metalloproteinase-1 is mediated by macrophage migration inhibitory factor (MIF) in human dermal fibroblasts. *J Biol Chem* **279**: 1676–1683.
69. Thelu J, Rossio P, Favier B. (2002). Notch signalling is linked to epidermal cell differentiation level in basal cell carcinoma, psoriasis and wound healing. *BMC Dermatol* **2**: 7.
70. Esworthy RS, Aranda R, Martin MG, *et al.* (2001) Mice with combined disruption of Gpx1 and Gpx2 genes have colitis. *Am J Physiol Gastrointest Liver Physiol* **281**: G848–855.
71. Selvi E, Tripodi SA, Catenaccio M, *et al.* (2003) Expression of macrophage migration inhibitory factor in diffuse systemic Sclerosis. *Ann Rheum Dis* **62**: 460–464.
72. Hagemann T, Wilson J, Kulbe H, *et al.* (2005) Macrophages induce invasiveness of epithelial cancer cells via NF-kappaB and JNK. *J Immunol* **175**: 1197–1205.
73. Lue H, Kapurniotu A, Fingerle-Rowson G, *et al.* (2006) Rapid and transient activation of the ERK MAPK signalling pathway by macrophage migration inhibitory factor (MIF) and dependence on JAB1/CSN5 and Src kinase activity. *Cell Signal* **18**: 688–703.
74. Morand EF (2005) New therapeutic target in inflammatory disease: macrophage migration inhibitory factor. *Intern Med J* **35**: 419–426.
75. Onodera S, Kaneda K, Mizue Y, *et al.* (2000) Macrophage migration inhibitory factor upregulates expression of matrix metalloproteinases in synovial fibroblasts of rheumatoid arthritis. *J Biol Chem* **275**: 444–450.
76. Leech M, Santos LL, Metz C, *et al.* (2000) Control of macrophage migration inhibitory factor (MIF) by endogenous glucocorticoids in rat adjuvant arthritis. *Arthritis Rheum* **43**: 827–833.
77. Ichiyama H, Onodera S, Nishihira J, *et al.* (2004) Inhibition of joint inflammation and destruction induced by anti-type II collagen antibody/lipopolysaccharide (LPS)-induced arthritis in mice due to deletion of macrophage migration inhibitory factor (MIF). *Cytokine* **26**: 187–94.
78. Lacey DC, Sampey AV, Mitchell R, *et al.* (2003) Control of fibroblast-like synoviocyte proliferation by macrophage migration inhibitory factor (MIF). *Arthritis Rheum* **48**: 103–109.
79. Gregory JL, Leech MT, David JR, *et al.* (2004) Reduced leukocyte–endothelial cell interactions in the inflamed microcirculation of macrophage migration inhibitory factor-deficient mice. *Arthritis Rheum* **50**: 3023–3034.
80. Baugh JA, Chitnis S, Donnelly SC, *et al.* A functional promoter polymorphism in the macrophage migration inhibitory factor (MIF) gene associated with disease severity in rheumatoid arthritis. *Genes Immun* **3**: 170–176.

81. Radstake TR, Sweep FC, Welsing P, *et al.* (2005) Correlation of rheumatoid arthritis severity with the genetic functional variants and circulating levels of macrophage migration inhibitory factor. *Arthritis Rheum* **52**: 3020–3029.
82. Van Doornum S, McColl G, Wicks IP. (2002) Accelerated atherosclerosis: an extraarticular feature of rheumatoid arthritis? *Arthritis Rheum* **46**: 862–873.
83. Burger-Kentischer A, Goebel H, Seiler R, *et al.* (2002) Expression of macrophage migration inhibitory factor in different stages of human atherosclerosis. Circulation **105**: 1561–1566.
84. Burger-Kentischer A, Gobel H, Kleemann R, *et al.* (2006) Reduction of the aortic inflammatory response in spontaneous atherosclerosis by blockade of macrophage migration inhibitory factor (MIF). *Atherosclerosis* **184**: 28–38.
85. Ashworth JJ, Smyth JV, Pendleton N, *et al.* (2005) The dinucleotide (CA) repeat polymorphism of estrogen receptor beta but not the dinucleotide (TA) repeat polymorphism of estrogen receptor alpha is associated with venous ulceration. *J Steroid Biochem Mol Biol* **97**: 266–270.

12

MIF and Atherosclerosis

Christian Weber and Andreas Schober†*

INTRODUCTION

Cardiovascular disease has turned out to be the most prevalent chronic disease in many parts of the world, and it is predicted to become the main cause of disability and death worldwide in the early 21st century.[1] The pathologic basis for cardiovascular diseases, such as myocardial infarction or stroke, is the growth of atherosclerotic plaques in the vascular wall over a period of many years, leading to stenosis that impedes blood flow or plaque disruption with acute thrombotic occlusion.[2–4] In early lesion formation, leukocyte adhesion on arterial endothelial cells is the hallmark of the disease. Since normal arterial endothelial cells generally do not support interactions with leukocytes, activation of endothelium predominantly by the subendothelial accumulation of modified lipoprotein particles, is required. The expression of leukocyte adhesion molecules on the endothelial surface, such as the vascular cell adhesion molecule (VCAM)-1, is known to be an important prerequisite for the firm adhesion of leukocytes to the

*,†Institute of Molecular Cardiovascular Research, RWTH Aachen University, 52074 Aachen, Germany.
*Corresponding author: Christian Weber.
*Institut für Kardiovaskuläre Molekularbiologie, Universitätsklinikum der RWTH Aachen, Pauwelsstrasse 30, 52074 Aachen, Germany. E-mail: cweber@ukaachen.de
†Present address: Medizinische Poliklinik, Ludwig Maximilans-Universität München, Germany.

arterial wall.[2–4] Interestingly, monocytes and T-cells, which accumulate in early atherosclerotic lesions, express the $\beta1$ integrin VLA-4, mediating the interactions of leukocytes with VCAM-1 on endothelial cells. Another group of chemoattractant cytokines, or chemokines, is essential for the rapid increase of the integrin-binding activity and other steps of leukocyte adhesion.[5,6] In animal models, the chemokine receptor CXCR2 and its CXC chemokine ligand CXCL1 (growth-related oncogene-α or keratinocyte-derived chemokine) have been shown to enhance VLA-4-mediated binding of monocytes to the luminal surface of carotid arteries.[7] Furthermore, the CC chemokine CCL2 (monocyte chemotactic protein-1) and its receptor CCR2 are involved in atherogenesis, presumably by directing the migration of leukocytes through the endothelial cell layer.[5,6] Once recruited to the intima of arteries, monocytes are transformed into foam cells as a result of excessive lipid accumulation, which is a characteristic of the nascent atheroma. Moreover, the predominance of monocyte-derived foam cells in the later stages of plaque growth indicates a higher inflammatory activity and a greater risk for acute plaque rupture with subsequent thrombotic vessel occlusion, mostly a devastating complication of atherosclerosis.[2–4]

In certain subtypes of arteriosclerosis, neointima formation is the response to different kinds of injuries. The mechanical injury of percutaneous arterial interventions by balloon angioplasty or stent implantation results in luminal re-narrowing in up to one-third of cases within 6 months. In contrast to native atherosclerosis, neointimal smooth muscle cell (SMC) accumulation predominates in restenosis.[8]

EXPRESSION OF MIF DURING ATHEROSCLEROTIC PLAQUE DEVELOPMENT

Eponymously discovered as a macrophage migration inhibitory factor, the pleiotropic cytokine MIF has been implicated in many acute and chronic inflammatory diseases such as septic shock, colitis, glomerulonephritis, or arthritis via a proinflammatory recruitment of mononuclear cells.[9] With inflammatory processes in the vessel wall widely acknowledged to give rise to atherosclerosis,[2–4] the first evidence for an involvement of MIF in atherogenesis was unveiled in hypercholesterolemic rabbits, which develop extensive lesions of all stages throughout the abdominal aorta.[10] MIF mRNA and protein expression was upregulated in vascular endothelial cells in early fatty-streak and advanced lesions compared with non-lesion-associated areas. In contrast, SMCs in the media exhibit increased MIF expression only in early lesions with subendothelial macrophage accumulation.[10] Since macrophages activated by lipopolysaccharide, tumor necrosis factor (TNF)-α and interferon-γ have been shown to secrete substantial amounts of preformed MIF, it is reasonable to assume that macrophage-derived MIF could be involved in atherogenesis. Indeed, in early fatty-streak lesions, monocytes adhering to the endothelial cells and subendothelial cells exhibit strong MIF expression, which was clearly diminished in foam cells and macrophages of advanced plaques.[10] These results were compatible with the hypothesis that MIF is predominantly involved in early lesion development, presumably by supporting monocyte recruitment. A mechanism of MIF-dependent activation of endothelial cells was proposed, since *in vitro* up-regulation of intercellular adhesion molecule-1 (ICAM-1) on endothelial cells was observed after treatment with MIF.[10]

These data from a rabbit model were largely confirmed in a study using human specimens from patients undergoing heart transplantation, carotid endarterectomy or from autopsied individuals. Although MIF was detected only in a small number of endothelial cells and SMCs by immunohistochemistry in arteries with normal histology or type I (initial) lesions without macrophage infiltration the number of MIF-expressing cells was significantly higher in fatty-streak (type II) lesions.[11] About 24% of intimal cells in these foam cell-rich lesions produce MIF, including endothelial cells, SMCs, macrophages and T-lymphocytes. In the medial vessel wall layer of fatty-streak lesions, roughly 21% of SMCs expressed MIF. MIF was found to be co-localized with its cellular co-activator Jab-1.[11] In contrast to the findings in hypercholesterolemic rabbits, a further increase in the number of MIF-expressing cells was reported in advanced lesions of human arteries.[11] A causal role of oxidized low-density lipoprotein (oxLDL) in MIF up-regulation was observed *in vitro* in endothelial cells and macrophages. The peak expression of MIF protein in human endothelial cells was observed after 24 hours and with 10 μg/ml of oxLDL.[11]

Furthermore, oxLDL-induced MIF secretion by macrophages at concentrations above 45 μg/ml. Therefore, an important role of MIF in the progression of human atherosclerosis was suggested, contributing to the regulation of plaque initiation and advanced lesion growth.

MIF AND NEOINTIMAL HYPERPLASIA AFTER VASCULAR INJURY

Initial insights into the functional role of MIF in vascular disease were derived from vascular-injury models using hypercholesterolemic low density lipoprotein receptor-deficient (LDL-R$^{-/-}$) or apolipoprotein E-deficient (ApoE$^{-/-}$) mice, where the MIF activity was inhibited with a blocking antibody. In LDL-R$^{-/-}$ mice, endothelial denudation and controlled stretching of the carotid artery was performed by using an angioplasty inflation device (2.5 atm) with subsequent air-drying of the luminal surface through an air exit hole at the proximal end of the common carotid artery. A monoclonal anti-MIF antibody or isotype control was injected intraperitoneally five times (500 μg each injection) up to 11 days after the carotid injury.[12] At 7 days after carotid injury, enhanced MIF expression was detected by immunostaining in the media, which was grossly diminished 28 days after injury. The anti-MIF antibody treatment had no effect on the neointimal area 7 days after carotid injury as compared with the isotype control, but reduced the proliferation of medial cells, while the apoptosis of neointimal and medial cells, although a rare event at that point of time, was increased in anti-MIF antibody-treated mice.[12] The inhibition of apoptosis appears to be related to the function of MIF as a disulfide reductase, as seen for cellular redox homeostasis and MIF-mediated monocyte/macrophage activation.[13] Moreover, the neointimal content of CD45-positive inflammatory cells was reduced from approximately 50% in the isotype control-treated to 36% in anti-MIF antibody-treated mice 7 days after injury. After 28 days, however, a 54% reduction of the neointimal area was found in LDL-R$^{-/-}$ mice treated with anti-MIF antibody without a significant difference in the neointimal CD45-positive cells compared with the isotype control. The proliferation rate of neointimal cells 28 days after injury was 5.9% in control mice, which appears much lower than in medial cells after 7 days (24%). Anti-MIF antibody treatment reduced the neointimal proliferation rate by 56%.[12] Unfortunately, it remains unclear whether a diminished accumulation of SMCs accounts for the reduction in the neointimal area. It is conceivable that the marked inhibition of the inflammatory response by neutralizing MIF at 7 days may affect the neointimal growth at advanced stages. The changes of the low-baseline neointimal proliferation rate at 28 days and in the small number of apoptotic cells at 7 days appear to hardly explain the differences in the neointimal area. Whether the inhibition of medial cell proliferation by the anti-MIF antibody 7 days after injury affects the neointimal area after 28 days remains at least partially elusive. *In vitro*, the addition of serum or platelet-derived growth factor (PDGF) to cultured SMCs induced the release of MIF and incubation with the neutralizing anti-MIF antibody inhibited the serum-induced proliferation of SMCs, suggesting a possible autocrine response SMC-derived MIF, which contributes to the mitogenic effect of serum stimulation.[12]

In a model of wire-induced carotid artery injury using hypercholesterolemic ApoE$^{-/-}$ mice, the role of MIF in neointima formation was evaluated by injection of a blocking monoclonal anti-MIF antibody (100 μg twice weekly for 3 weeks) or isotype control antibody.[14] Plaque size tended to be lower in MIF antibody-treated mice, but the difference was not significant. A comparison of the plaque composition in the two groups revealed a marked decrease of neointimal macrophage content in anti-MIF antibody-treated ApoE$^{-/-}$ mice.[14] Moreover, neointimal foam-cell accumulation, which is a characteristic of this model of vascular remodeling, in relation to the content of macrophages was also attenuated by treatment with anti-MIF antibody. Therefore, MIF appears to be centrally involved in neointimal macrophage accumulation and may also promote the transformation of macrophages into foam cells *in vivo*. Conversely, a compensatory increase in neointimal SMC and collagen type I content in ApoE$^{-/-}$ mice was evident, when MIF was inhibited by antibody treatment, resulting in a more stable plaque phenotype.[14] How this relates to the possibly subtype-specific short-term enhancement of PDGF-induced vascular SMC migration by MIF stimulation[15] remains to be determined. This constituted the first evidence of a pro-inflammatory role of MIF in atherosclerotic plaque destabilization. Interestingly, monocyte adhesion on oxLDL-treated human aortic endothelial cells depends almost completely on endothelial MIF and stimulation of aortic endothelial cells with MIF for 2 hours induced monocyte adhesion under flow conditions, providing preliminary hints at a chemokine-like function of MIF in leukocyte recruitment.[14] In conjunction with this, the data infer that blocking MIF in manifest atherosclerosis may serve to mediate a conversion of vulnerable plaques into a more stable condition less susceptible to rupture.

It is notable that differences in neointima formation in the common carotid artery after partial ligation of the external and internal carotid artery in four different mouse strains were associated with higher baseline and neointimal levels of MIF gene expression in a mouse strain with increased neointimal hyperplasia as determined by microarray profiling.[16]

ROLE OF MIF IN NATIVE ATHEROSCLEROSIS

The important involvement of MIF in atherosclerotic plaque progression was corroborated in animal studies using MIF-deficient ($MIF^{-/-}$) LDL-$R^{-/-}$ double knockout mice on an atherogenic diet for 12 and 26 weeks.[17] Hyperlipidemic LDL-$R^{-/-}$ mice develop typical atherosclerotic plaques of various stages in the aortic arch and the abdominal aorta. In MIF-deficient $LDL^{-/-}$ mice, the lesion area in the abdominal aorta and the intimal thickness in the aortic arch were significantly reduced.[17] Further characterization revealed less advanced lesions in $MIF^{-/-}$ LDL-$R^{-/-}$ mice compared with those in MIF+/+ $LDL^{-/-}$ mice. This finding may be explained by a diminished proliferation rate of neointimal cells in MIF-deficient mice. Of note, the total serum cholesterol and LDL-cholesterol were also reduced in MIF-deficient mice.[17] It cannot entirely be excluded that differences in the genetic background of the 5th generation backcrossed $MIF^{-/-}$ $LDL^{-/-}$ C57/Bl6 mice and LDL-$R^{-/-}$ C57/Bl6 mice may in part account for the reduced lesion area in $MIF^{-/-}$, since different strains of mice have a great variability in the susceptibility to atherogenic diet and development of atherosclerosis. MIF has not been shown to affect cholesterol metabolism; therefore, the reduced total cholesterol and LDL concentrations in $MIF^{-/-}$ mice may be attributable to the mixed genetic background. *In vitro*, MIF-deficient murine SMCs exhibited impaired proteolytic activity as evidenced by reduced elastin and collagen degradation.[17]

In another study using an $ApoE^{-/-}$ mouse model of spontaneous atherosclerosis, mice on normal chow diet for 14 weeks were treated with a monoclonal anti-MIF antibody (200 μg twice weekly) or isotypic control antibody.[18] MIF expression was detected abundantly in all atheroslerotic plaques, predominantly in endothelial cells and foam cells. Interestingly, the inhibition of MIF resulted in markedly reduced intimal macrophage content, which is in agreement with the findings for wire-induced neointima formation.[18] In addition, the blood levels of inflammatory proteins were reduced in MIF antibody-treated $ApoE^{-/-}$ mice. In the aortic wall local expression of ICAM-1, TNF-α and CD40 ligand was decreased by anti-MIF antibody treatment, whereas matrix metalloprotease (MMP)-2 and interleukin (IL)-12-p40 mRNA levels were upregulated.[18] Possible transcriptional regulators of plaque development and inflammation, such as P-c-jun and C-EBP-b, were downregulated in anti-MIF antibody-treated mice.[18] In contrast to the results obtained in $MIF^{-/-}$ LDL-$R^{-/-}$ mice, plaque volumes in total aortas and in the aortic root did not significantly differ between anti-MIF antibody-treated and isotype control antibody-treated mice.[17] Therefore, MIF appears to be involved in the inflammatory component of atherogenesis, which is considered to be responsible for plaque vulnerability and the subsequent risk for thrombotic arterial occlusion.

In a recent study (Bernhagen and Weber, *personal communication*), the molecular mechanisms of MIF-mediated atherogenesis have been evaluated in more detail. In a series of experiments, MIF was shown to induce firm arrest of monocytes on stimulated endothelial cells under flow conditions, and therefore a chemokine-like function of MIF has been clearly established. This effect is dependent on the MIF-binding protein CD74 and the chemokine receptor CXCR2 on monocytes, as shown by inhibition with blocking antibodies. These results were confirmed in *ex vivo* perfused murine carotid arteries, where anti-MIF antibody blockade reduced monocyte arrest in $ApoE^{-/-}$ or LDL-$R^{-/-}$ mice to a similar extent as neutralizing CXCR2. In $MIF^{-/-}$ LDL-$R^{-/-}$ mice, the inhibition of monocyte arrest after CXCR2 or anti-MIF antibody blockade was only observed after loading with exogenous MIF. Importantly, monocyte arrest on endothelial cells after short-term stimulation with MIF depended on a rapid activation of leukocyte integrins. Furthermore, MIF evokes monocyte chemotaxis, since transmigration *in vitro* was induced by an MIF gradient with comparable efficacy as that of CCL2. Similarly, a function of MIF in stimulating migration has been reported for SMCs.[15] MIF-triggered monocyte chemotaxis was dependent on $G_{\alpha i}$ protein and phosphoinositide-3-kinase, typical of chemokine-mediated transmigration. As for adhesion, monocyte migration towards MIF was mediated via CD74 and CXCR2 on monocytes. This provides a novel pathway conferring proatherogenic actions of MIF.

MIF AND PLAQUE DESTABILIZATION

Additional evidence for the contribution of MIF to plaque destabilization was provided by a study of human iliac arteries with obstructive atherosclerosis.[19] Up-regulation of MIF and MMP-9 was found predominantly in vulnerable atherosclerotic plaques, but not in fibrous lesions, concomitant with an increased macrophage accumulation.[19] Double-immunohistochemistry further demonstrated a cellular co-localization of MIF and MMP-9, which was detected in serial sections to CD68-positive macrophages and a subtype of SMCs. In a macrophage cell line and primary aortic SMCs, mRNA and protein expression for MMP-9 was found to be enhanced by MIF treatment in a dose-dependent manner.[19] Since MMP-9 can degrade most components of the extracellular matrix in atherosclerotic plaques, a major role of this proteinase has been suggested in plaque destabilization.[2–4] Therefore, MIF-dependent MMP-9 up-regulation may be involved in the process of plaque stabilization observed in ApoE$^{-/-}$ mice treated with a neutralizing MIF antibody.

In another study by the same group, increased MMP-1 expression was described in MIF-positive macrophages and SMCs in vulnerable atherosclerotic plaques, which was associated with enhanced collagenolysis in plaque tissue, as demonstrated by immunostaining for the collagenase-cleaved type I collagen fibrils.[20] The hypothesis that MIF can induce MMP-1 in plaque cells was further evaluated in cultured SMCs. Indeed, the stimulation of SMCs with MIF resulted in increased MMP-1 mRNA, protein expression and collagenase activity.[20] Although the role of MMP-1 in atherosclerotic plaque development and destabilization is unclear, these results are consistent with an MIF-induced collagen degradation via MMP-1, which may be a relevant mechanism leading to plaque vulnerability.

Similar findings were obtained in human carotid endatherectomy samples, where MIF expression was shown to be progressively increased in the advanced plaque stages.[20] In particular, CD40 ligand- and angiotensin II-positive mononuclear cells in complicated atherosclerotic plaques exhibited enhanced MIF staining. In addition to CD68-expressing intimal macrophages, endothelial cells in plaque microvessels were positive for MIF,[21] while CD40 ligand-expression was absent in microvascular endothelial cells. To identify the molecular pathways involved in MIF up-regulation, *in vitro* studies that employed the monocytic cell line THP-1 and involving treatment with CD40 ligand, angiotensin II, glucose, insulin and hypoxia, where carried out. Although hyperglycemia and insulin had no effect on MIF expression, angiotensin II induced MIF secretion in THP-1 cells via the angiotensin I-receptor.[21] Furthermore, CD40 ligand stimulation of THP-1 cells enhanced the MIF mRNA expression and MIF secretion, which could provide an explanation for the co-localization of MIF and CD40 ligand demonstrated in human plaque tissue. Thus, CD40 ligand and angiotensin II stimulation appears to represent an important pathway for MIF up-regulation and plaque instability.

ROLE OF MIF IN ANEURYSM FORMATION

Abdominal aortic aneurysm is another manifestation of atherosclerosis tightly linked to the inflammatory component of the disease. In a clinical study, not only the local expression in aneurysmal lesions but also MIF serum levels were associated with inflammatory vascular disease and found to specifically correlate with the size and expansion of abdominal aortic aneurysm.[22] This was substantiated by findings that the aneurysmal expression of MIF was paralleled by an enhanced expression of MMP-1, MMP-9 and MMP-12; a co-localization with MMPs in ruptured aneurysms; and by a decreased expression of their inhibitors.[23] In line with the well-documented inflammatory pathogenesis of atherosclerosis[2–4] and the association of MIF expression with intima-media thickening,[16] this could imply that MIF levels may serve as a prognostic biomarker for clinical manifestations of atherosclerosis, as demonstrated for other and partly related inflammatory mediators, e.g., for the arrest chemokine CCL5/RANTES.[2–6,24]

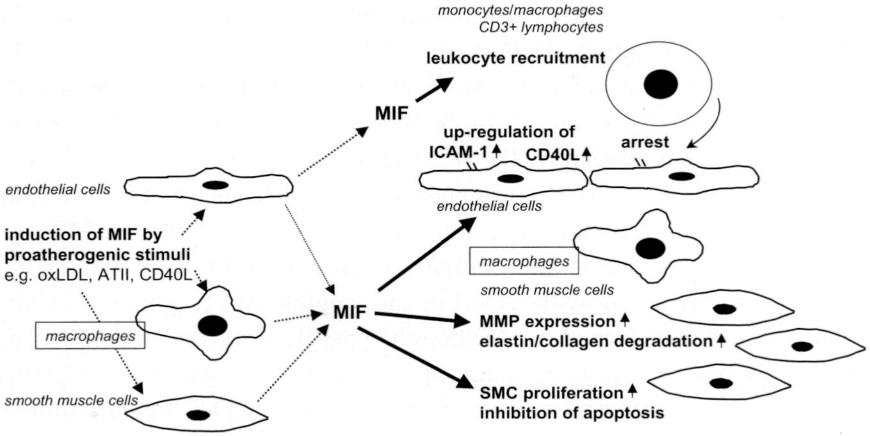

Fig. 1 Cellular effects of MIF in atherosclerotic arteries.

Table 1 Cellular Mechanisms of MIF in Atherosclerosis and Vascular Remodeling

Vascular Effects of MIF	Ref. No.
Expression of MIF in intimal macrophages, T-cells, endothelial cells, SMCs	10, 11
Up-regulation of ICAM-1 on endothelial cells by macrophage-derived MIF	10
Up-regulation of MIF in endothelial cells or macrophages by oxLDL	11
Reduced proliferation of medial and neointimal cells by blocking MIF	12
Increased apoptosis of medial cells after vascular injury by blocking MIF	12
Autocrine contribution to SMC proliferation	12
Neointimal $CD45^{+}$-positive leukocyte and macrophage infiltration	12, 14
Induction of monocyte arrest on activated vascular endothelium	14
Short-term enhancement of PDGF-stimulated SMC migration by MIF	15
Impaired proteolytic degradation of elastin or collagen in MIF-deficient SMCs	17
Decreased expression of aortic ICAM-1, TNF-α and CD40L by blocking MIF	18
Down-regulation of transcriptional regulators (e.g. P-c-jun) by blocking MIF	18
Induction of MMP expression and collagenolysis in SMCs and macrophages	19, 20
Stimulation of MIF secretion in macrophages by angiotensin II and CD40L	21
Co-localization of MIF and MMP up-regulation during aneurysmal expansion	22, 23

As summarized in Table 1, a considerable body of data has collectively helped to establish the remarkable activity of MIF as an inflammatory mediator in the atherogenic recruitment of mononuclear cells and as a regulator of vascular cell functions. These effects of MIF explain its importance in the pathogenesis of unstable atherosclerosis and possibly its role in risk assessment of atherosclerotic disease.

REFERENCES

1. Fuster V. (1999) Epidemic of cardiovascular disease and stroke: the three main challenges. Presented at the 71st Scientific Sessions of the American Heart Association, Dallas, Texas. *Circulation* **99**: 1132–1137.
2. Ross R. (1993) The pathogenesis of atherosclerosis: a perspective for the 1990s. *Nature* **362**: 801–809.
3. Libby P. (2002) Inflammation in atherosclerosis. *Nature* **420**: 868–874.
4. Hansson GK. (2005) Inflammation, atherosclerosis, and coronary artery disease. *N Engl J Med* **352**: 1685–1695.
5. Weber C, Schober A, Zernecke A. (2004) Chemokines: key regulators of mononuclear cell recruitment in atherosclerotic vascular disease. *Arterioscler Thromb Vasc Biol* **24**: 1997–2008.
6. Weber C. (2003) Novel mechanistic concepts for the control of leukocyte transmigration: specialization of integrins, chemokines, and junctional molecules. *J Mol Med* **81**: 4–19.
7. Huo Y, Weber C, Forlow SB, *et al.* (2001) The chemokine KC, but not monocyte chemoattractant protein-1, triggers monocyte arrest on early atherosclerotic endothelium. *J Clin Invest* **108**: 1307–1314.
8. Bittl JA. (1996) Advances in coronary angioplasty. *N Engl J Med* **335**: 1290–1302.
9. Calandra T, Roger T. (2003) Macrophage migration inhibitory factor: a regulator of innate immunity. *Nat Rev Immunol* **3**: 791–800.
10. Lin SG, Yu XY, Chen YX, *et al.* (2000) De novo expression of macrophage migration inhibitory factor in atherogenesis in rabbits. *Circ Res* **87**: 1202–1208.
11. Burger-Kentischer A, Goebel H, Seiler R, *et al.* (2002) Expression of macrophage migration inhibitory factor in different stages of human atherosclerosis. *Circulation* **105**: 1561–1666.
12. Chen Z, Sakuma M, Zago AC, *et al.* (2004) Evidence for a role of macrophage migration inhibitory factor in vascular disease. *Arterioscler Thromb Vasc Biol* **24**: 709–714.

13. Thiele M, Bernhagen J. (2005) Link between macrophage migration inhibitory factor and cellular redox regulation. *Antioxid Redox Signal* **7**: 1234–1248.
14. Schober A, Bernhagen J, Thiele M, *et al.* (2004) Stabilization of atherosclerotic plaques by blockade of macrophage migration inhibitory factor after vascular injury in apolipoprotein E-deficient mice. *Circulation* **109**: 380–385.
15. Schrans-Stassen BH, Lue H, Sonnemans DG, *et al.* (2005) Stimulation of vascular smooth muscle cell migration by macrophage migration inhibitory factor. *Antioxid Redox Signal* **7**: 1211–1216.
16. Korshunov VA, Nikonenko TA, Tkachuk VA, *et al.* (2006) Interleukin-18 and macrophage migration inhibitory factor are associated with increased carotid intima-media thickening. *Arterioscler Thromb Vasc Biol* **26**: 295–300.
17. Pan JH, Sukhova GK, Yang JT, *et al.* (2004) Macrophage migration inhibitory factor deficiency impairs atherosclerosis in low-density lipoprotein receptor-deficient mice. *Circulation* **109**: 3149–3153.
18. Burger-Kentischer A, Gobel H, Kleemann R, *et al.* (2006) Reduction of the aortic inflammatory response in spontaneous atherosclerosis by blockade of macrophage migration inhibitory factor (MIF). *Atherosclerosis* **184**: 28–38.
19. Kong YZ, Yu X, Tang JJ, *et al.* (2005) Macrophage migration inhibitory factor induces MMP-9 expression: implications for destabilization of human atherosclerotic plaques. *Atherosclerosis* **178**: 207–215.
20. Kong YZ, Huang XR, Ouyang X, *et al.* (2005) Evidence for vascular macrophage migration inhibitory factor in destabilization of human atherosclerotic plaques. *Cardiovasc Res* **65**: 272–282.
21. Schmeisser A, Marquetant R, Illmer T, *et al.* (2005) The expression of macrophage migration inhibitory factor 1alpha (MIF 1alpha) in human atherosclerotic plaques is induced by different proatherogenic stimuli and associated with plaque instability. *Atherosclerosis* **178**: 83–94.
22. Pan JH, Lindholt JS, Sukhova GK, *et al.* (2003) Macrophage migration inhibitory factor is associated with aneurysmal expansion. *J Vasc Surg* **37**: 628–635.
23. Verschuren L, Lindeman JH, van Bockel JH, *et al.* (2005) Up-regulation and coexpression of MIF and matrix metalloproteinases in human abdominal aortic aneurysms. *Antioxid Redox Signal* **7**: 1195–1202.
24. von Hundelshausen P, Koenen RR, Sack M, *et al.* (2005) Heterophilic interactions of platelet factor 4 and RANTES promote monocyte arrest on endothelium *Blood* **105**: 924–930.

13

Inflammatory Arthritis

Eric F. Morand[*,†] *and Michelle Leech*[*]

INTRODUCTION

Inflammatory arthritis is a common cause of pain and disability in adults. There are over 100 different clinical syndromes of inflammatory arthritis, of which rheumatoid arthritis (RA) is the most common, affecting approximately 1% of adults worldwide. RA is characterized clinically by chronic inflammatory polyarthritis, which is highly destructive in nature and leads to significant permanent damage to critical anatomical structures in the joint. Joint involvement is characterized by pain, swelling, stiffness and loss of function even preceding the onset of structural damage. Structural damage in the form of cartilage and bone destruction is seen within 1–2 years of disease onset. RA is characterized by systemic complications, including non-specific effects of chronic inflammation such as fatigue and anemia, and more severe extraarticular manifestations of disease such as vasculitis, ocular disease, pulmonary alveolitis, and soft tissue granulomatous nodules. RA is associated with significantly increased mortality, with a reduction of life expectancy of the order of 10 years for patients with severe disease. Contributing significantly to this is the marked increase in the prevalence

*Centre for Inflammatory Diseases, Monash Institute for Medical Research/Monash University, Department of Medicine, 246 Clayton Road, Clayton 3168 VIC Australia.
†Corresponding author. E-mail: eric.morand@med.monash.edu.au

and severity of atherosclerotic vascular disease in patients with RA. Finally, glucocorticoids (GC), of which the first report of human use was in the treatment of RA, are still used in up to 70% of patients despite major dose-dependent toxicity. The unique relationship between MIF and GC is therefore of particular relevance in the case of RA.

In recent years the critical role of cytokines in the pathogenesis of RA has become clear, with the demonstration of clinical efficacy of specific cytokine antagonism in human subjects with RA. Although many patients with RA are responsive to existing cytokine antagonists, at least 50% of subjects do not achieve a sufficient clinical response, and treatments directed at alternative therapeutic targets are therefore required. Considerable evidence now supports the macrophage migration inhibitory factor (MIF) as a potential therapeutic target in RA, and this evidence will be reviewed in this chapter.

PATHOGENESIS OF RHEUMATOID ARTHRITIS

The pathogenesis of RA has been the subject of several excellent recent reviews,[21,74] and a full dissection of the field is beyond the scope of this chapter. Nonetheless, information regarding the key elements of RA pathogenesis is required in order to interpret the evidence that MIF is involved in this disease.

The Pathology of RA

The normal synovium is a relatively hypocellular, loosely assembled tissue, with a lining layer of macrophage-like and fibroblast-like synoviocytes (FLS).[18] The lining layer is normally only 1–2 cells thick (Fig. 1), and although it has some functions of an epithelium, such as the secretion of hyalouronic acid, it is not truly an epithelium and has no basement membrane. The key pathological feature of RA is a marked expansion of the synovium, characterized both by infiltration of leukocytes (especially monocytes/macrophages and Th_1 CD4+ T lymphocytes) and by hyperplasia of resident cells such as fibroblast-like synoviocytes (FLS).[74] (Fig. 1). The lining layer therefore becomes markedly thickened, up to 6–10 cells

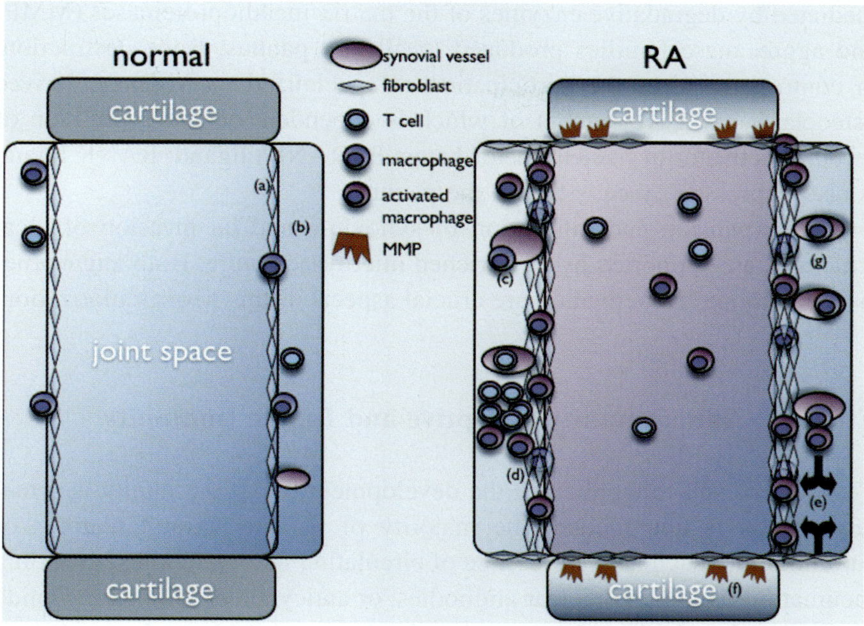

Fig. 1 Actions of MIF in RA. The left panel shows a representation of a normal joint, with a thin lining layer (a) and relatively hypocellular and hypovascular sublining tissue (b). The right panel shows the changes associated with RA, including recruitment and activation of blood-borne leukocytes (c), hyperplasia of the lining layer fibroblasts and macrophages (d), increased production of cytokines by macrophages and fibroblasts (e), degradation of cartilage by enzymes such as matrix metalloproteinases (MMP) (f), and increased numbers of sublining tissue blood vessels (g). Each of these phenomena can be initiated and/or regulated by MIF.

thick, and the sparsely populated sublining tissue becomes densely infiltrated with leukocytes. Although macrophages and CD4+ T cells predominate, the RA synovial lesion is also rich in mast cells, dendritic cells, B cells and plasma cells. Interestingly, while neutrophils are the dominant cell type in the free synovial fluid, they are comparatively rare in the synovial tissue itself.

Established RA is characterized by the development of pannus, an invasive outgrowth of synovial tissue over the surface of articular cartilage. Pannus is composed of macrophages and FLS and is capable of directly invading cartilage and promoting bone destruction. Cartilage damage is mediated by degradative enzymes of the matrix metalloproteinases (MMP) and aggrecanase families produced locally by pannus. Bone destruction, in contrast, requires the participation of specialized macrophage-derived osteoclasts, the development of which is dependent on the expression of receptor activator of nuclear factor kappa B (RANK) ligand; RANK ligand is also expressed by cells within the pannus.

The expansion and infiltration of synovium and its invasion of local structures are supported by an enriched microvasculature. Both angiogenesis and endothelial activation are crucial aspects of this neovascularization.

Etiology: Autoimmunity, Adaptive and Innate Immunity

That loss of self-tolerance and the development of autoimmunity is a feature of RA is unarguable. The majority of patients exhibit features of autoimmunity such as the presence of circulating autoantibodies, including rheumatoid factor, antinuclear antibodies, or anticyclic citrullinated peptide antibodies. In addition, RA synovium is characterized by a dense infiltration of CD4+ T cells, which often organize into pseudo-lymphoid follicles. Involvement of the adaptive immune response is also supported by genetic studies, which demonstrate a linkage between the prevalence and severity of RA and expression of certain HLA class II haplotypes. A sequence which has come to be known as the "shared epitope" is located in amino acids 70 to 74 (glutamine-leucine-arginine-alanine-alanine (QKRAA) or QRRAA) of the third hypervariable region of DRβ chains, and is found in RA-associated DR genes, including DR4, DR14 and DR1. Moreover, many animal models of RA are T cell dependent and characterized by autoimmunity. The best evidence for the importance of monocyte/macrophage-derived cytokines (monokines) monokines such as tumor necrosis factor (TNF) in RA comes from clinical trials of specific cytokine antagonists. Similarly, the best evidence in support of the role of the adaptive immune in the pathology of RA also comes from clinical studies. Therapies directed at T cells (such as cyclosporin A), B cells (such as the anti-CD20 B cell depleting antibody, rituximab) or co-stimulation molecules (such as recombinant CTLA4-Ig, abatacept) have been shown to have clinical benefit in RA and are approved for use in this disease in some countries.[20,35,90]

Evidence for the activation of innate immune pathways in RA is, however, also strong. In addition to indirect evidence of the involvement of "monokines" (see below), activation of intracellular signal transduction pathways known to be activated by innate immune signals is well described in RA. Innate immune system activators, such as bacterial endotoxins like lipopolysaccharide, activate cells through Toll-like receptors (TLR) such as TLR-4,[67] and TLR-4 expression has been reported in RA synovial tissue.[60] Expression of other TLRs, such as TLR-2, -3, and -7, has also been reported in human RA and modulation of these molecules is beneficial in animal models of RA.[31,32,65] Intracellular signal transduction pathways activated by TLR ligation are strikingly similar to those identified as responsive to cytokines, such as interleukin (IL)-1 and to a lesser extent TNF, and these are highly upregulated in RA synovial tissue.[22] It is not yet clear whether TLR ligands have an important role in human RA, or whether the sharing of signalling pathways with IL-1 and TNF is dominant in the innate immune activation phenotype of RA.

Pathology of RA: The Pivotal Role of Cytokines

Cytokines have been shown in many preclinical studies to be pivotal to the pathology of synovial inflammation. Indeed, cytokines have proven to be excellent therapeutic targets in RA, with antagonists directed at TNF and IL-1 already a routine part of the therapeutic armamentarium in RA and antagonism of IL-6 in late clinical development. Targeting cytokines has been successful at least in part because this approach allows RA pathology to be treated even though the etiology of the disease remains unknown.

Cytokines such as TNF and IL-1 contribute to pain and swelling via the upregulation of mediators such as prostaglandin E_2 (PGE_2), and to joint damage via activation of matrix metalloproteinases (MMPs), aggrecanases, and proteins involved in osteoclast activation such as RANKL. TNF-mediated activation of the NF-κB pathway has also been shown to be important in the regulation of synovial hyperplasia by mediating an anti-apoptotic influence on synovial cells.[46]

Human and animal studies also demonstrate the participation of cytokines such as granulocyte macrophage-colony stimulating factor (GM-CSF), and IL-8, IL-15, IL-18 and IL-17 in human RA and animal models. Many of these cytokines are the targets of therapeutic development programs for RA.

EXPRESSION AND FUNCTION OF MIF IN RHEUMATOID ARTHRITIS

Human RA

Leech *et al.* first reported the expression of MIF in RA.[41] MIF was reported to be overexpressed in the serum, synovial fluid, and cultured synovial fibroblasts of patients with RA compared with healthy or osteoarthritic controls. In these studies, MIF was localized to macrophages, endothelial cells and fibroblast-like synoviocytes (FLS) in RA synovial tissue, and was less readily demonstrable in CD3-positive synovial lymphoid aggregates.[41] The expression of MIF by cultured synovial fibroblasts was also confirmed in this study, with intracellular and released protein readily detected and MIF mRNA demonstrated by RT-PCR. Onodera *et al.* also reported on the expression of MIF in RA tissues.[57] The overexpression of MIF in RA synovial compared with control samples was again demonstrated. In this study, T cell expression of MIF was reported, but the authors cited as a marker of T cell phenotype CD45, which is in fact a pan-leukocyte marker.

Morand *et al.* have described an association between RA disease activity, as measured by serum CRP, and synovial MIF content in patients with RA.[50] Significantly increased synovial MIF was detected in synovial biopsies from patients with active RA, compared with paired samples obtained following treatment. Moreover, successful induction of remission with conventional anti-rheumatic drugs was associated with significant reduction in synovial MIF content. Direct associations between synovial MIF and other markers of disease activity, such as joint counts, were not demonstrated in this small study.

Significant associations of polymorphisms in the *MIF* gene with the incidence and severity of inflammatory arthropathies including RA have been reported, and are reviewed extensively elsewhere in this text. Worth mentioning here is the report in 2005 of Radstake *et al.* who detected associations between two major polymorphisms associated with increased MIF expression and disease progression (as measured by radiographic assessment of joint damage).[61] If this was confirmed in other populations, a strong predictive role for MIF measurement and genotyping in RA will have been established.

Animal Models of RA

Mikulowska *et al.* first explored the role of MIF in an animal model of RA, in the collagen-induced arthritis (CIA) model.[47] In this study, treatment with neutralizing anti-MIF Ab prior to immunization with type II collagen led to delayed onset and lowered frequency of arthritis. This was associated with decreased IgG2a responses to type II collagen, with no differences observed in the production of IgG1. This was interpreted as consistent with the idea that anti-MIF treatment may be associated with a decrease in Th1 responsiveness. Surprisingly, however, the overall T cell proliferative responses to type II collagen were increased with anti-MIF treatment.

Leech *et al.* subsequently reported a role for MIF in the evolution of adjuvant arthritis in the rat.[42] MIF expression was detected in the synovial tissues of rats with arthritis, and in their serum, at greater levels than in healthy rats. In rats developing adjuvant arthritis, MIF was almost undetectable in normal (pre-disease) joints but the level of MIF significantly increased over time in diseased joints.[42] Anti-MIF mAb administration during the evolution of adjuvant arthritis led to a dramatic reduction in clinical and histologic disease parameters. At a high dose of antiMIF mAb, there were virtually no signs of clinical or histological arthritis in the treated group. This is an impressive result as the rat adjuvant arthritis model is one of the most severe. Of note, the doses of mAb used in this study were considerably higher than in the earlier CIA studies, suggesting that a more significant effect on CIA might have been achieved with a higher dose of mAb.

Rat adjuvant arthritis can be further exacerbated by adrenalectomy (ADX), removing the tonic suppressive effects of endogenous glucocorticoids. Indeed, ADX results in up to 80% fatality in this model.[43] Anti-MIF mAb treatment of ADX rats not only reduced arthritis severity, but the lethality observed in control-treated mice was also completely prevented by MIF neutralization.

A further model of RA, antigen-induced arthritis, was also profoundly inhibited by anti-MIF mAb administration,[71] as shown in a study which also confirmed inhibition of delayed-type hypersentivity by anti-MIF mAb treatment. The more profound effects of anti-MIF treatment detected in these studies when compared with those in collagen-induced arthritis may reflect the differences in the models, but more likely relate to the fact that much lower doses of neutralizing mAb were used in the CIA studies.[47] This study also confirmed the glucocorticoid-antagonist role of MIF *in vivo*, by showing that exogenous MIF reversed the therapeutic effects of dexamethasone.[71]

$MIF^{-/-}$ mice have been recently used to confirm the role of MIF in models of RA. A recent pair of studies re-confirmed the suppression of CIA by anti-MIF antibodies and in $MIF^{-/-}$ mice,[29] and demonstrated suppression of a passive-transfer model of CIA by anti-MIF antibodies.[54] In the antigen-induced arthritis model, $MIF^{-/-}$ mice were found to exhibit decreased histological evidence of arthritis severity, compared with wild-type controls, including evidence of reduced cartilage damage.[39] Interestingly, and in line with human data on the effect of MIF on the cell-cycle regulatory protein p53 (see below), reduced arthritis in $MIF^{-/-}$ mice was associated with increased synovial p53 expression and increased synovial apoptosis.

ACTIONS OF MIF IN RHEUMATOID ARTHRITIS

MIF and Innate and Adaptive Immunity

MIF has an important role in the regulation of innate immune responses, as demonstrated by numerous studies in which exogenous MIF exacerbates, or MIF neutralization or MIF deficiency ameliorates, innate immune system activation by bacterial endotoxin.[8,9] MIF has been shown to regulate the expression of TLR-4,[67] which as noted above has been detected in RA tissues. MIF deficiency has been shown to be associated with defective activation of intracellular signal transduction events downstream of innate immune response receptors, including NF-κB and MAP kinases, which are known to be activated in RA tissues.[44,67]

In the first study of the role of MIF in an animal model of RA, it was shown that titers of anti-collagen IgG2a antibodies were inhibited by MIF neutralization, implying an effect on Th1 immunity. This effect of MIF neutralization is supported by findings in the Th-1-dependent antigen-induced model of arthritis, in which inhibition of arthritis by anti-MIF mAb was accompanied by reduction in skin DTH.[71] Recent studies by the author and co-workers have extended this to demonstrate that antigen-specific IgG titers and antigen-specific T-cell activation are impaired in MIF$^{-/-}$ mice in the context of antigen-induced arthritis (unpublished observations). These findings are in keeping with the reported role of MIF in T cell activation. Human peripheral blood and mouse splenic T cells are known to express MIF mRNA and release MIF protein following stimulation by specific antigen, mitogen and anti-CD-3 antibody.[3,76] Anti-MIF mAb treatment inhibits T cell proliferation and IL-2 production *in vivo*.[6]

Synovial Hyperplasia

As outlined above, hypercellularity of the synovium is a hallmark of RA pathology, and indeed is required for the development of pannus, the invasive front of proliferative cells which invades cartilage and bone. Contributing to hypercellularity in RA is a balance of the influx of bone-marrow derived leukocytes, local proliferation of cells, and impaired apoptosis. MIF has been shown to contribute to each of these phenomena in RA, and is therefore shown to be a major factor in this aspect of RA pathology.

Leukocyte recruitment

A number of studies now suggest a key role for MIF in leukocyte recruitment in inflammatory diseases including RA. MIF has been identified in human vascular endothelial cells by RT-PCR, Southern blotting, Western blotting and immunohistochemistry.[34] MIF protein was localized to the cytoplasm of unstimulated cells and MIF expression was profoundly and dose-dependently upregulated by LPS stimulation. A bell-shaped dose-response curve for MIF expression in response to LPS was similar to that demonstrated for macrophages,[22] and in both endothelial cells and monocyte-macrophages, this could be postulated as a protective response to prevent an excessive or life-threatening immune response to endotoxin.

Using intravital microscopy in $MIF^{-/-}$ mice, it was recently demonstrated that MIF is required for normal basal and inflammatory leukocyte-endothelial interactions in the microcirculation of living mice.[38] $MIF^{-/-}$ mice demonstrated reduced leukocyte rolling, adherence and transmigration in response to both LPS and TNF. The findings of reduced leukocyte adhesion and transmigration were confirmed in the joints of living mice with arthritis induced by carrageenin injection. These findings are consistent with the reduced leukocyte infiltrate observed in the synovium of mice treated with anti-MIF mAb or of $MIF^{-/-}$ mice.[21,48] This is a key discovery in terms of the contribution of MIF to inflammation, as leukocyte-endothelial interactions are a critical prerequisite of virtually all inflammatory responses. The induction by MIF of IL-8 expression may be highly relevant to this observation.[45,46] Using the same technique, the author and co-workers have also recently demonstrated that administration of recombinant MIF induces leukocyte-endothelial interactions *in vivo* (Gregory *et al.*, *J Immunol* 2006, in press). The administration of MIF was shown to induce leukocyte adhesion and emigration *in vivo*, and these effects were partly dependent on the induction by MIF of monocyte chemoattractant protein 1 (MCP-1, also known as CCL-2). It has previously been shown that MIF deficiency reduces homing of CD4+ T cells to the brain in a model of multiple sclerosis.[17] Amin *et al.* have recently reported the upregulation of adhesion molecules VCAM and ICAM-1 by MIF *in vitro*.[4]

Synoviocyte proliferation and apoptosis

In vivo, it is unresolved whether increases in local proliferation or reduction in apoptosis contribute most to synovial hyperplasia, with many authors suggesting that impairment of apoptosis is the dominant process. *In vitro*, however, it is well known that cytokines can induce synoviocyte proliferation. While fibroblast outgrowth and proliferation is readily demonstrable from joint tissue of wild-type mice, dramatic reduction in these phenomena was demonstrated in $MIF^{-/-}$ mice.[39] *In vitro*, it has been shown that recombinant MIF induces proliferation of RA FLS.[38] This effect of MIF is dependent on the activation by MIF of ERK MAP kinase in these cells.[38] In addition to a direct effect, MIF appears to be required for the pro-proliferative effects of other cytokines. Anti-MIF mAb completely prevents increased FLS proliferation induced by IL-1,[38] despite the magnitude of the proliferative effect of exogenous IL-1 being significantly greater than that of exogenous MIF.

Even under basal conditions, RA-derived FLS exhibit substantially increased *in vitro* proliferation than cells obtained from patients with non-inflammatory disease such as osteoarthritis (OA). Given that MIF expression is greater in RA-derived than OA-derived cells,[41] it is tempting to hypothesize that MIF expression is at least in part responsible for this profile. This remains to be established, and will require experiments using MIF neutralization and overexpression in RA and OA cells, respectively.

MIF also has effects on the tumor suppressor protein p53. Recombinant MIF inhibits human RA FLS p53 expression *in vitro*.[39] By downstream effects on p21, reductions in p53 can "unlock" cells to allow entry into cell cycle and therefore support proliferation. MIF inhibition of p53 was also accompanied by inhibition of apoptosis.[39] Leech *et al.* also reported that reduced inflammation in arthritic $MIF^{-/-}$ mice was accompanied by increased synovial apoptosis and p53 expression.[39] This is in line with observations of an inhibitory effect of MIF on p53 expression and function in macrophages.[48] Impairment of synoviocyte apoptosis is a hallmark of synovial pathology in RA, and has been related to a loss of normally functioning p53 in RA tissues, via either effects on p53 expression or somatic mutations.[36,58,88] Together, these data suggest a net deleterious effect of MIF on RA synovial hyperplasia.

Cytokine and Inflammatory Mediator Production

Monokines and fibroblast mediators

Cytokines such as TNF and IL-1 are known to be important in RA pathology. As noted above, TNF production by monocyte/macrophage lineage cells is reduced in the absence of MIF.[9] Leech *et al.* also showed that induction of TNF expression by blood monocytes, induced by culturing with RA synovial fibroblast conditioned media, was suppressed by anti-MIF mAb.[41] This is consistent with the high levels of MIF in RA contributing to the expression of TNF.

Sampey *et al.* have demonstrated a role for MIF in human RA synoviocyte activation, as expressed in phospholipase A_2 (PLA_2) and cyclooygenase-2 (COX-2) expression and activation.[69] In these studies, MIF induced activation of human RA FLS *in vitro*, and moreover, anti-MIF mAb prevented the activating effects of the pro-inflammatory cytokine IL-1. MIF upregulates FLS release of prostaglandin E_2.[70] This prostaglandin is the end-product of PLA_2 and COX-2 activation, and is responsible for pain and edema in RA.

In terms of the intracellular signal transduction pathways utilized by MIF in exerting its effects, it has been shown that MIF upregulates ERK and p38 MAP kinase activation.[37,72] Interestingly, the authors have not been able to demonstrate MIF activation of the NF-κB pathway in RA FLS. At and above concentrations that activate MAP kinases in these cells, no evidence of NF-κB activation could be detected.[37] Moreover, anti-MIF mAb treatment, which completely prevented IL-1-mediated activation of FLS, had no effect on IL-1-induced NF-κB activation. These results suggest selective utililzation of MAP kinase pathways by MIF, which may be implicated in the relationship between MIF and glucocorticoids. Another group has reported the upregulation of synoviocyte IL-8 and IL-1 by MIF, suggesting activation of NF-κB as well as MAP kinases.[54] In the latter study, however, only supraphysiological concentrations of MIF (1 μg/ml) were associated with NF-κB activation, and the relevance of this to RA is not clear.

Cartilage and Bone Damage in Rheumatoid Arthritis

In studies on human RA synoviocytes, Onodera *et al.* have reported the activation of expression of MMP-1 and -3 by MIF *in vitro*.[52] As noted above, MMPs are involved in tissue remodelling and may be critical to cartilage injury in RA.

Onodera *et al.*, reporting on bone healing, reported on MIF expression in chondrocytes,[55] and a recent microarray study confirmed MIF expression by chondrocytes.[14] In a study in a collagen-induced model of RA in mice, reduced MMP-13 expression was observed in anti-MIF mAb-treated mice.[29] $MIF^{-/-}$ mice were also found to exhibit decreased histological evidence of cartilage damage in the antigen-induced model of RA.[39]

To date, only very limited studies of MIF on bone biology have been undertaken, but these also favor an important role for MIF. MIF is expressed in calvarial osteoblasts[56]; in osteoblasts, it is present at sites of fracture healing,[55] and upregulates osteoblast expression of matrix metalloproteinases (MMP)-9 and –13.[53] There have been no studies of the effects of MIF on osteoclast differentiation or function, or on the expression of the key regulatory molecules of osteoclast function, receptor-activator of NF-κB (RANK), RANK-ligand (RANKL) and osteoprotegerin (OPG). These data do not suggest any potential osteotoxicity of MIF antagonism.

There is additional indirect evidence for a role of MIF in bone biology. Cytokine-dependent communication between cells of the immune system and cells of the osseous system is important in physiological and pathological regulation of bone biology. The effects of RANKL, IL-1, IL-6, TNF and IL-11 on bone cell biology are well described.[28,64] Expression of many of these osteogenic factors can be regulated by MIF.[28,54,64,69,70] Recent studies have also brought attention to the role of immune responses in postmenopausal bone loss. Macrophage MIF expression is directly inhibited by oestrogen, in an oestrogen-receptor-dependent fashion, and MIF is overexpressed in oestrogen-deficient mice.[5] Therefore, considering the immunological actions of MIF and its relationship with oestrogen, the data indicate that MIF may directly or indirectly influence both physiologic and pathologic bone loss, particularly in postmenopausal osteoporosis.

MIF AND GLUCOCORTICOIDS IN RHEUMATOID ARTHRITIS

Glucocorticoids in RA

Glucocorticoids (GCs) represent the most important and frequently used class of anti-inflammatory drugs in routine clinical practice.[73] Since the successful use in RA of cortisol (the principal GC of the adrenal cortex) in 1948, GC have become so important that some authors divide the history of rheumatology into "BC and AC" (before and after cortisol).[33] Based on community survey data, the prevalence of oral GC use has been estimated to be 0.5% of the general population and 1.4% of those older than 55 years.[62,86] Despite the use of new biological disease modifying drugs in RA, between 56% and 68% of patients continue to require GC therapy.[23,51]

Although GCs are clearly beneficial,[84] substantial dose-dependent and often irreversible side-effects are recognized, which limit their use.[59] Side-effects of most concern include hypertension, obesity, osteoporosis, myopathy, edema, and immune compromise. Premature death related to atherosclerosis is also increasingly recognized in inflammatory diseases, including RA and systemic lupus erythematosus,[16,19,45,75,85] and the potential contribution of GC therapy to atherosclerosis in this setting is suggested by a number of studies.[77] A recent study in RA patients found that GC exposure was associated with increased carotid plaque incidence.[15]

While exogenous GC are important in the treatment of RA, endogenous GC produced by the hypothalamo-pituitary-adrenal (HPA) axis are also critical modulators of the inflammatory response. Inflammation can powerfully activate the HPA axis, leading to increased production of GC. Adrenalectomy, which removes the source of endogenous GC, dramatically amplifies experimental inflammation.[26,43] These observations have led to the concept that the production of endogenous glucocorticoids constitutes a built-in anti-inflammatory regulatory loop.[12] In the context of arthritis, the protective effect of endogenous GC is very well established, with numerous studies demonstrating the exacerbation of RA models in response to adrenalectomy.[26,40,89] The clinical severity of adjuvant arthritis in the rat, for example, is exacerbated by adrenalectomy.[78] Lastly, susceptibility to the streptococcal cell wall (SCW) model of inflammatory arthritis is observed in a normally disease-resistant Fischer rat strain after administration of the GC-receptor antagonist compound RU486.[79] These findings and those of many similar studies confirm that endogenous GC not only limit the severity of inflammatory disease but may also regulate disease susceptibility.

Relationship between MIF and Glucocorticoids in RA

The induction of MIF release by glucocorticoids was among the key findings which led to the "rediscovery" of MIF as a pivotal proinflammatory cytokine.[10] GC are clearly anti-inflammatory in the setting of RA and MIF is a cytokine which mediates proinflammatory processes in the same setting. The unique relationship between GC and MIF may, therefore, at first seem counterintuitive. It is important to appreciate that GC induction of MIF expression takes place at low GC concentrations in the physiological range, with a bell-shaped dose-response associated with inhibition of MIF at high pharmacological concentrations of GC. The concept that MIF directly regulates GC sensitivity has recently been strongly supported by experimental data. Two groups have independently reported that MIF has a negative effect on GC sensitivity of endotoxin-induced macrophage TNF release, with a left-shift of the GC dose-response curve of approximately 3-fold in the absence of MIF.[2,66] If a reduction in GC requirement of 3-fold could be achieved in the clinical application of MIF inhibition, this would result in a very significant clinical benefit in terms of reduced GC toxicity.

Leech *et al.* first reported the regulation of synoviocyte MIF by GC.[41] In cultured primary RA-derived FLS, both intracellular and released MIF protein were induced by dexamethasone, with a bell-shaped dose-response curve peaking at 10^{-10}–10^{-11} M. Confirmation of MIF upregulation by dexamethasone in FLS at the transcriptional level was obtained with increases in MIF mRNA demonstrated by semi-quantitative RT-PCR.

Leech *et al.* also first reported the regulation of MIF by GC *in vivo*.[43] In this study, adrenalectomized (ADX) or sham-ADX control rats underwent induction of adjuvant arthritis. MIF expression in synovial tissue was lower in ADX arthritic rats, indicating that endogenous GC contribute positively to the synovial expression of MIF. Lethality in ADX rats undergoing adjuvant arthritis, which is well described due to the severity of the model in the absence of endogenous GC, was completely prevented by anti-MIF mAb administration, which was also associated with reduced arthritis severity.[43] These data suggested that even the lower levels of synovial MIF observed could drive synovial inflammation in the absence of GC.

Mechanisms of Interaction between MIF and Glucocorticoids

The mechanism through which MIF antagonizes the effects of GC is not fully elucidated. Cell activation via NF-κB is highly GC-sensitive, such that inhibition of NF-κB represents a dominant means by which GC inhibit the inflammatory response.[7] Daun *et al.* reported that while MIF did not directly induce NF-κB activation, it impaired the effects of GC on the NF-κB inhibitory molecule I-κB.[13] As noted above, however, we and others have found no evidence that MIF acts through direct NF-κB activation.[34]

Evasion of GC-mediated suppression could depend on the preferential utilization by MIF of MAP kinase signalling pathways, which are relatively glucocorticoid-insensitive.[80] As noted above, MIF activates the phosphorylation of MAP kinases, including ERK and p38 MAP kinase in RA cells.[37,72] There are no direct effects of the glucocorticoid-receptor complex on MAP kinases. Glucocorticoids do, however, induce the expression of MAP kinase phosphatase (MKP)-1, a natural inhibitory regulator of MAP kinase activation, thereby contributing indirectly to modulation of MAP kinases, and glucocorticoids also influence the interaction of transcription factors at the level of DNA unfolding and binding.[30] MKP-1 has emerged as a major regulatory protein in innate immune responses[11] and in models of RA.[68] The author has previously reported on the expression, function, and regulation by glucocorticoids of MKP-1 in RA tissues; RA FLS MKP-1 expression is increased by glucocorticoids and this is associated with inhibition of MAP kinase activation.[83]

Recent observations suggest that MIF is indeed a direct negative regulator of MKP-1 expression.[2] MIF$^{-/-}$ cells exhibited a greater sensitivity to induction of MKP-1 by dexamethasone, and exogenous MIF directly inhibited dexamethasone-induced MKP-1 expression. These observations were confirmed in independent studies using MIF anti-sense.[66] This is consistent with the idea that MIF present in inflammatory disease permits over activation of MAP kinases via inhibition of MKP-1, and that MIF reversal of glucocorticoid-induced MKP1 expression is a key mechanism in MIF's role in controlling glucocorticoid sensitivity. The direct relationship between MIF and MKP-1 in RA or animal models of RA has not been investigated. Interestingly, however, MKP-1 expression is regulated by p53, and this may represent a mechanism through which MIF downregulation of MKP-1 is achieved.[1]

POTENTIAL FOR MIF ANTAGONISM IN RA

Summary: MIF as a Therapeutic Target in RA

MIF antagonism might be first applied clinically in RA. Many aspects of the biology of MIF in RA have been established, and suggest MIF antagonism could be a very attractive therapeutic pathway in this disease.

Multiple interacting processes are involved in the pathogenesis and pathology of RA. These include aspects of the innate and adaptive immune responses, leukocyte recruitment, cytokine production and responses, hyperplasia and disordered apoptosis, intracellular signal transduction events, inflammatory mediator and proteolytic enzyme synthesis, and ultimately synovitis, cartilage damage and bone erosion. Evidence exists for a potential role of MIF in all of these phenomena in RA, and studies in animal models of RA strongly suggest the potential clinical value of MIF inhibition in this disease. MIF inhibition in RA, therefore, has the potential to positively influence disease onset, inflammatory activity and disease perpetuation, and damage-related outcomes.

Despite this evidence, it must be stated that the ultimate test of the role of MIF in RA will rely on the development of specific MIF inhibitory therapeutic agents and their application in clinical trials. The successful development of TNF antagonist therapy in RA follows the unsuccessful application of similar technology against ostensibly validated targets that did not prove valuable in clinical trials.

The MIF-GC relationship is perhaps the most intriguing aspect of the biology of this protein. The clear evidence for the role of endogenous GC in controlling inflammatory responses *in vivo*, and the persistent dependence on GC therapy for RA even in the presence of anti-TNF biological therapy, presents the most persuasive argument for trials of MIF inhibition in RA. Reducing or abolishing the use of exogenous GC in the treatment of RA is a near-universal aim of physicians treating inflammatory diseases, including RA. The evidence that MIF directly modulates GC sensitivity suggests a specific "steroid-sparing" role for MIF inhibition that strongly argues for the testing of MIF inhibitor therapies in this disease. This requires that MIF blockade does not cause similar metabolic, glycemic, pressor and bone resorptive effects to GC. Although this is currently unknown, there is no mechanistic data which would support GC-like toxicity of MIF-directed therapies.

Finally, emerging evidence of mechanistic links between the pathogenesis of "traditional" inflammatory disease such as RA, and co-morbidities more recently recognized to have an inflammatory basis such as atherosclerosis, provides a compelling case for addressing molecules implicated in both disease processes. As reviewed elsewhere in this text, there is little doubt that the evidence base for the role of MIF in atherosclerosis is at least as strong as that in RA, and the possibility that MIF over-expression might actually explain the increased incidence of atherosclerosis in RA and other inflammatory diseases, such as systemic lupus erythematosus (SLE), has recently been proposed.[49] This argues for the testing of MIF inhibitor therapies in clinical trials with both disease activity and co-morbidity outcome measurements.

Routes to MIF antagonism in arthritis

As will be reviewed extensively in other chapters, several potential routes to develop inhibitors of the biological function of MIF exist. Other soluble cytokines have been successfully antagonized using mAb, and anti-MIF mAb have proven effective in a wide range of animal models of RA as well as in inhibiting the function of cells derived from human RA synovium. Anti-MIF mAb powerfully suppresses human RA synoviocyte PLA_2, COX-2 expression and proliferation induced by IL-1, and suppresses TNF production by monocytes exposed to FLS-conditioned media. As reviewed elsewhere, however, MIF has significant intracellular functions that may be less amenable to targeting by macromolecular approaches such as mAb.[49] Small molecule MIF antagonists, which could be engineered to cross the cell membrane, might have the advantage of blocking the function of both extracellular and intracellular MIF. Although small molecule antagonists have been shown to be effective in models of endotoxic shock and diabetes mellitus, no published reports exist on the potential effects in models of RA. Recently, however, the author and co-workers have demonstrated the efficacy of a small molecule MIF antagonist compound in the antigen-induced arthritis model (unpublished observations). Both parenteral and oral administration of this compound were effective in reducing joint swelling and, importantly, histological evidence of joint inflammation, cartilage damage and bone erosion.

MIF Inhibitor Profile for Treatment of RA

Clearly, RA is a chronic disease, and evidence from trials of TNF antagonists suggest that, at least in the case of treating established disease, treatment cessation is associated with rapid relapse.[87] An MIF inhibitory therapy would therefore be required to have a sufficient safety profile to permit long-term therapy. Leaving to one side possible "off-target" toxicity related to non-selectivity, chronic MIF inhibition may still be associated with various risks, not least of which would be immunosuppression. Although it is difficult to conceive of a therapy for the inflammation of RA which was not associated with at least some immunosuppression, it may be that MIF inhibition is relatively attractive in this regard. For example, $MIF^{-/-}$ mice have normal breeding, reproduction, and growth, and are disease-free in non-specific pathogen free housing without succumbing to opportunistic infections. Moreover, the selective activation by MIF of MAP kinase pathways as opposed to NF-κB pathways differentiates it from antagonists of TNF, or indeed direct antagonists of the NF-κB pathway such as GC or NF-κB pathway kinase inhibitors.

The long-term risk of tumors in immunosuppressed patients is a concern for many established therapies of RA. Fukushima *et al.* reported studies in which MIF-transgenic (Tg) mice were inoculated with a murine sarcoma cell line. Although tumor growth rates were initially enhanced in Tg mice, subsequent tumor regression was significantly better in Tg mice.[24] These data suggest the possibility of differential effects of MIF on early, compared to late, events in tumor development *in vivo*. Against this, however, it may be that in fact cancer rates are reduced, as abundant preclinical data suggests that MIF inhibition is in fact protective from models of cancer onset and metastasis.[25,27,63,81,82]

The potential toxicity of MIF inhibitory therapy would of course have to be balanced against any such therapy it replaced, such as GC or non-specific immunosuppressants. Certainly, a specific route to "steroid-sparing" therapy through MIF inhibition would be desirable because of the ability to reduce the very well known toxicity of exogenous GC. There is no theoretical reason to suspect that MIF inhibition would produce Cushingoid side effects, as the GC-receptor dependence of these toxicities of GC are not related to the mechanisms of action of MIF. Even highly toxic drugs such as azathioprine are routinely used for "steroid-sparing" in disease such as SLE, and there is no evidence that specific MIF inhibitory therapy would have any such toxicity.

Conclusions

Evidence from a wide variety of *in vitro*, animal and human experiments now suggests that MIF has an important and unique role in the pathology of RA. Effects on all nearly all aspects of RA pathology have been reported, including leukocyte recruitment, adaptive and innate immune responses, cytokine and mediator expression, tissue hyperplasia and cartilage damage. The field lacks exploration only in the area of bone damage. Therapeutic MIF inhibition in RA, therefore, has a rational basis in terms of the role of MIF in mechanisms of disease. Compared with existing immunosuppressive approaches, MIF inhibition may have a favorable profile in regard to co-morbidities such as atherosclerosis and neoplasia. The unique relationship of MIF with GC additionally suggests MIF inhibition as providing the first targeted route to "steroid-sparing." Progress towards clinical trials of MIF inhibitor biological or small molecule therapies will permit testing of the hypothesis that MIF is important in the pathology of RA in clinical trials in the near future.

REFERENCES

1. Aeberli D, Leech M, Morand EF. (2006) Macrophage migration inhibitory factor and glucocorticoid sensitivity. *Rheumatology* **45**: 937–943.
2. Aeberli D, Yang YH, Mansell A, *et al.* (2006) Macrophage migration inhibitory factor modulates glucocorticoid sensitivity in macrophages via effects on MAP kinase phosphatase-1 and p38 MAP kinase. *FEBS Lett* **580**: 974–981.
3. Altboum I, Pick E. (1979) Antigen and mitogen induced production of macrophage migration inhibitory factor in the mouse. *Int Arch Allergy Appl Immunol* **60**: 29–43.

4. Amin MA, Haas CS, Zhu K, *et al.* (2006) Migration inhibitory factor up-regulates vascular cell adhesion molecule-1 and intercellular adhesion molecule-1 via Src, PI3 kinase, and NFkappaB. *Blood* **107**: 2252–2261.
5. Ashcroft GS, Mills SJ, Lei K, *et al.* (2003) Estrogen modulates cutaneous wound healing by downregulating macrophage migration inhibitory factor. *J Clin Invest* **111**: 1309–1318.
6. Bacher M, Metz CN, Calandra T, *et al.* (1996) An essential regulatory role for macrophage migration inhibitory factor in T-cell activation. *Proc Natl Acad Sci USA* **93**: 7849–7854.
7. Barnes PJ. (1998) Anti-inflammatory actions of glucocorticoids — molecular mechanisms. *Clin Sci* **94**: 557–572.
8. Bernhagen J, Calandra T, Mitchell RA, *et al.* (1993) MIF is a pituitary-derived cytokine that potentiates lethal endotoxaemia. *Nature* **365**: 756–759.
9. Bozza M, Satoskar AB, Lin G, *et al.* (1999) Targeted disruption of migration inhibitory factor gene reveals its critical role in sepsis. *J Exp Med* **189**: 341–346.
10. Calandra T, Bernhagen J, Metz CN, *et al.* (1995) MIF as a glucocorticoid-induced modulator of cytokine production. *Nature* **377**: 68–71.
11. Chi H, Barry SP, Roth RJ, *et al.* (2006) Dynamic regulation of pro- and anti-inflammatory cytokines by MAPK phosphatase 1 (MKP-1) in innate immune responses. *Proc Natl Acad Sci USA* **103**: 2274–2279.
12. Chrousos GP. (1995) The hypothalamic-pituitary-adrenal axis and immune-mediated inflammation. *N Eng J Med* **332**: 1351–1362.
13. Daun JM, Cannon JG. (2000) Macrophage migration inhibitory factor antagonizes hydrocortisone-induced increases in cytosolic IκBα. *Am J Physiol* **279**: R1043–R1049.
14. De Ceuninck F, Dassencourt L, Anract P. (2004) The inflammatory side of human chondrocytes unveiled by antibody microarrays. *Biochem Biophys Res Commun* **323**: 960–969.
15. del Rincon I, O'Leary DH, Haas RW, Escalante A. (2004) Effect of glucocorticoids on the arteries in rheumatoid arthritis. *Arthritis Rheum* **50**: 3813–3822.
16. del Rincon ID, Williams K, Stern MP, *et al.* (2001) High incidence of cardiovascular events in a rheumatoid arthritis cohort not explained by traditional cardiac risk factors. *Arthritis Rheum* **44**: 2737–2745.
17. Denkinger CM, Denkinger M, Kort JJ, *et al.* (2003) *In vivo* blockade of macrophage migration inhibitory factor ameliorates acute experimental autoimmune encephalomyelitis by impairing the homing of encephalitogenic T cells to the central nervous system. *J Immunol* **170**: 1274–1282.
18. Edwards JCW. (1994) Synovium. In: Klippel JH, Dieppe PA (eds.), *Rheumatology*. Mosby-Year Book, London, pp. 371–378.
19. El-Magadmi M, Bodill H, Ahmad Y, *et al.* (2004) Systemic lupus erythematosus: an independent risk factor for endothelial dysfunction in women. *Circulation* **110**: 399–404.

20. Emery P, Fleischmann R, Filipowicz-Sosnowska A, *et al.* (2006) The efficacy and safety of rituximab in patients with active rheumatoid arthritis despite methotrexate treatment: results of a phase IIB randomized, double-blind, placebo-controlled, dose-ranging trial. *Arthritis Rheum* **54**: 1390–1400.
21. Firestein GS. (2003) Evolving concepts of rheumatoid arthritis. *Nature* **423**: 356–361.
22. Firestein GS, Manning AM. (1999) Signal transduction and transcription factors in rheumatic disease. *Arthritis Rheum* **42**: 609–621.
23. Fleischmann RM, Schechtman J, Bennett R, *et al.* (2003) Anakinra, a recombinant human interleukin-1 receptor antagonist (r-metHuIL-1ra), in patients with rheumatoid arthritis: a large, international, multicenter, placebo-controlled trial. *Arthritis Rheum* **48**: 927–934.
24. Fukushima T, Nishihira J, Yoshiki T, *et al.* (2005) Evidence of dual function of macrophage migration inhibitory factor relevant to tumor progression and regression. *Int J Mol Med* **16** 119–126.
25. Hagemann T, Wilson J, Kulbe H, *et al.* (2005) Macrophages induce invasiveness of epithelial cancer cells via NF-kappaB and JNK. *J Immunol* **175**: 1197–1205.
26. Harbuz MS, Rees RG, Lightman SL. (1993) HPA axis responses to acute stress and adrenalectomy during adjuvant-induced arthritis in the rat. *Am J Physiol* **264**: R179–R185.
27. Hira E, Ono T, Dhar DK, *et al.* (2004) Overexpression of macrophage migration inhibitory factor induces angiogenesis and deteriorates prognosis after radical resection for hepatocellular carcinoma. *Cancer* **103**: 588–598.
28. Hofbauer LC, Heufelder AE. (2001) Role of receptor activator of nuclear factor-kappaB ligand and osteoprotegerin in bone cell biology. *J Mol Med* **79**: 243–253.
29. Ichiyama H, Onodera S, Nishihira J, *et al.* (2004) Inhibition of joint inflammation and destruction induced by anti-type II collagen antibody/lipopolysaccharide (LPS)-induced arthritis in mice due to deletion of macrophage migration inhibitory factor (MIF). *Cytokine* **26**: 187–194.
30. Ito K, Yamamura S, Essilfie-Quaye S, *et al.* (2006) Histone deacetylase 2-mediated deacetylation of the glucocorticoid receptor enables NF-kappaB suppression. *J Exp Med* **203**:7–13.
31. Iwahashi M, Yamamura M, Aita T, *et al.* (2004) Expression of Toll-like receptor 2 on CD16+ blood monocytes and synovial tissue macrophages in rheumatoid arthritis. *Arthritis Rheum* **50**: 1457–1467.
32. Joosten LA, Koenders MI, Smeets RL, *et al.* (2003) Toll-like receptor 2 pathway drives streptococcal cell wall-induced joint inflammation: critical role of myeloid differentiation factor 88. *J Immunol* **171**: 6145–6153.
33. Kirwan JR, Balint G, Szebenyi B. (1999) Anniversary: 50 years of glucocorticoid treatment in rheumatoid arthritis. *Rheumatology* **38**: 100–102.

34. Kleemann R, Hausser A, Geiger G, *et al.* (2000) Intracellular action of the cytokine MIF to modulate AP-1 activity and the cell cycle through Jab1. *Nature* **408**: 211–216.
35. Kremer JM, Westhovens R, Leon M, *et al.* (2003) Treatment of rheumatoid arthritis by selective inhibition of T-cell activation with fusion protein CTLA4Ig. *N Engl J Med* **349**: 1907–1915.
36. Kullmann F, Judex M, Neudecker I, *et al.* (1999) Analysis of the p53 tumor suppressor gene in rheumatoid arthritis synovial fibroblasts. *Arthritis Rheum* **42**: 1594–1600.
37. Lacey D, Sampey A, Mitchell R, *et al.* (2003) Control of fibroblast-like synoviocyte proliferation by macrophage migration inhibitory factor. *Arthritis Rheum* **48**: 103–109.
38. Lacey DC, Sampey AV, Mitchell R, *et al.* (2003) Control of fibroblast-like synoviocyte proliferation by macrophage migration inhibitory factor (MIF). *Arthritis Rheum* **48**: 103–109.
39. Leech M, Lacey DC, Xue JR, *et al.* (2003) Macrophage migration inhibitory factor (MIF) regulates p53 in inflammatory arthritis. *Arthritis Rheum* **48**: 1881–1889.
40. Leech M, Metz C, Bucala R, Morand EF. (2000) Regulation of macrophage migration inhibitory factor by endogenous glucocorticoids in rat adjuvant-induced arthritis. *Arthritis Rheum* **43**: 827–833.
41. Leech M, Metz C, Hall P, *et al.* (1999) Macrophage migration inhibitory factor in rheumatoid arthritis: evidence of proinflammatory function and regulation by glucocorticoids. *Arthritis Rheum* **42**: 1601–1608.
42. Leech M, Metz CN, Santos LL, *et al.* (1998) Involvement of macrophage migration inhibitory factor in the evolution of rat adjuvant arthritis. *Arthritis Rheum* **41**: 910–917.
43. Leech M, Santos LL, Metz C, *et al.* (2000) Control of macrophage migration inhibitory factor (MIF) by endogenous glucocorticoids in rat adjuvant arthritis. *Arthritis Rheum* **43**: 827–833.
44. Liao H, Bucala R, Mitchell RA. (2003) Adhesion-dependent signaling by macrophage migration inhibitory factor (MIF). *J Biol Chem* **278**: 76–81.
45. Manzi S, Selzer F, Sutton-Tyrrell K, *et al.* (1999) Prevalence and risk factors of carotid plaque in women with systemic lupus erythematosus. *Arthritis Rheum* **42**: 51–60.
46. Miagkov AV, Kovalenko DV, Brown CE, *et al.* (1998) NF-kappa-B activation provides the potential link between inflammation and hyperplasia in the arthritic joint. *Proc Natl Acad Sci USA* **95**: 13859–13864.
47. Mikulowska A, Metz CN, Bucala R, Holmdahl R. (1997) Macrophage migration inhibitory factor is involved in the pathogenesis of collagen type II-induced arthritis in mice. *J Immunol* **158**: 5514–5517.

48. Mitchell RA, Liao H, Chesney J, *et al.* (2002) Macrophage migration inhibitory factor (MIF) sustains macrophage proinflammatory function by inhibiting p53: regulatory role in the innate immune response. *Proc Natl Acad Sci USA* **99**: 345–350.
49. Morand EF, Leech M, Bernhagen J. (2006) MIF: a new cytokine link between rheumatoid arthritis and atherosclerosis. *Nat Rev Drug Discov* **5**: 399–411.
50. Morand EF, Leech M, Weedon H, *et al.* (2002) Macrophage migration inhibitory factor in rheumatoid arthritis: clinical correlations. *Rheumatology* **41**: 558–562.
51. Moreland LW, Schiff MH, Baumgartner SW, *et al.* (1999) Etanercept therapy in rheumatoid arthritis. A randomized, controlled trial. *Ann Intern Med* **130**: 478–486.
52. Onodera S, Kaneda K, Mizue Y, *et al.* (2000) Macrophage migration inhibitory factor up-regulates expression of matrix metalloproteinases in synovial fibroblasts of rheumatoid arthritis. *J Biol Chem* **275**: 444–450.
53. Onodera S, Nishihira J, Iwabuchi K, *et al.* (2001) Macrophage migration inhibitory factor up-regulates matrix metalloproteinase-9 and -13 in rat osteoblasts. Relevance to intracellular signaling pathways. *J Biol Chem* **20**: 20.
54. Onodera S, Nishihira J, Koyama Y, *et al.* (2004) Macrophage migration inhibitory factor up-regulates the expression of interleukin-8 messenger RNA in synovial fibroblasts of rheumatoid arthritis patients: common transcriptional regulatory mechanism between interleukin-8 and interleukin-1beta. *Arthritis Rheum* **50**: 1437–1447.
55. Onodera S, Nishihira J, Yamazaki M, *et al.* (2004) Increased expression of macrophage migration inhibitory factor during fracture healing in rats. *Histochem Cell Biol.*
56. Onodera S, Suzuki K, Matsuno T, *et al.* (1996) Identification of macrophage migration inhibitory factor in murine neonatal calvariae and osteoblasts. *Immunology* **89**: 430–435.
57. Onodera S, Tanji H, Suzuki K, *et al.* (1999) High expression of macrophage migration inhibitory factor in the synovial tissues of rheumatoid joints. *Cytokine* **11**: 163–167.
58. Pap T, Aupperle KR, Gay S, *et al.* (2001) Invasiveness of synovial fibroblasts is regulated by p53 in the SCID mouse *in vivo* model of cartilage invasion. *Arthritis Rheum* **44**: 676–681.
59. Pisu M, James N, Sampsel S, Saag KG. (2005) The cost of glucocorticoid-associated adverse events in rheumatoid arthritis. *Rheumatology* **44**: 781–788.
60. Radstake TR, Roelofs MF, Jenniskens YM, *et al.* (2004) Expression of toll-like receptors 2 and 4 in rheumatoid synovial tissue and regulation by proinflammatory cytokines interleukin-12 and interleukin-18 via interferon-gamma. *Arthritis Rheum* **50**: 3856–3865.

61. Radstake TR, Sweep FC, Welsing P, *et al.* (2005) Correlation of rheumatoid arthritis severity with the genetic functional variants and circulating levels of macrophage migration inhibitory factor. *Arthritis Rheum* **52**: 3020–3029.
62. Ramsey-Goldman R. (2002) Missed opportunities in physician management of glucocorticoid-induced osteoporosis? *Arthritis Rheum* **46**: 3115–3120.
63. Ren Y, Law S, Huang X, *et al.* (2005) Macrophage migration inhibitory factor stimulates angiogenic factor expression and correlates with differentiation and lymph node status in patients with esophageal squamous cell carcinoma. *Ann Surg* **242**: 55–63.
64. Rho J, Takami M, Choi Y. (2004) Osteoimmunology: interactions of the immune and skeletal systems. *Mol Cells* **17**: 1–9.
65. Roelofs MF, Joosten LA, Abdollahi-Roodsaz S, *et al.* (2005) The expression of toll-like receptors 3 and 7 in rheumatoid arthritis synovium is increased and costimulation of toll-like receptors 3, 4, and 7/8 results in synergistic cytokine production by dendritic cells. *Arthritis Rheum* **52**: 2313–2322.
66. Roger T, Chanson AL, Knaup-Reymond M, Calandra T. (2005) Macrophage migration inhibitory factor promotes innate immune responses by suppressing glucocorticoid-induced expression of mitogen-activated protein kinase phosphatase-1. *Eur J Immunol* **35**: 3405–3413.
67. Roger T, David J, Glauser MP, Calandra T. (2001) MIF regulates innate immune responses through modulation of Toll-like receptor 4. *Nature* **414**: 920–924.
68. Salojin KV, Owusu IB, Millerchip KA, *et al.* (2006) Essential role of MAPK phosphatase-1 in the negative control of innate immune responses. *J Immunol* **176**: 1899–1907.
69. Sampey AV, Hall PH, Mitchell RA, *et al.* (2001) Regulation of synoviocyte phospholipase A2 and cyclooxygenase 2 by macrophage migration inhibitory factor. *Arthritis Rheum* **44**: 1273–1280.
70. Santos L, Lacey DC, Yang Y, *et al.* (2004) Activation of synovial cell p38 MAP kinase by macrophage migration inhibitory factor (MIF). *J Rheumatol* **31**: 1038–1043.
71. Santos LL, Hall P, Metz CN, *et al.* (2001) Role of macrophage migration inhibitory factor (MIF) in murine antigen-induced arthritis: interaction with glucocorticoids. *Clin Exp Immunol* **123**: 309–314.
72. Santos LL, Lacey D, Yang Y, *et al.* (2004) Activation of synovial cell p38 MAP kinase by macrophage migration inhibitory factor. *J Rheumatol* **31**: 1038–1043.
73. Schacke H, Docke WD, Asadullah K. (2002) Mechanisms involved in the side effects of glucocorticoids. *Pharmacol Ther* **96**: 23–43.
74. Smolen JS, Steiner G. (2003) Therapeutic strategies for rheumatoid arthritis. *Nat Rev Drug Discov* **2**: 473–488.

75. Solomon DH, Karlson EW, Rimm EB, *et al.* (2003) Cardiovascular morbidity and mortality in women diagnosed with rheumatoid arthritis. *Circulation* **107**: 1303–1307.
76. Sorg C. (1980) Characterization of murine macrophage migration inhibitory activities (MIF) released by concanavalin A stimulated thymus or spleen cells. *Mol Immunol* **17**: 565–569.
77. Souverein PC, Berard A, Van Staa TP, *et al.* (2004) Use of oral glucocorticoids and risk of cardiovascular and cerebrovascular disease in a population based case-control study. *Heart* **90**: 859–865.
78. Stephanou A, Sarlis NJ, Knight RA, *et al.* (1992) Glucocorticoid-mediated responses of plasma ACTH and anterior pituitary pro-opiomelanocortin, growth hormone and prolactin mRNAs during adjuvant-induced arthritis in the rat. *J Mol Endocrinol* **9**: 273–281.
79. Sternberg EM, Hill JM, Chrousos GP, *et al.* (1989) Inflammatory mediator-induced hypothalamic-pituitary-adrenal axis activation is defective in streptococcal cell wall arthritis susceptible Lewis rats. *Proc Nat Acad Sci USA* **86**: 2374–2378.
80. Swantek JL, Cobb MH, Geppert TD. (1997) Jun N-terminal kinase/stress-activated protein kinase (JNK/SAPK) is required for lipopolysaccharide stimulation of tumor necrosis factor alpha (TNF) translation: glucocorticoids inhibit TNF translation by blocking JNK/SAPK. *Mole Cell Biol* **17**: 6274–6282.
81. Takahashi K, Kato Y, Sugiyama T, *et al.* (1999) Production of murine collagen-induced arthritis using Klebsiella pneumoniae O3 lipopolysaccharide as a potent immunological adjuvant. *Microbiol Immunol* **43**: 795–801.
82. Takahashi N, Nishihira J, Sato Y, *et al.* (1998) Involvement of macrophage migration inhibitory factor (MIF) in the mechanism of tumor cell growth. *Mol Med* **4**: 707–714.
83. Toh M-L, Yang Y, Leech M, Morand EF. (2004) Expression of MAP kinase phosphatase-1, a negative regulator of MAP kinases, in: *Rheumatoid Arthritis: Upregulation by IL-1 and Glucocorticoids. Arthritis Rheumatism* **50**: 3118–3128.
84. van Everdingen AA, Jacobs JW, Siewertsz Van Reesema DR, Bijlsma JW. (2002) Low-dose prednisone therapy for patients with early active rheumatoid arthritis: clinical efficacy, disease-modifying properties, and side effects: a randomized, double-blind, placebo-controlled clinical trial. *Ann Intern Med* **136**: 1–12.
85. Wallberg-Jonsson S, Ohman ML, Dahlqvist SR. (1997) Cardiovascular morbidity and mortality in patients with seropositive rheumatoid arthritis in Northern Sweden. *J Rheumatol* **24**: 445–451.
86. Walsh LJ, Wong CA, Pringle M, Tattersfield AE. (1996) Use of oral corticosteroids in the community and the prevention of secondary osteoporosis: a cross sectional study. *BMJ* **313**: 344–346.

87. Weinblatt ME, Kremer JM, Bankhurst AD, *et al.* (1999) A trial of Etanercept, a recombinant tumor necrosis factor receptor: Fc fusion protein, in patients with rheumatoid arthritis receiving methotrexate. *New Engl J Medi* **340**: 253–259.
88. Yamanishi Y, Boyle DL, Pinkoski MJ, *et al.* (2002) Regulation of joint destruction and inflammation by p53 in collagen-induced arthritis. *Am J Pathol* **160**: 123–130.
89. Yang YH, Hutchinson P, Leech M, Morand EF. (1997) Exacerbation of adjuvant arthritis by adrenalectomy is associated with reduced leukocyte lipocortin 1. *J Rheumatol* **24**: 1758–1764.
90. Yocum DE, Klippel JH, Wilder RL, *et al.* (1988) Cyclosporin A in severe, treatment-refractory rheumatoid arthritis. A randomized study. *Ann Intern Med* **109**: 863–869.
91. Gregory JL, Morand EF, McKeown SJ, *et al.* (2006) Macrophage migration inhibitory factor induces macrophage recruitment via CC chemokine ligand 2. *J Immunol* **177**: 8072–8079.

14

Gene Structure and Functional MIF Polymorphisms in Respiratory Disease

*John Baugh**

INTRODUCTION

Macrophage migration inhibitory factor (MIF) was originally described in 1966 as a T-lymphocyte-derived "activity" that inhibited the random migration of macrophages.[1,2] Over the subsequent 20 years, MIF expression was found to correlate with general macrophage activation functions, including adherence, spreading, phagocytosis and enhanced tumoricidal activity.[3–5] It was not until the cloning of the MIF cDNA in 1989, the subsequent expression of recombinant protein and the generation of specific anti-MIF antibodies, however, that the full extent of MIF actions began to become unravelled. Since this time MIF has been well described as a potent proinflammatory mediator, and has emerged as a key integrator of the immuno-neuroendocrine interface and a potential regulator of tumorigenesis (reviewed in Refs. 6–8).

Early in the history of MIF research it was demonstrated that MIF expression increased in inflammatory tissues and in the serum of patients with inflammatory disease. Interestingly, it also became apparent that

*The Conway Institute of Biomolecular and Biomedical Research, Department of Medicine and Medical Science, University College Dublin, Belfield, Dublin 4, Ireland. Tel.: +353 1 7166729; Mobile: +353 863841221; Fax: +353 1 7166701.

within both normal and diseased patient populations there was a large range in serum MIF levels. These data raised the postulate that MIF is regulated both locally and systemically and that genetic variations may render individuals prone to exaggerated MIF expression. The cloning of the human and mouse MIF genes in 1994 and 1995, respectively, allowed the initiation of more detailed studies into the transcriptional regulation of MIF expression and the starting place for analysis of genetic variation. This chapter will highlight the advances that have been made in the understanding of MIF gene expression and the potential role played by promoter polymorphisms. Particular emphasis will be placed on the potential relationship between functional MIF polymorphisms and respiratory disease.

MIF GENE STRUCTURE

Human

A cDNA encoding MIF was isolated in 1989 from a lectin-stimulated T cell hybridoma library using functional expression cloning in Cos-1 cells.[9] A cDNA library representing approximately 60,000 cloned mRNA's was screened by transfection of pooled plasmid DNA into Cos-1 cells, followed by functional analysis of Cos-1 supernatants in an *in vitro* macrophage migration assay. Having identified a pool of plasmid that caused the secretion of a factor from transfected Cos-1 cells that inhibited macrophage migration, the pooled plasmid was subdivided until a single clone was identified. Sequencing of the identified plasmid revealed a 345-base pair open reading frame encoding a 115-amino acid polypeptide with a calculated molecular weight of 12,650 Da, which was in agreement with the apparent molecular weight of the MIF-specific protein band observed in metabolic labeling experiments. Analysis of this sequence also revealed that MIF did not contain the usual hydrophobic leader sequence normally associated with secreted proteins and implied a non-classical method of protein secretion.

Following the cloning of MIF from a T cell cDNA library, it became apparent that MIF was expressed in various non-lymphoid tissues, suggesting a role for MIF beyond the immune system. Furthermore, in order to gain a greater understanding of the tissue- and stimulus-specific expression of MIF, a more thorough analysis of the MIF gene was required. In 1994 Paraklar and Wistow were the first to clone the human MIF gene (GeneBank Accession number: L19686).[10] Restriction digested human genomic DNA revealed a single band when probed with human MIF cDNA, implying a single MIF gene. Screening of a human placental genomic library with the same MIF cDNA then identified a single phage which was isolated and subcloned for sequencing. Sequence analysis revealed the full MIF coding sequence and comparison with human MIF cDNA sequence identified three exons of 204, 172, and 182 base pairs separated by introns of 189 and 95 base pairs (see Fig. 1). In addition, the isolated genomic DNA also revealed 1 kb of 5′-flanking sequence and about 250 bp of 3′-flanking sequence. Primer extension analysis from a primer that was complementary to a region in exon 1 produced a single faint product consistent with a start site 97 base pairs upstream of the initiator methionine. This start site was confirmed independently using 5′RACE and indicated that transcription of MIF started from a region that was flanked by a cAMP response element (CRE) and multiple SP1 sites but not a classical TATA box. Indeed analysis of the 5′-flanking region revealed putative CCAAT and TATA boxes that were 1000 base pairs upstream of the transcriptional start site but their role in MIF expression remains unclear.

Initial genetic mapping using PCR screening of a human chromosome panel suggested that the human MIF gene lay on chromosome 19.[11] Subsequent detailed analysis by the same group, however, showed the localization to be actually on chromosome 22.[12] Further PCR screening of a chromosome 22 somatic cell hybrid panel, which divides the chromosome into 24 separate regions, localized the MIF gene to region 11 in 22q11.2. This location was confirmed using fluorescence *in situ* hybridisation (FISH).[12]

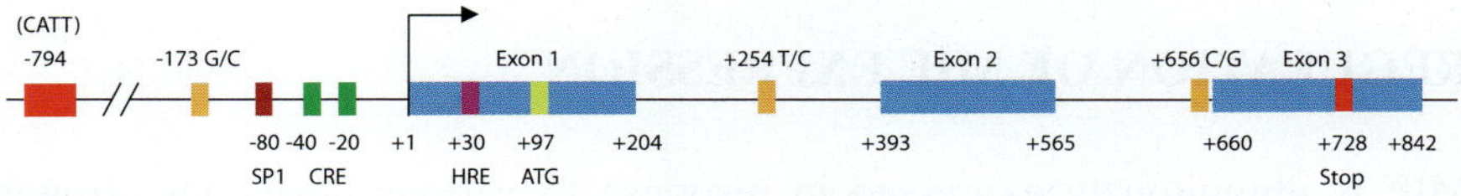

Fig. 1 Schematic representation of human MIF gene structure including promoter region. GeneBank Accession number: L19686, chromosomal localization 22q11.2.

Mouse

A 543bp cDNA encoding a murine homolog to human MIF was cloned in 1992 as part of a study analyzing growth factor-induced delayed early response genes in mouse fibroblasts.[13] The cloned sequence revealed a single open reading frame for a 115-amino acid polypeptide that shared an 88% sequence identity to human MIF and yielded a product of approximately 12 kDa upon *in vitro* transcription and translation. At the time, however, as the cDNA probe used to identify MIF hybridized to multiple rat, mouse, and human genomic restriction fragments, it was concluded that the identified cDNA belonged to a member of an MIF-related family. Independent studies cloned a cDNA for MIF from mouse pituitary in 1993, but it was not until 1995 that the full mouse MIF gene was cloned.[14,15] Southern blotting of a mouse liver genomic DNA with a probe derived from amplification of mouse MIF cDNA revealed multiple bands, indicative of either a closely homologous gene family or multiple pseudogenes. In contrast, high stringency Southern analysis with an apparent intronic probe that was generated by PCR amplification of mouse genomic DNA with MIF-specific exon primers yielded a more simple hybridization pattern indicative of a single gene copy.[15] Screening of a genomic DNA phage library ultimately identified the full mouse MIF gene and two MIF pseudogenes. Independent studies using a PCR-based approach also identified the functional full length mouse MIF gene as well as nine processed pseudogenes.[11] The structure of the mouse MIF gene is found to be highly similar to that of humans, with three exons separated by two short introns (201 and 145 nucleotides) and maps to chromosome 10.[11,15] 5′RACE and primer extension indicated that the transcriptional start site for mouse MIF was 85 bases upstream of the start methionine, and similar to the human MIF promoter, there is no classical TATA box in the immediate 5′-flanking region.[15]

REGULATION OF MIF EXPRESSION

MIF is ubiquitously expressed in response to various hormones, growth factors, and inflammatory stimuli such as lipopolysaccharide. Furthermore, MIF is expressed at high levels in many resting inflammatory cells and is stored preformed, ready for immediate secretion. Specific release of MIF is seen from macrophages in response to LPS, T cells in response to anti-CD3/CD28 stimulation, and fibroblasts in response to adhesion, integrin ligation and growth factor stimulation.

Elevated expression of MIF has been described in many inflammatory pathologies, including acute respiratory distress syndrome,[16] asthma,[17,18] pulmonary fibrosis,[19] septic shock,[20,21] rheumatoid arthritis,[22] and glomerulonephritis.[23] Interestingly, elevated MIF expression is also found in a large variety of human malignancies. Prostate, breast, colon, brain, skin and lung-derived tumors have all been shown to contain significantly higher levels of MIF mRNA and protein than their non-cancerous cell counterparts.[24–26] Despite the compelling evidence that MIF is over expressed in, and contributes to, the pathology of so many inflammatory and malignant diseases, the mechanisms that contribute to exaggerated expression of MIF have been poorly described. Indeed, very little is known about tissue- and stimulus-specific signalling events that lead to the upregulation of MIF expression.

Promoter scanning software suggests several putative transcription factor binding sites in the immediate 5′-flanking region that are conserved between the murine and human MIF promoters. To date, however, the only response elements that have been functionally tested are a cyclic adenosine monophosphate response element (CRE) in the mouse promoter and a hypoxia inducible factor (HIF-1α) response element (HRE) in the 5′UTR of the human MIF gene and the proximal promoter region in the mouse MIF gene.

Studies analyzing the regulation of MIF expression in the pituitary identified a proximal CRE (-41 to -48 nucleotides upstream of the transcriptional start site) that is essential for corticotrophin releasing factor (CRF)-induced transcription from the murine MIF promoter in rat pituitary cells.[27] The human MIF promoter also contains a proposed CRE at around the same location, but it has not been determined whether this site is involved in human pituitary expression of MIF.

Several recent studies have implicated MIF as a hypoxia-responsive gene in cancer, cardiac ischemic injury, and inflammatory responses. Gene array studies have shown MIF to be upregulated by hypoxia in squamous cell carcinoma.[28] and in the breast cancer cell line MCF-7.[29] In colon cancer, hypoxic induction of MIF has been shown to correlate with sensitivity to apoptosis.[30] Furthermore, a role for MIF in ischemic responses in cardiac tissues was proposed by Takahashi *et al.*, who described elevated MIF expression in patients with acute myocardial infarction and demonstrated that hypoxia induced the secretion of MIF from rat cardiac myocytes.[31] More recently, Bacher *et al.* have shown that MIF is highly expressed in glioblastoma multiforme and that hypoxic stimulation of a glioblastoma cell line resulted in increased expression of MIF mRNA and protein.[32] The mechanism by which hypoxia induces MIF expression has recently been shown to be dependent upon a functional HRE in the 5′UTR of the human MIF gene but is also influenced by CREB activity.[33] Under conditions of hypoxia, HIF-1α regulated MIF promoter activity is blocked by overexpression of CREB, implying a role for CREB in modulating hypoxia-induced MIF expression. Similar responses have also been confirmed in the mouse MIF promoter where a HRE in the 5′UTR is essential for hypoxia-induced promoter activity.[34]

In the context of an inflammatory response NF-κB has been implicated in the expression of MIF in response to IL-1β and TNFα in human endometrial cells.[35,36] IL-1β and TNFα were convincingly shown to induce MIF mRNA expression and protein secretion in a dose-dependent manner in parallel with an increase in NF-κB translocation and IκB phosphorylation but a direct interaction between NF-κB and the MIF promoter was not shown.

Interestingly, NF-κB has also been implicated in the regulation of MIF expression by estrogen. Studies addressing the delay in wound healing in the elderly associated decreased estrogen levels with increased serum and wound MIF levels.[37] Extending from these data, it was shown that estrogen inhibits basal and LPS-induced MIF promoter activity and that the effect of estrogen was abrogated by the NF-κB inhibitor PDCT. In contrast to the proposed role of NF-κB in driving IL-1β and TNFα-induced MIF expression.[35,36] in collaboration with an active estrogen, receptor NF-κB may also be essential in downregulating MIF expression.[37]

Over recent years there has been great interest in hormonal regulation of MIF expression, particularly the relationship between MIF and glucocorticoid biology. Studies in the early 1990s rediscovered MIF as a pituitary-derived mediator of systemic stress responses (reviewed in

Refs. 38 and 39). In addition to being secreted from the pituitary, MIF was found to be released from immune/inflammatory cells as a consequence of glucocorticoid stimulation.[40,41] Further analysis in rodents suggested that organ MIF protein content was induced by parenteral glucocorticoid administration, suggesting a direct effect on MIF production, as well as on MIF secretion. Interestingly, however, there was no concordant increase in MIF mRNA, suggesting an effect on protein translation without any modulation of transcriptional activity.[42]

MIF GENE POLYMORPHISMS

With the advent of sensitive assays for quantifying serum MIF levels, it became apparent that significantly elevated MIF levels were associated with many inflammatory and malignant pathologies. Interestingly, it also became apparent that within both normal and diseased populations there was a broad range in serum MIF levels. Such observed variation in expression could be partly contributed to by the circadian regulation of MIF[43] but raised the question of whether genetic variation determined susceptibility to exaggerated MIF expression.

The concept that polymorphisms in immune response genes contribute to the pathogenesis of certain human autoimmune/inflammatory diseases has received increasing interest over the last decade. At present, very few gene polymorphisms have been shown to be functionally significant and to be of prognostic value in specific disease states. Previously defined examples include polymorphisms in TNFα and IL-1ra that have been shown to have certain prognostic significance in malaria and ischaemic heart disease, respectively.[44,45] Similarly, a number of polymorphisms in TNFα and IL-1β have been reported to be associated with rheumatoid arthritis severity.[46–48]

Sequencing of the MIF gene in six individuals highlighted a tetranucleotide CATT-repeat at position −794, that was originally described by Paraklar and Wistow,[49] and revealed a variation in the number of repeats.[50] Further screening of a larger population identified individuals that were homozygous or heterozygous for 5-, 6-, 7- or 8-CATT repeats. Independent studies using high performance denaturing liquid chromatography also revealed the existence of a G- to C-single nucleotide polymorphism (SNP) at position −173 of the human MIF promoter.[51] as well as a T- to C-SNP at position +254 (intron 1), and a C- to G-SNP at position +656 (intron 2)[52] (see Fig. 1). It is currently unclear whether the intronic SNPs play any role in regulating MIF expression, but it has become well established that both the −794 CATT-repeat polymorphism and the −173 SNP affect transcriptional activity of the human MIF promoter.

In vitro analysis has shown that MIF promoter activity is proportional to the number of CATT repeats at position −794 with the 5-CATT allele conferring significantly lower basal as well as serum- and forskolin-induced transcriptional activity.[50] Similarly, it has been shown that the −173 C variant also increases MIF promoter activity.[52] Multiple studies have correlated carriage of non-5 CATT alleles and −173 C alleles with exaggerated inflammatory disease, supporting the hypothesis that these variants are responsible for elevated MIF expression, but the precise mechanism by which these polymophisms affect MIF promoter activity is currently unknown. Promoter scanning software shows that the pituitary-specific transcription factor Pit-1 could potentially bind to the CATT-repeat region. Such an interaction has been confirmed *in vitro* using electromobility shift assays (Baugh unpublished), but the functional significance of this interaction is unclear. Given the proposed role of pituitary-derived MIF in determining circulating MIF levels, it is attractive to propose that Pit-1 may play a role in determining an individual's genetic susceptibility to enhanced MIF expression and exaggerated inflammatory responses, but further studies are required to address this.

Analysis of the −173 position of the MIF promoter reveals that the mutation to a cytosine residue generates a putative activator protein (AP)-4 binding site. Despite clear evidence that this site is involved in exaggerated MIF expression, the transcriptional complexes involved have not been identified and the role of AP-4 is unproven.

MIF POLYMORPHISMS AND DISEASE

Since the description of the two MIF promoter polymorphic regions, these have been linked with several inflammatory diseases. In support of the proposed role of MIF in promoting exaggerated inflammatory responses, the high expressing non-5 CATT and −173 C alleles have both been associated with disease progression and severity in multiple conditions. The primary focus of this chapter is to review the role of MIF promoter polymorphisms in respiratory disease.

MIF Polymorphisms in Sarcoidosis

Sarcoidosis is a multisystemic granulomatous disorder of unknown etiology, which can affect any organ system; however, lung involvement is seen in more than 90% of patients. Common extrathoracic sites include lymph nodes, skin, and eyes.[53,54] The general concept has emerged that sarcoidosis results from the exposure of genetically susceptible individuals to specific environmental agents as yet unknown.[53,54] Genetic susceptibility to sarcoidosis is supported by familial clustering[54,55] and a higher prevalence of disease in first-degree relatives.[56] In Ireland, the prevalence of sarcoidosis cases among siblings is 2.4%.[57]

Macrophage migration inhibitory factor (MIF) is secreted by a variety of pulmonary cells, including monocyte/macrophages,[58] endothelial cells and various epithelial cells including bronchial epithelium.[16] Elevated serum MIF levels have been reported in sarcoidosis[59] and culture supernatants of cutaneous granulomas from patients with sarcoidosis contain MIF.[60] MIF potentially contributes towards an excessive inflammatory response in the lung both directly via an induction of pro-inflammatory cytokine secretion, including TNF-α and IL-8[21] and indirectly through its ability to override the anti-inflammatory activity of glucocorticoids.[16] As well as driving innate responses, MIF is also an important regulator of adaptive immunity. T cells constitutively express MIF and expression studies using Th1 and Th2 subsets revealed that while both subsets of T cells express MIF, secretion is predominantly in activated Th2 clones.[61] These data would favor a role for MIF in the development of Th2-driven antibody production. MIF is implicated in a wide number of acute and chronic immune and inflammatory diseases, including ARDS (acute respiratory distress syndrome),[16] asthma,[18] septic shock,[21] and rheumatoid arthritis.[62] Anti-MIF neutralizing antibodies reduce the production of TNF-α and IL-8,[16,21] highlighting the potential role of an anti-MIF strategy as a therapeutic target in inflammatory lung disease. The potential role of MIF in granulomatous disease was highlighted by the ability of an anti-MIF monoclonal antibody strategy to attenuate granulomatous inflammatory bowel disease in an animal model of Crohns disease.[63]

The passage from granuloma to fibrosis is not well understood in sarcoidosis. Th1 cytokines (IL-2, IL-12, and IFN-γ) are likely to favor granuloma formation.[64] A shift from a Th1 to a Th2 phenotype (IL-4, IL-5, and IL-10) may be important for persistent and progressive fibrotic disease.[65] Given MIF's proinflammatory role in innate immunity, its ability both to over-ride the anti-inflammatory action of glucocorticoids and to drive a Th2 phenotype, it was proposed that the MIF gene may potentially modify the clinical course in sarcoidosis.

In a case-control study of an Irish population with sarcoidosis, 176 patients were specifically radiologically classified into two groups: those with milder radiological disease and those with interstitial or fibrotic changes. Patients and a healthy control population of 166 were genotyped in regard to the −794 CATT repeat. No significant association was found between the presence of the low expressing 5-CATT MIF-promoter polymorphism and milder clinical disease.[66] In particular, no association was demonstrated between those patients heterozygous for the 5-CATT polymorphism and disease activity, in contrast to other studies of patients with chronic pulmonary and non-pulmonary inflammatory diseases (e.g. asthma,[17] cystic fibrosis[67] and rheumatoid arthritis[50]). With reference to a possible association between 5-CATT heterozygosity and disease activity as the 5/5 genotype considered as quite rare, a larger population of patients would need to be studied to definitively address this question.

Interestingly, however, the G- to C-single nucleotide polymorphism (SNP) at position −173 of the MIF gene was found to be associated with the development of sarcoidosis in patients with erythema nodosum (EN) in a Spanish cohort.[26] A significantly increased frequency of the C mutant allele was observed in patients with EN secondary to sarcoidosis as compared to controls ($p = 0.0016$; p (corr)$=0.0032$; OR, 2.78; 95% CI 1.45–5.35) and also compared to patients with EN unrelated to sarcoidosis ($p = 0.0004$; p (corr) $= 0.0008$; OR, 3.72; 95% CI 1.75–7.87). Patients with EN carrying an MIF 173 C allele were found to have an increased risk of sarcoidosis (57% in EN secondary to sarcoidosis vs. 24% in patients with EN related to other etiologies; $p = 0.002$; p (corr) $= 0.004$; OR, 4.16;

95% CI 1.64–10.50). In contrast to these data, the possession of a 5-CATT repeat allele was not associated with a decreased risk of an Irish sarcoidosis patient presenting with EN.[66]

MIF Polymorphisms in Asthma

Although MIF is most well known as a key contributor to acute lung injury,[16] it is also apparent that MIF plays a role in the development of allergic inflammatory disease. Human eosinophils, derived from atopic donors are potent sources of MIF.[18] Both C5a and IL-5, physiologically relevant stimuli in allergic inflammatory disease, cause significant immediate release of MIF from these *ex-vivo* circulating cells. Eosinophils have been implicated as a key cell type in the pathogenesis of allergic inflammatory diseases such as bronchial asthma, allergic rhinitis and atopic dermatitis. Analysis of bronchoalveolar lavage fluid samples from a cohort of stable asthmatics revealed an elevated level of MIF protein expression and it is suggested that the elevated numbers of eosinophils present in the alveolar airspaces are responsible for its production.[18] More recently, in animal models of sepsis, C5a has also been shown to induce MIF expression and release from neutrophils.[68] The identification of eosinophils and notably, the type II alveolar epithelial cells as additional cellular sources of MIF, and the correlation of MIF expression with the development of a Th2 phenotype highlights the potential importance of MIF in allergic pulmonary inflammatory disease.

Studies in OVA-primed and inhalation-challenged mice support a role for MIF in the immune response to OVA and the development of inflammatory airway disease analogous to human asthma.[17] When compared to wild-type mice, $MIF^{-/-}$ mice showed a reduction in total and OVA-specific secondary response antibodies, especially IgE. These findings suggest that MIF deficiency results in a selective defect in Th2-priming, or in the B cell response that is necessary for isotype switching. Supporting the potential role for MIF in Th2 skewing, it was also demonstrated that the reduced level of IgE in challenged $MIF^{-/-}$ mice was associated with a lower level of IL-5 and eotaxin in bronchiolar lavage fluid; a decrease in IL-4, IL-5, and IL-13 in the lung; and a decrease in the secretion of these cytokines by splenocytes.[17] Furthermore, the decrease in Th2 cytokine production in challenged MIF$^{-/-}$ mice was paralleled by decreased peribronchial infiltration and eosinophil content in alveoli, decreased mucus production, and decreased airway hyperresponsiveness.[17]

Given the role of MIF as a proinflammatory mediator and counter-regulator of glucocorticoid function, as well as the clear evidence linking MIF with disease progression in the mouse OVA-challenge model, great interest has been generated in MIF as a potential therapeutic target in asthma. Also, as increased MIF expression has been documented in asthma, it was a logical progression to address the potential association between MIF promoter polymorphisms and asthmatic disease. Two independent studies screening an ethnically homogenous (white) population and a Japanese population, respectively, revealed no differences between the distribution of the CATT alleles in the asthmatic vs. the control groups.[17,69] However, further stratification of the asthmatic group by disease severity, as defined by the Global Initiative for Asthma criteria[70] revealed significant associations between the 5-CATT allele and the less severe clinical disease (5/5 vs. X/X alleles, $P = 0.007$; and 5/5 and 5/X vs. X/X alleles, $p = 0.05$).[17] A trend was also described between the 5/5 and combined 5/5 + 5/X groups for both a) higher recorded FEV1 ($P = 0.06$ and $P = 0.13$, respectively); and (b) fewer hospital admissions for asthma ($P = 0.09$ and $P = 0.08$, respectively). These data support a protective role for the low expressing 5-CATT MIF allele in the development of severe asthma. In combination with the evidence highlighting a pathophysiological role for MIF in the development of asthma, these data may also prove useful in guiding MIF neutralizing therapy to individuals who are genetically susceptible to exaggerated MIF expression.

MIF Polymorphisms in Cystic Fibrosis

Cystic fibrosis (CF), caused by mutations in the CF transmembrane receptor (CFTR) gene,[71] is the most common life-limiting autosomal recessive disease among people of European heritage, with a frequency of 1 in 1,461 live births in Ireland.[72] In CF lung disease, an exaggerated, sustained response to gram-negative bacteria characterized by neutrophil-dominated airway inflammation is well recognized.[73–75] CF lung disease renders patients susceptible to colonization by the opportunistic gram-negative pathogen *Pseudomonas aueriginosa* (PA) and responses to PA are one of the major causes of morbidity and mortality in these patients.[73–77]

MIF has been shown to contribute towards an exaggerated host response to gram-negative infection via its ability to augment Toll-like receptor 4 (TLR4) expression.[78] Given the reported role for MIF in sepsis,[20,21] it was proposed that exaggerated MIF expression due to promoter polymorphic variation may be associated with disease severity. A case-control study of an Irish CF population analyzed distribution of the various CATT-repeat genotypes in 167 patients with CF and 166 age- and sex-matched control subjects. MIF genotype frequencies did not differ significantly between cases and control subjects (χ^2 test, $P > 0.48$), but a significant association was found between the low expressing 5-CATT allele and both decreased PA colonization and pancreatic insufficiency.[67] Specifically, the 5/X allele group displayed a PA colonization of 75% compared to 91% in the X/X allele group (OR, 0.21; 95% CI 0.09–0.65; $P = 0.004$) and a reduced 86% risk of pancreatic insufficiency compared to 96% in the X/X allele group (OR, 0.27; 95% CI 0.07–1.00; $P = 0.05$). Although pancreatic insufficiency and diabetes mellitus in CF are believed to occur as a result of reduced volume of pancreatic secretions with low bicarbonate concentrations, resulting in the retention and premature activation of digestive proenzymes and leading to tissue destruction and fibrosis,[73,79] it is worth noting that the 5-CATT MIF allele may attenuate this process.[67] It is thus proposed that exaggerated MIF expression from the non-5 CATT genotypes promotes pancreatic tissue destruction and fibrosis. In support of this hypothesis, it has recently been reported that serum MIF levels in patients with pancreatitis directly correlated with severity of disease and that anti-MIF treatment attenuated disease progression in a rat model of acute pancreatitis.[80]

SUMMARY

Since the cloning of MIF cDNA, significant advances have been made in the understanding of MIF biology. Characterization of the MIF gene and promoter sequence has further added to our knowledge in terms of the tissue- and stimulus-specific regulation of MIF expression. Extending from these data, the identification of polymorphic variation in the MIF gene, and in particular in the promoter region has highlighted the possibility of genetic susceptibility to exaggerated MIF expression. This chapter has highlighted the association between MIF promoter polymorphisms and inflammatory lung disease. Specifically, to date it has been shown that the low expressing 5-CATT allele appears to be protective of disease progression in cystic fibrosis and asthma. In addition, the −173 C allele has been linked with the development of sarcoidosis in a Spanish EN population, despite the lack of association between the CATT alleles and sarcoidosis in an Irish population. Although the CATT and −173 SNP variants have individually been linked with regulation of MIF expression and disease severity in many inflammatory diseases, it is important to note that several studies have also made an association between the 7-CATT allele and the −173 C allele.[81–87] Taken together, these data highlight the possibility that genetic susceptibility to enhanced MIF expression may be a critical determinant in the outcome of inflammatory diseases, including asthma and cystic fibrosis, and strongly influence the efficacy of glucocorticoid treatment. Furthermore, identification of patients susceptible to exaggerated MIF expression may play an important role in targeting focused anti-MIF therapeutic strategies to those patients most likely to benefit.

REFERENCES

1. Bloom BR, Bennett B. (1966) Mechanism of a reaction *in vitro* associated with delayed-type hypersensitivity. *Science* **153**: 80–82.
2. David JR. (1966) Delayed hypersensitivity *in vitro*: its mediation by cell-free substances formed by lymphoid cell-antigen interaction. *Proc Natl Acad Sci USA* **56**: 72–77.
3. Churchill WHJ, Piessens WF, Sulis CA, David JR. (1975) Macrophages activated as suspension cultures with lymphocyte mediators devoid of antigen become cytotoxic for tumor cells. *J Immunol* **115**: 781–786.
4. Nathan CF, Karnovsky ML, David JR. (1971) Alterations of macrophage functions by mediators from lymphocytes. *J Exp Med* **133**: 1356–1376.
5. Nathan CF, Remold HG, David JR. (1973) Characterization of a lymphocyte factor which alters macrophage functions. *J Exp Med* **137**: 275–290.

6. Baugh JA, Bucala R. (2002) Macrophage migration inhibitory factor. *Crit Care Med* **30**: S27–S35.
7. Baugh JA, Donnelly SC. (2003) Macrophage migration inhibitory factor: a neuroendocrine modulator of chronic inflammation. *J Endocrinol* **179**: 15–23.
8. Mitchell RA. (2004) Mechanisms and effectors of MIF-dependent promotion of tumourigenesis. *Cell Signal* **16**: 13–19.
9. Weiser WY, Temple PA, Witek-Giannotti JS, *et al.* (1989) Molecular cloning of a cDNA encoding a human macrophage migration inhibitory factor. *Proc Natl Acad Sci USA* **86**: 7522–7526.
10. Paralkar V, Wistow G. (1995) Mif — an essential intermediate for growth-factor induced mitogenesis. *Faseb Journal* **9**: A1332.
11. Kozak CA, Adamson MC, Buckler CE, *et al.* (1995) Genomic cloning of mouse mif (macrophage inhibitory factor) and genetic-mapping of the human and mouse expressed gene and 9 mouse pseudogenes. *Genomics* **27**: 405–411.
12. Budarf M, McDonald T, Sellinger B, *et al.* (1997) Localization of the human gene for macrophage migration inhibitory factor (MIF) to chromosome 22q11.2. *Genomics* **39**: 235–236.
13. Lanahan A, Williams JB, Sanders LK, Nathans D. (1992) Growth factor-induced delayed early response genes. *Mol Cell Biol* **12**: 3919–3929.
14. Kozak CA, Adamson MC, Buckler CE, *et al.* (1995) Genomic cloning of mouse MIF (macrophage inhibitory factor) and genetic mapping of the human and mouse expressed gene and nine mouse pseudogenes. *Genomics* **27**: 405–411.
15. Mitchell R, Bacher M, Bernhagen J, *et al.* (1995) Cloning and characterization of the gene for mouse macrophage migration inhibitory factor (MIF). *J Immunol* **154**: 3863–3870.
16. Donnelly SC, Haslett C, Reid PT, *et al.* (1997) Regulatory role for macrophage migration inhibitory factor in acute respiratory distress syndrome. *Nat Med* **3**: 320–323.
17. Mizue Y, Ghani S, Leng L, *et al.* (2005) Role for macrophage migration inhibitory factor in asthma. *Proc Natl Acad Sci USA* **102**: 14410–14415.
18. Rossi AG, Haslett C, Hirani N, *et al.* (1998) Human circulating eosinophils secrete macrophage migration inhibitory factor (MIF). Potential role in asthma. *J Clin Invest* **101**: 2869–2874.
19. Magi B, Bini L, Perari MG, *et al.* (2002) Bronchoalveolar lavage fluid protein composition in patients with sarcoidosis and idiopathic pulmonary fibrosis: a two-dimensional electrophoretic study. *Electrophoresis* **23**: 3434–3444.
20. Bernhagen J, Calandra T, Mitchell RA, *et al.* (1993) Mif is a pituitary-derived cytokine that potentiates lethal endotoxemia. *Nature* **365**: 756–759.
21. Calandra T, Echtenacher B, Roy DL, *et al.* (2000) Protection from septic shock by neutralization of macrophage migration inhibitory factor. *Nat Med* **6**: 164–170.

22. Morand EF, Leech M, Weedon H, *et al.* (2002) Macrophage migration inhibitory factor in rheumatoid arthritis: clinical correlations. *Rheumatology* **41**: 558–562.
23. Lan HY, Yang NS, Nikolic-Paterson DJ, *et al.* (2000) Expression of macrophage migration inhibitory factor in human glomerulonephritis. *Kidney Int* **57**: 499–509.
24. Takahashi N, Nishihira J, Sato Y, *et al.* (1998) Involvement of macrophage migration inhibitory factor (MIF) in the mechanism of tumor cell growth. *Molecul Med* **4**: 707–714.
25. Meyer-Siegler K, Fattor RA, Hudson PB. (1998) Expression of macrophage migration inhibitory factor in the human prostate. *Diagn Molecul Pathol* **7**: 44–50.
26. Tomiyasu M, Yoshino I, Suemitsu R, *et al.* (2002) Quantification of macrophage migration inhibitory factor mRNA expression in non-small cell lung cancer tissues and its clinical significance. *Clin Cancer Res* **8**: 3755–3760.
27. Waeber G, Thompson N, Chautard T, *et al.* (1998) Transcriptional activation of the macrophage migration-inhibitory factor gene by the corticotropin-releasing factor is mediated by the cyclic adenosine 3′,5′-monophosphate responsive element-binding protein CREB in pituitary cells. *Mol Endocrinol* **12**: 698–705.
28. Koong AC, Denko NC, Hudson KM, *et al.* (2000) Candidate genes for the hypoxic tumor phenotype. *Cancer Res* **60**: 883–887.
29. Bando H, Toi M, Kitada K, Koike M. (2003) Genes commonly upregulated by hypoxia in human breast cancer cells MCF-7 and MDA-MB-231. *Biomed Pharmacother* **57**: 333–340.
30. Yao K, Shida S, Selvakumaran M, *et al.* (2005) Macrophage migration inhibitory factor is a determinant of hypoxia-induced apoptosis in colon cancer cell lines. *Clin Cancer Res* **11**: 7264–7272.
31. Takahashi M, Nishihira J, Shimpo M, *et al.* (2001) Macrophage migration inhibitory factor as a redox-sensitive cytokine in cardiac myocytes. *Cardiovasc Res* **52**: 438–445.
32. Bacher M, Schrader J, Thompson N, *et al.* (2003) Up-regulation of macrophage migration inhibitory factor gene and protein expression in glial tumor cells during hypoxic and hypoglycemic stress indicates a critical role for angiogenesis in glioblastoma multiforme. *Am J Pathol* **162**: 11–17.
33. Baugh JA, Gantier M, Li L, *et al.* (2006) Dual regulation of macrophage migration inhibitory factor (MIF) expression in hypoxia by CREB and HIF-1. *Biochem Biophys Res Commun* **347**: 895–903.
34. Welford SM, Bedogni B, Gradin K, *et al.* (2006) HIF1alpha delays premature senescence through the activation of MIF. *Genes Dev* **20**: 3366–3371.

35. Cao WG, Morin M, Sengers V, *et al.* (2006) Tumour necrosis factor-alpha up-regulates macrophage migration inhibitory factor expression in endometrial stromal cells via the nuclear transcription factor NF-κB. *Hum Reprod* **21**: 421–428.
36. Cao WG, Morin M, Metz C, *et al.* (2005) Stimulation of macrophage migration inhibitory factor expression in endometrial stromal cells by interleukin 1, beta involving the nuclear transcription factor NF-κB. *Biol Reprod* **73**: 565–570.
37. Hardman MJ, Waite A, Zeef L, *et al.* (2005) Macrophage migration inhibitory factor: a central regulator of wound healing. *Am J Pathol* **167**: 1561–1574.
38. Bucala R. (1996) Mif rediscovered: cytokine, pituitary hormone, and glucocorticoid- induced regulator of the immune response. *Faseb J* **10**: 1607–1613.
39. Flaster H, Bernhagen J, Calandra T, Bucala R. (2007) The MIF-Glucocorticoid Dyad: Regulation of Inflammation and Immunity. *Mol Endocrinol.*
40. Calandra T, Bernhagen J, Metz CN, *et al.* (1995) Mif as a glucocorticoid-induced modulator of cytokine production. *Nature* **377**: 68–71.
41. Leech M, Metz C, Bucala R, Morand EF. (2000) Regulation of macrophage migration inhibitory factor by endogenous glucocorticoids in rat adjuvant-induced arthritis. *Arthritis Rheum* **43**: 827–833.
42. Fingerle-Rowson G, Koch P, Bikoff R, *et al.* (2003) Regulation of macrophage migration inhibitory factor expression by glucocorticoids *in vivo*. *Am J Pathol* **162**: 47–56.
43. Petrovsky N, Socha L, Silva D, *et al.* (2003) Macrophage migration inhibitory factor exhibits a pronounced circadian rhythm relevant to its role as a glucocorticoid counter-regulator. *Immunol Cell Biol* **81**: 137–143.
44. McGuire W, Hill AV, Allsopp CE, *et al.* (1994) Variation in the TNF-alpha promoter region associated with susceptibility to cerebral malaria. *Nature* **371**: 508–510.
45. Wilson AG, Symons JA, McDowell TL, *et al.* (1997) Effects of a polymorphism in the human tumor necrosis factor alpha promoter on transcriptional activation. *Proc Natl Acad Sci USA* **94**: 3195–3199.
46. Tarlow JK, Cork MJ, Clay FE, *et al.* (1997) Association between interleukin-1 receptor antagonist (IL-1ra) gene polymorphism and early and late-onset psoriasis. *Br J Dermatol* **136**: 147–148.
47. Francis SE, Camp NJ, Dewberry RM, *et al.* (1999) Interleukin-1 receptor antagonist gene polymorphism and coronary artery disease. *Circulation* **99**: 861–866.
48. Cox A, Camp NJ, Cannings C, *et al.* (1999) Combined sib-TDT and TDT provide evidence for linkage of the interleukin-1 gene cluster to erosive rheumatoid arthritis. *Hum Mol Genet* **8**: 1707–1713.

49. Paralkar V, Wistow C. (1994) Cloning the human gene for macrophage-migration inhibitory factor (Mif). *Genomics* **19**: 48–51.
50. Baugh JA, Chitnis S, Donnelly SC, *et al.* (2002) A functional promoter polymorphism in the macrophage migration inhibitory factor (MIF) gene associated with disease severity in rheumatoid arthritis. *Genes Immun* **3**: 170–176.
51. Donn RP, Shelley E, Ollier WE, Thomson W. (2001) A novel 5′-flanking region polymorphism of macrophage migration inhibitory factor is associated with systemic-onset juvenile idiopathic arthritis. *Arthritis Rheum* **44**: 1782–1785.
52. Donn R, Alourfi Z, De Benedetti F, *et al.* (2002) Mutation screening of the macrophage migration inhibitory factor gene: positive association of a functional polymorphism of macrophage migration inhibitory factor with juvenile idiopathic arthritis. *Arthritis Rheum* **46**: 2402–2409.
53. Costabel U. (2001) Sarcoidosis: clinical update. *Eur Respir J Suppl* **32**: 56s–68s.
54. Statement on sarcoidosis. (1999) Joint Statement of the American Thoracic Society (ATS), the European Respiratory Society (ERS) and the World Association of Sarcoidosis and Other Granulomatous Disorders (WASOG) adopted by the ATS Board of Directors and by the ERS Executive Committee, February 1999. *Am J Respir Crit Care Med* **160**: 736–755.
55. Rybicki BA, Iannuzzi MC, Frederick MM, *et al.* (2001) Familial aggregation of sarcoidosis. A case-control etiologic study of sarcoidosis (ACCESS). *Am J Respir Crit Care Med* **164**: 2085–2091.
56. Rybicki BA, Kirkey KL, Major M, *et al.* (2001) Familial risk ratio of sarcoidosis in African-American sibs and parents. *Am J Epidemiol* **153**: 188–193.
57. Brennan NJ, Crean P, Long JP, FitzGerald MX. (1984) High prevalence of familial sarcoidosis in an Irish population. *Thorax* **39**: 14–18.
58. Bernhagen J, Bacher M, Calandra T, *et al.* (1996) An essential role for macrophage migration inhibitory factor in the tuberculin delayed-type hypersensitivity reaction. *J Exper Med* **183**: 277–282.
59. Yoshida T, Siltzbach LE, Masih N, Cohen S. (1979) Serum-migration inhibitory activity in patients with sarcoidosis. *Clin Immunol Immunopathol* **13**: 39–46.
60. Campbell PB, Kataria YP, Tolson TA. (1984) *In vitro* production of inhibitors of monocyte locomotion by the granuloma of sarcoidosis. *Am Rev Respir Dis* **130**: 417–423.
61. Bacher M, Metz CN, Calandra T, *et al.* (1996) An essential regulatory role for macrophage migration inhibitory factor in t-cell activation. *Proc Nat Acad Sci USA* **93**: 7849–7854.

62. Morand EF, Leech M, Weedon H, *et al.* (2002) Macrophage migration inhibitory factor in rheumatoid arthritis: clinical correlations. *Rheumatology* **41**: 558–562.
63. de Jong YP, Abadia-Molina AC, Satoskar AR, *et al.* (2001) Development of chronic colitis is dependent on the cytokine MIF. *Nat Immunol* **2**: 1061–1066.
64. Moller DR. (1999) Cells and cytokines involved in the pathogenesis of sarcoidosis. *Sarcoidosis Vasc Diffuse Lung Dis* **16**: 24–31.
65. Kunkel SL, Lukacs NW, Strieter RM, Chensue SW. (1996) Th1 and Th2 responses regulate experimental lung granuloma development. *Sarcoidosis Vasc Diffuse Lung Dis* **13**: 120–128.
66. Plant BJ, Ghani S, O'Mahony MJ, *et al.* (2007) Sarcoidosis and MIF gene polymorphism: a case-control study in an Irish population. *Eur Respir J* **29**: 325–329.
67. Plant BJ, Gallagher CG, Bucala R, *et al.* (2005) Cystic fibrosis, disease severity, and a macrophage migration inhibitory factor polymorphism. *Am J Respir Crit Care Med* **172**: 1412–1415.
68. Riedemann NC, Guo RF, Gao H, *et al.* (2004) Regulatory role of C5a on macrophage migration inhibitory factor release from neutrophils. *J Immunol* **173**: 1355–1359.
69. Hizawa N, Yamaguchi E, Takahashi D, *et al.* (2004) Functional polymorphisms in the promoter region of macrophage migration inhibitory factor and atopy. *Am J Respir Crit Care Med* **169**: 1014–1018.
70. Global Inititative for Asthma (GINA): Global Strategy for Asthma Management and Prevention. (2002) National Heart, Lun, and Blood Institute/World Health Organisation Workshop Report (Natl. Inst. of Health, Bethesda).
71. Collins FS. (1992) Cystic fibrosis: molecular biology and therapeutic implications. *Science* **256**: 774–779.
72. Cashman SM, Patino A, Delgado MG, *et al.* (1995) The Irish cystic fibrosis database. *J Med Genet* **32**: 972–975.
73. Ratjen F, Doring G. (2003) Cystic fibrosis. *Lancet* **361**: 681–689.
74. Gibson RL, Burns JL, Ramsey BW. (2003) Pathophysiology and management of pulmonary infections in cystic fibrosis. *Am J Respir Crit Care Med* **168**: 918–951.
75. Yankaskas JR, Marshall BC, Sufian B, *et al.* (2004) Cystic fibrosis adult care: consensus conference report. *Chest* **125**: 1S–39S.
76. Kerem E, Corey M, Gold R, Levison H. (1990) Pulmonary function and clinical course in patients with cystic fibrosis after pulmonary colonization with Pseudomonas aeruginosa. *J Pediatr* **116**: 714–719.
77. Steinkamp G, Wiedemann B. (2002) Relationship between nutritional status and lung function in cystic fibrosis: cross sectional and longitudinal analyses from the German CF quality assurance (CFQA) project. *Thorax* **57**: 596–601.

78. Roger T, David J, Glauser MP, Calandra T. (2001) MIF regulates innate immune responses through modulation of Toll-like receptor 4. *Nature* **414**: 920–924.
79. Kopelman H, Durie P, Gaskin K, *et al.* (1985) Pancreatic fluid secretion and protein hyperconcentration in cystic fibrosis. *N Engl J Med* **312**: 329–334.
80. Sakai Y, Masamune A, Satoh A, *et al.* (2003) Macrophage migration inhibitory factor is a critical mediator of severe acute pancreatitis. *Gastroenterology* **124**: 725–736.
81. Nunez C, Rueda B, Martinez A, *et al.* (2007) Involvement of macrophage migration inhibitory factor gene in celiac disease susceptibility. *Genes Immun* **8**: 168–170.
82. Wu SP, Leng L, Feng Z, *et al.* (2006) Macrophage migration inhibitory factor promoter polymorphisms and the clinical expression of scleroderma. *Arthritis Rheum* **54**: 3661–3669.
83. Sanchez E, Gomez LM, Lopez-Nevot MA, *et al.* (2006) Evidence of association of macrophage migration inhibitory factor gene polymorphisms with systemic lupus erythematosus. *Genes Immun* **7**: 433–436.
84. Radstake TR, Sweep FC, Welsing P, *et al.* (2005) Correlation of rheumatoid arthritis severity with the genetic functional variants and circulating levels of macrophage migration inhibitory factor. *Arthritis Rheum* **52**: 3020–3029.
85. Donn RP, Plant D, Jury F, *et al.* (2004) Macrophage migration inhibitory factor gene polymorphism is associated with psoriasis. *J Invest Dermatol* **123**: 484–487.
86. Donn R, Alourfi Z, Zeggini E, *et al.* (2004) A functional promoter haplotype of macrophage migration inhibitory factor is linked and associated with juvenile idiopathic arthritis. *Arthritis Rheum* **50**: 1604–1610.
87. Barton A, Lamb R, Symmons D, *et al.* (2003) Macrophage migration inhibitory factor (MIF) gene polymorphism is associated with susceptibility to but not severity of inflammatory polyarthritis. *Genes Immun* **4**: 487–491.

15

MIF Alleles in Inflammatory Arthritis

*Rachelle Donn** *and David Ray*†

Many aspects of the macrophage migration inhibitory factor (MIF) make it an important candidate to investigate, at the genetic level, for association with rheumatic diseases. Much of this background information is detailed in Chapter 13 — Morand's "Inflammatory Arthritis" of this book. Overall, strong evidence from animal models, *in vitro* studies and human studies have shown MIF to have a central role in the inflammatory phenotype of chronic rheumatic diseases.[1–3]

INTRODUCTION

MIF Alleles and Arthritis

This chapter focuses on the identification and subsequent association of MIF gene polymorphisms and arthritis. To date MIF gene polymorphisms have been investigated in only two arthritis conditions: juvenile idiopathic arthritis (JIA) and rheumatoid arthritis (RA).

JIA is the commonest chronic rheumatic disease of childhood and is characterized by persistent idiopathic inflammation of more than one joint,

*ARC/EU & Centre for Molecular Medicine, Stopford Building, Oxford Road, Manchester M13 9PT, UK. E-mails: Rachelle.Donn@manchester.ac.uk; †David.W.Ray@manchester.ac.uk

with an onset before the age of 16 years and a duration of six weeks or greater. Other diseases such as juvenile polymyositis, mixed connective tissue disease, acute rheumatic fever and post-infective arthropathies must be excluded before a diagnosis of JIA can be made.[4] JIA encompasses seven distinct subgroups. The subgroups of JIA differ in their clinical manifestations, prognosis and specific autoimmune characteristics, but all share chronic idiopathic synovial joint inflammation.[5] Systemic onset JIA (So-JIA) accounts for ∼15–20% of all JIA. It has features distinct from the other subgroups, such as a daily spiking fever, macular erythematous rash, serositis, hepatosplenomegaly, lymphadenopathy and laboratory evidence of profound systemic inflammation. There is substantial morbidity and associated mortality with this subgroup of patients.

The incidence rate of juvenile arthritis in the UK has been estimated to be 10 per 100,000 children per year[6] and the prevalence, 65 per 100,000.[7] The incidence rates in America are similar to the incidence rates observed in European studies of JIA.[8]

RA is a chronic, systemic inflammatory disease primarily affecting the synovial joints. Joint involvement is usually symmetrical with the small joints of the hands and feet most commonly affected. With the progression of the disease, any synovial joint may be affected and in 80% of RA cases, inflammation can result in erosion of the cartilage and eventually the bone. Extraarticular manifestations may also develop, such as fatigue, weight-loss, vasculitis, lymphadenopathy, splenomegaly, Raynauds phenomenon, scleritis and lung and cardiac involvement.

RA is a common complex disease with an estimated prevalence of ∼1% in the UK, based on a Norfolk population study.[9] RA has been identified throughout the world with distinct variation in prevalence between populations. Native American populations (Pima Indians) have the highest prevalence (5.3%), and low prevalence rates are found in rural China and rural Africa. Prevalence is always much higher in females when compared with that in males, with almost three times more females affected.[10]

The MIF Gene

The genomic organization of the human MIF gene was first reported by Paralkar and Wistow.[11] MIF is a remarkably small gene and comprises three exons of 205, 173 and 183 base pairs (bp), separated by two introns of 189 and 95 bps (*EMBL ID: HSMIF* http://www.ebi.ac.uk/htbin/emblfetch?L19686).

Only a single functional MIF gene exists in humans (chrom 22q11.2), unlike in mice, where several processed (intronless) pseudogenes have been described.[12–14] The clones sequenced by Paralkar and Wistow identified 250 bps of 3′ untranslated region (3′UTR) and 1 Kb of 5′ flanking region. Primer extension and 5′ rapid amplification of cDNA ends (RACE) were used to map the transcription start site, and identified a single site 97 bp upstream from the initiator methionine, in a TATA-less promoter.[11] These approaches cannot exclude the presence of additional transcription start sites, and the anatomy of the MIF promoter, consisting of abundant glucocorticoid (Gc) content, and no TATA box, would suggest the presence of multiple transcription start sites. However, analysis of mRNA harvested from multiple tissues, heart, brain, placenta, lung, liver, skeletal muscle, kidney and pancreas, by northern analysis identified a single transcript (of approx ~800 nucleotides).

MIF Gene Polymorphisms

The presence of raised MIF levels within the serum, plasma or synovial tissue of JIA and RA cases suggests a role for MIF in the pathogenesis. The increased expression of MIF in these sites could be consequential, rather than causative. In an attempt to address this conundrum, sequence variation of the MIF gene was determined. The identification of any polymorphic variants of MIF would allow future comparisons of specific combinations of these genetic changes (or genotypes) with diseased and nondiseased individuals to be undertaken. If a genetic basis to altered endogenous MIF protein production were found to underlie the disease states, this would become an important evidence for a central contributory role of MIF in the pathogenesis of the disease.

Polymorphism Identification

Donn *et al.*[15] used denaturing high-performance liquid chromatography (dHPLC, Wave, Transgenomic) to look for variation across the MIF gene and 1 Kb of the 5′ flanking region. This screen was performed on 32 healthy, normal UK Caucasian individuals and 96 Caucasian children with JIA, a chronic inflammatory disease in which raised MIF protein concentrations had been described.[16] Four polymorphic positions were found. These were all seen in both the healthy volunteers and the subjects with JIA. The CATT repeat element, originally documented by Paralkar and Wistow,[11] was found to be polymorphic with 5–8 alleles existing. The 8-allele repeat is found at only very low allele frequency in the UK Caucasians. Also, within the 5′ flanking region, a single nucleotide polymorphism (SNP) at nucleotide position −173 (G to C) (relative to HSMIF) was seen (rs755622). Two intronic polymorphisms at nucleotide positions +254 (T to C) (rs2096525) and at +656 (C to G) (rs2070766) were also identified. Baugh *et al.*,[17] having sequenced the 5′ flanking region of MIF in six normal individuals and six RA patients, also described the tetranucleotide CATT repeat element, which begins at position −794.

The genomic organization of the MIF gene and the positions of these polymorphisms are shown in Fig. 1.

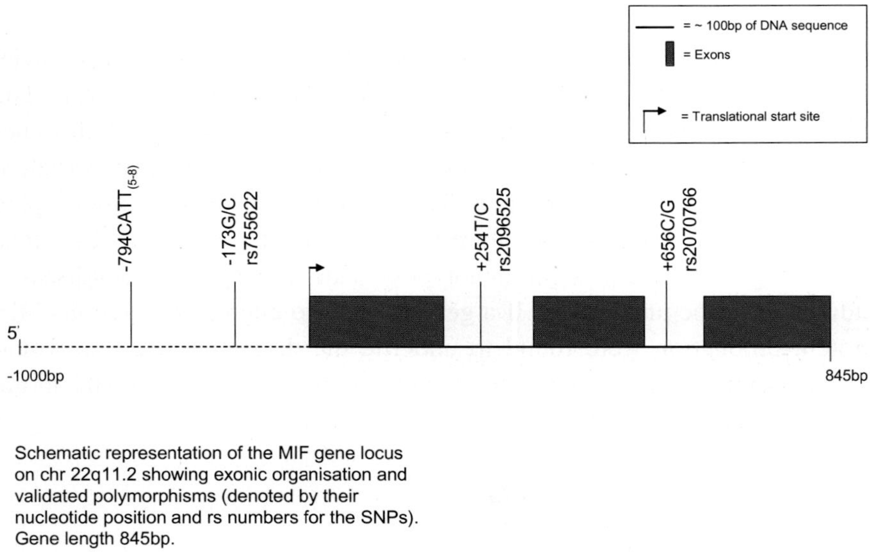

Fig. 1 Genomic organisation of the MIF gene and polymorphic variants

STUDY DESIGNS FOR THE INVESTIGATION OF CANDIDATE GENES FOR COMPLEX RHEUMATIC DISEASES

The lack of consistent data for any genetic locus hinders the belief in that locus being a critical element in disease etiology. Before describing the limited studies that we have looked at and MIF gene polymorphisms in arthritis, it is worth reviewing the common approaches used to study genetic variation in complex diseases and to document their relative advantages and disadvantages.

Methods Appropriate for Investigating the Genetic Basis of Complex Diseases

Two types of study designs are widely employed to determine whether nucleotide variations within a gene contribute to the expression of a complex genetic disease. These are linkage- and association-based studies. Linkage and association studies are distinct approaches, each with advantages and limitations in the context of studying and are best viewed as complementary methods.

Association Studies

The commonest form of association analysis is a case-control study. Here, assessment is made regarding whether the frequency of a particular allele (i.e., difference in DNA sequence in or near the gene of interest), genotype (the set of alleles on both chromosomes in an individual), or haplotype (a series of alleles on one chromosome) is enriched in patients compared with that observed in unaffected controls. The controls should be individuals who are not related to the cases, but are of the same genetic ancestry. Genetic markers are usually, but not necessarily, polymorphic sites within a gene of interest. Much recent interest has focussed on single nucleotide polymorphisms (SNPs). Allelic enrichment (or association) implies that:

a) The marker is causally involved in disease susceptibility or severity *or*

b) The pathogenic gene(s) is in *linkage disequilibrium* (LD) (i.e. nonrandomly mixed during meiosis) with alleles of the associated marker.

Polymorphic changes conferring only a marginal increase in disease risk (defined as *neither necessary nor sufficient for disease to manifest*) can be identified by association studies. Cumulatively, or individually, these changes may have functional consequences that cause the disease, or alter the disease severity and course.

Association studies are advantageous in terms of statistical power as each affected case and control subject is utilized in the analysis, leading to high sensitivity in the test. However, they are plagued by the concern of population stratification. This occurs where cases and controls differ, not only with respect to the expressed phenotype (i.e., observed clinical outcome) and its underlying genetic risk components, but also with respect to their overall population genetic ancestry. This can occur despite efforts of researchers to genetically match cases and controls. As a result, many genetic markers can spuriously appear to be associated with disease. To circumvent this problem, case-control studies have to be replicated in different populations, or confirmed by the transmission disequilibrium test (see next Section).

Linkage Studies

Linkage studies employ the use of *family* material, such as affected sibling pairs and multicase families. Linkage compares the segregation pattern of genetic markers across the genome and the disease state. Since stretches of DNA that are physically close to each other are less likely to be separated during meiotic recombination than distantly spaced sequences, markers that cosegregate with the disease (in family studies) are close to the chromosomal localization of the underlying disease gene. For a positive result from linkage analysis, the mathematics dictates that the mutations/polymorphism under investigation must be strong predictors of the disease (*actually necessary and sufficient for the expression of the disease to occur*). The comparison of different alleles of the tested markers in different families can provide cumulative evidence for positive linkage to the disease.

To ensure that linkage studies have adequate statistical power to identify the susceptibility loci, sufficient numbers of either multicase families or affected sibling pairs must be studied. Unfortunately, this is largely an unfeasible undertaking for rare diseases such as JIA, since multicase families are exceedingly rare and sibling pairs uncommon. A valuable alternative is the transmission disequilibrium test (TDT).[18] The advantage of the TDT is that it requires only one affected (child) and one, or both, of their parents (simplex or TDT-families). The analysis is dependent on the genotype information from a parent heterozygous for the marker locus. It considers whether one allele is preferentially transmitted to the affected child more often than would be expected by chance. If a heterozygous parent has one copy of a marker allele, then there would be a 50:50 chance that they would transmit it to a child. However, if a marker is both linked to and associated with a disease gene, then the total number of transmission of this allele as compared with that of the nonassociated allele in a cohort of simplex families, will be greater than 50:50. Usually a large cohort of simplex families is needed to have sufficient statistical power to assess this distortion in transmission. A distinct advantage of the TDT method is that even if only one parent is available, the test can be informative. It does, however, depend on that parent being a heterozygote. Hence, the usefulness of the TDT is maximized when using genetic markers with high heterozygosity.

Replication Studies

With any positive association found in a case-control study, *replication* is critical. This can be done by:

(a) Looking at the same factors in a disease population of different genetic ancestry.
(b) Studying the disease in one population, but taking an initial result from an association study and attempting to confirm the result by linkage analysis, or vice versa.

MIF POLYMORPHISMS AND JIA

Association of MIF-173G/C and Systemic Onset JIA

Following the identification of the novel SNP polymorphism (G to C) at position −173 of the upstream region of MIF, Donn *et al.*[15] showed significant differences in allele and genotype frequencies between 117 UK Caucasian systemic onset JIA cases and 172 unrelated healthy controls. Individuals possessing an MIF-173*C allele have an increased risk of susceptibility to So-JIA (OR 2.3, 95% CI 1.34–3.86, $p = 0.0005$). De Benedetti *et al.*[19] then showed an association of MIF-173G/C SNP and severity of outcome of So-JIA. In this cohort of So-JIA patients, those carrying the MIF-173*C allele had serum and synovial fluid MIF levels significantly higher than did those patients with the GG genotype. This strongly suggest MIF polymorphisms to be driving MIF production, which in turn was related to the disease status. Other measures of severity were assessed in this group of patients, and the MIF-173*C allele was found overall to be a predictor of poor outcome in So-JIA. Using data collected from Italian patients who had been followed up for a minimum of five years, De Benedetti analyzed retrospectively: 1) the number of joints with active arthritis and the number of joints with a limited range of motion; 2) the score at the last visit on the Italian version of the Childhood Health Assessment Questionnaire (C-HAQ); and 3) data concerning the treatment regimens during the disease course. Their findings showed that at the last visit, the number of joints with a limited range of motion were significantly higher in patients carrying the MIF-173*C allele. An interesting relationship with the MIF SNP and glucocorticoid (Gc) usage was also found. MIF polymorphism was associated with Gc requirement with patients who carry the MIF-173*C allele needing significantly longer treatment with Gc on a daily regime (median 42 months, range 26–108 months) compared with that of the MIF-173*GG homozygotes (median 19 months, range 0–94 months) ($p = 0.0025$). Furthermore, when correlation between MIF-173 polymorphism and intraarticular injection of triamcinolone hexacetonide (TXA) was examined, the duration of clinical response (months with no clinical evidence of synovitis in the injected joint) was significantly shorter in patients carrying the MIF-173*C allele. Taking a cutoff of six months of complete clinical response (i.e., no evidence of synovitis in the injected joint for >6 months after the injection), this remission was seen in only 22 of 52 injected joints for MIF-173*C carriers compared with 52 of 60 injected joints of GG homozygotes (OR from population averaged model, 0.14, 95% CI 0.04–0.48).[19]

These results were consistent with earlier findings from the same group showing that high MIF levels were associated with a shorter clinical response to TXA[16] and with the association of the MIF-173*C allele with higher synovial fluid MIF levels.[20] The data also supports the well-known ability of MIF to override the immunosuppressive and anti-inflammatory properties of Gc.[2,21] The greater requirement for longer treatment with Gcs supports the concept of a functional role of the MIF-173 SNP in so-JIA, and it appears that carriage of an MIF-173*C allele predicts a poorer response to treatment with Gc.

Following the observations in so-JIA, Donn *et al.*[22] reported MIF-173 association with all JIA, not just with the so-JIA subgroup.[22] The finding again was that carriage of the MIF-173*C allele conferred the risk. In this large study of 526 UK Caucasian JIA cases and 259 UK controls, individuals with an MIF-173*C allele had an increased risk of JIA susceptibility (34.8% vs. 21.6%) (OR 1.9, 95%CI 1.4–2.7, $p = 0.002$). This paper also demonstrated that the MIF-173**G and C variants resulted in altered expression of MIF in a cell type-specific manner. The result of transient transfections in two human cell lines, a T-lymphoblast cell line CEM C7A and A549s (lung epithelial cells) showed cell type-specific results. In the A549 cell line, a construct of MIF-173*G and a luciferase reporter (MIF-173G-luc) had significantly greater luciferase activity compared with the MIF-173C-luc, whereas in the CEM C7A, the reverse was found with the MIF-173*C-luc, showing a consistent and significant increase in promoter activity as compared with the MIF-173*G-luc reporter.[22]

Association and Linkage of MIF with JIA

Multiple false positive associations exist in the literature for a large number of candidate genes studied in complex rheumatic diseases. To specifically address such a possibility, Donn *et al.*[20] then went on to replicate their initial findings of MIF association with JIA, but using a cohort of 321 UK

Caucasian simplex families (one affected per family).[20] These JIA families were independent of the JIA cases that had been studied earlier by Donn *et al.*[15,22] This study design allowed the use of the transmission disequilibrium test and overcomes any population stratification of a control population, which can distort association studies. Using this study design, Donn *et al.* showed that MIF was linked and associated with JIA ($p = 0.0016$) and that a 2-point promoter haplotype CATT7-MIF-173*C was found to be overtransmitted to the JIA cases (38 transmitted; 21 not transmitted). If no linkage in the presence of association existed, the frequency of transmissions would not deviate from Mendelian segregation.

Donn *et al.* also showed that JIA patients carrying the MIF-173*C allele had significantly higher plasma MIF. This was found to be independent of disease activity as measured by CRP level, and was not specific to any one JIA subgroup.[20] Having previously shown cell-type-specific differences of the MIF-173G/C,[22] an assessment was made of the potential interactive effect of the MIF-173 and CATT polymorphisms by using luciferase reporter gene constructs. Again cell type-specific differences were observed.[20]

Only a limited number of additional studies have been carried out for MIF polymorphisms and JIA. In an approach similar to that of De Benedetti *et al.*,[19] Berdeli *et al.* evaluated the association of the MIF-173G/C polymorphism with outcome in Turkish juvenile arthritis cases who were followed up for >3 years. Eleven cases with so-JIA, 27 with pauci-JIA (≤ 4 joints affected) and 29 with poly-JIA (>5 joints affected) were included in the study. Data concerning the treatment regimen during the disease course, duration of glucocorticoid treatment, C-HAQ score, and the number of joints with active arthritis at the last visit were determined. It was shown that for patients carrying the MIF-173*C allele, the number of disease modifying antirheumatic drugs (DMARDS) required for the treatment was more, the duration of glucocorticoid treatment significantly longer, and at the last visit the C-HAQ scores and number of joints with active arthritis were significantly higher. These findings were observed across all three subgroups studied.[23] This extends De Benedetti's observations and suggests that the MIF-173*C allele is not only a predictor of poor outcome for so-JIA, but also in the pauciarticular and polyarticular types.

In a study of several different candidate genes, Miterski *et al.*[24] investigated the MIF-173 and the CATT repeat polymorphisms in German juvenile arthritis cases and controls. No association with the MIF-173 polymorphism was seen but an increase in the CATT5 allele was found (16.5% vs. 11.0% in the controls, $p = 0.01$). Overall, a body of evidence exists for the importance of MIF as a susceptibility locus for JIA. However, additional studies in diverse populations are still needed. Potentially more exciting is the emerging evidence of MIF-173 as a possible predictive indicator of severity of disease course in JIA. This needs further study, ideally in a prospective cohort of JIA cases.

MIF POLYMORPHISMS AND RA

Association with Disease Susceptibility or Severity?

In a similar way that observations of raised MIF in the serum of JIA patients had prompted investigation of the MIF gene by Donn *et al.*, Baugh *et al.* were aware of the substantial body of evidence implicating MIF in the etiopathogeneis of RA. Studying the CATT repeat element at position −794, Baugh *et al.*[17] showed that the shortest repeat length, the CATT-5 allele, had the lowest level of basal and stimulated MIF promoter activity *in vitro*. The distribution of the MIF CATT alleles was determined in normal controls, mild RA and severe RA cases from a rheumatology practice in Wichita. Comparing within cases, homozygosity for CATT5 appeared to be protective for the development of a severe disease.

Radstake *et al.*[25] studied an early RA inception cohort from The Netherlands.[25] Genotyping for both the MIF-173 and CATT polymorphisms was undertaken. No association with RA susceptibility was seen, but a trend towards an increase of the CATT7-MIF-173*C haplotype did occur in the RA patients (11.2%) vs. 8.8% in the controls. However, carriage of the MIF-173C allele, or the MIF CATT7 allele, was associated with significantly higher circulating MIF levels as compared with those individuals without these alleles. Furthermore, higher levels of circulating MIF correlated with more severe radiological joint damage ($r = 0.64$, $p = 0.001$).

This led Radstake *et al.* to conclude that MIF expression is genetically determined and may be used as a novel prognostic tool in RA.

In contrast to the studies showing MIF association with severity of outcome of RA, a study of a UK RA inception cohort failed to show any association of MIF promoter polymorphism with RA severity, but did find association with the RA susceptibility. In these adult cases, it was carriage of the CATT7-MIF-173*C, the same haplotype found to confer increased risk of JIA susceptibility, that was found to be associated with susceptibility to onset of RA.[26]

The possibility of the MIF-173 being a susceptibility locus for RA was not supported by a large study of >4,000 samples from North America and Sweden.[27] Designed specifically to attempt to replicate putative candidate gene associations with RA described in the literature, Plenge *et al.* tested 17 alleles from 14 genes in 2,370 RA cases and 1,757 controls from the North American Rheumatoid Arthritis Consortium and the Swedish Epidemiological Investigation of Rheumatoid Arthritis. The study provided evidence for protein tyrosine phosphatase, non-receptor-type, 22 (PTPN22), cytotoxic T-lymphocyte-associated 4 (CTLA4) and peptidylarginine deiminase, type IV (PADI4) as RA susceptibility genes. No significant association was found with the MIF-173 polymorphism. This study did not look at any genetic associations with RA severity/outcome.

From the limited studies that exist, a possible role for MIF in RA severity is suggested. Further studies are needed to replicate these initial observations and to address the critical question of whether MIF gene polymorphisms could act as prognostic indicators for disease outcome. At present, no assessment of MIF polymorphisms and Gc usage with RA has been undertaken. This is very relevant in this older population where steroid side effects are known to be more common.

FUTURE DIRECTIONS

Multiple human and animal studies have shown a role for MIF in the chronic inflammation that manifests as RA.[28–35] However, further evaluation is needed to address the role of MIF polymorphisms in RA, and to decipher to which aspects of the disease they are associated with. While the role for MIF and RA susceptibility seems limited, more studies are required to determine an MIF role in disease progression. The functional significance of the promoter polymorphisms needs evaluating and such an analysis may aid an understanding of the disease pathogenesis of the arthritic conditions where they have been shown to play a role.

Most excitingly, Morand *et al.* have developed a novel MIF antagonist small molecule compound (COR100140), which has been tested in a mouse model of RA. Following either IP or oral administration, significant inhibition of joint swelling and of a histological evidence of inflammation and damage was observed. Particularly significant is the finding that this molecule was effective in a model in which the anti-TNF agent etanercept was found to be ineffective.[36]

What is the role of MIF alleles in contributing to the potential use of such novel drugs? If endogenous high producers of MIF could be identified genetically, then perhaps these individuals would benefit most from such an anti-MIF therapy. MIF genotype has been shown to be correlated with MIF level, so prospective genotyping may be warranted. In JIA children, where the initial therapy is mainly intraarticular Gc injection, the work of De Benedetti *et al.*[19] shows that being aware of the individuals, MIF genotype could tailor better use of Gc injections. More specifically, perhaps it should be MIF-173*C cases that receive only or coadministered of the small molecule MIF antagonist. Tantalizing genetic data currently exists and what is now required is evaluation within a clinical trial setting to provide an evidence base to the value of MIF genotyping in contributing to the therapeutic decisions for RA and JIA.

REFERENCES

1. Calandra T, Roger T. (2003) Macrophage migration inhibitory factor: a regulator of innate immunity. *Nat Rev Immunol* **3**: 791–800.
2. Morand EF. (2005) New therapeutic target in inflammatory disease: macrophage migration inhibitory factor. *Intern Med J* **35**: 419–426.
3. Santos LL, Morand EF. (2006) The role of macrophage migration inhibitory factor in the inflammatory immune response and rheumatoid arthritis. *Wien Med Wochenschr* **156**: 11–18.
4. Chapel H, Haeney M, Misbah S, Snowden N. (1999) *Essentials of Clinical Immunology*, 4th ed. Blackwell Science, Oxford.

5. Glass DN, Giannini EH. (1999) Juvenile rheumatoid arthritis as a complex genetic trait. *Arthritis Rheum* **42**: 2261–2268.
6. Symmons DP, Jones M, Osborne J, *et al.* (1996) Pediatric rheumatology in the United Kingdom: data from the British Pediatric Rheumatology Group National Diagnostic Register. *J Rheumatol* **23**: 1975–1980.
7. Bywaters EGL. (1968) Report from the subcommittee on diagnostic criteria for juvenile rheumatoid arthritis (Still's disease). In: Bennett PH & Wood PHN (eds), *Population Studies of the Rheumatic Diseases: Proceedings of the Third International Symposium, New York, June 5–10, 1966*. p. 241. Excerpta Medica Foundation, Amsterdam.
8. Thomson W, Silman AJ. (2001) Juvenile idiopathic arthritis. In: *Epidemiology of Rheumatic Diseases*, pp. 72–84. Oxford University Press, Oxford.
9. Symmons DP. (2005) Looking back: rheumatoid arthritis–aetiology, occurrence and mortality. *Rheumatology (Oxford)* **44**(Suppl 4): iv14–iv17.
10. Hazes JM, Silman AJ. (1990) Review of UK data on the rheumatic diseases–2. Rheumatoid arthritis. *Br J Rheumatol* **29**: 310–312.
11. Paralkar V, Wistow G. (1994) Cloning the human gene for macrophage migration inhibitory factor (MIF). *Genomics* **19**: 48–51.
12. Bozza M, Kolakowski LF Jr, Jenkins NA, *et al.* (1995) Structural characterization and chromosomal location of the mouse macrophage migration inhibitory factor gene and pseudogenes. *Genomics* **27**: 412–419.
13. Kozak CA, Adamson MC, Buckler CE, *et al.* (1995) Genomic cloning of mouse MIF (macrophage inhibitory factor) and genetic mapping of the human and mouse expressed gene and nine mouse pseudogenes. *Genomics* **27**: 405–411.
14. Mitchell R, Bacher M, Bernhagen J, *et al.* (1995) Cloning and characterization of the gene for mouse macrophage migration inhibitory factor (MIF). *J Immunol* **154**: 3863–3870.
15. Donn RP, Shelley E, Ollier WE, Thomson W. (2001) A novel 5′-flanking region polymorphism of macrophage migration inhibitory factor is associated with systemic-onset juvenile idiopathic arthritis. *Arthritis Rheum* **44**: 1782–1785.
16. Meazza C, Travaglino P, Pignatti P, *et al.* (2002) Macrophage migration inhibitory factor in patients with juvenile idiopathic arthritis. *Arthritis Rheum* **46**: 232–237.
17. Baugh JA, Chitnis S, Donnelly SC, *et al.* (2002) A functional promoter polymorphism in the macrophage migration inhibitory factor (MIF) gene associated with disease severity in rheumatoid arthritis. *Genes Immun* **3**: 170–176.
18. Koeleman BP, Herr MH, Zavattari P, *et al.* (2000) Conditional ETDT analysis of the human leukocyte antigen region in type 1 diabetes. *Ann Hum Genet* **64**: 215–221.

19. De Benedetti F, Meazza C, Vivarelli M, *et al.* (2003) Functional and prognostic relevance of the −173 polymorphism of the macrophage migration inhibitory factor gene in systemic-onset juvenile idiopathic arthritis. *Arthritis Rheum* **48**: 1398–1407.
20. Donn RP, Alourfi Z, Zeggini E, *et al.* (2004) A functional promoter haplotype of macrophage migration inhibitory factor (MIF) is linked and associated with juvenile idiopathic arthritis. *Arthritis Rheum* **50**: 1604–1610.
21. Lolis E. (2001) Glucocorticoid counter regulation: macrophage migration inhibitory factor as a target for drug discovery. *Curr Opin Pharmacol* **1**: 662–668.
22. Donn R, Alourfi Z, De Benedetti F, *et al.* (2002) Mutation screening of the macrophage migration inhibitory factor gene: positive association of a functional polymorphism of macrophage migration inhibitory factor with juvenile idiopathic arthritis. *Arthritis Rheum* **46**: 2402–2409.
23. Berdeli A, Ozyurek AR, Ulger Z, *et al.* (2005) Association of macrophage migration inhibitory factor gene −173 G/C polymorphism with prognosis in turkish children with juvenile rheumatoid arthritis. *Rheumatol Int* 1–6.
24. Miterski B, Drynda S, Boschow G, *et al.* (2004) Complex genetic predisposition in adult and juvenile rheumatoid arthritis. *BMC Genet* **5**: 2.
25. Radstake TR, Sweep FC, Welsing P, *et al.* (2005) Correlation of rheumatoid arthritis severity with the genetic functional variants and circulating levels of macrophage migration inhibitory factor. *Arthritis Rheum* **52**: 3020–3029.
26. Barton A, Lamb R, Symmons D, *et al.* (2003) Macrophage migration inhibitory factor (MIF) gene polymorphism is associated with susceptibility to but not severity of inflammatory polyarthritis. *Genes Immun* **4**: 487–491.
27. Plenge RM, Padyukov L, Remmers EF, *et al.* (2005) Replication of putative candidate-gene associations with rheumatoid arthritis in >4,000 samples from North America and Sweden: association of susceptibility with PTPN22, CTLA4, and PADI4. *Am J Hum Genet* **77**: 1044–1060.
28. Leech M, Metz C, Hall P, *et al.* (1999) Macrophage migration inhibitory factor in rheumatoid arthritis: evidence of proinflammatory function and regulation by glucocorticoids. *Arthritis Rheum* **42**: 1601–1608.
29. Sampey AV, Hall PH, Mitchell RA, *et al.* (2001) Regulation of synoviocyte phospholipase A2 and cyclooxygenase 2 by macrophage migration inhibitory factor. *Arthritis Rheum* **44**: 1273–1280.
30. Toh ML, Yang Y, Leech M, *et al.* (2004) Expression of mitogen-activated protein kinase phosphatase 1, a negative regulator of the mitogen-activated protein kinases, in rheumatoid arthritis: up-regulation by interleukin-1beta and glucocorticoids. *Arthritis Rheum* **50**: 3118–3128.
31. Santos LL, Lacey D, Yang Y, *et al.* (2004) Activation of synovial cell p38 MAP kinase by macrophage migration inhibitory factor. *J Rheumatol* **31**: 1038–1043.

32. Mikulowska A, Metz CN, Bucala R, Holmdahl R. (1997) Macrophage migration inhibitory factor is involved in the pathogenesis of collagen type II-induced arthritis in mice. *J Immunol* **158**: 5514–5517.
33. Santos L, Hall P, Metz C, *et al.* (2001) Role of macrophage migration inhibitory factor (MIF) in murine antigen-induced arthritis: interaction with glucocorticoids. *Clin Exp Immunol* **123**: 309–314.
34. Leech M, Metz C, Santos L, *et al.* (1998) Involvement of macrophage migration inhibitory factor in the evolution of rat adjuvant arthritis. *Arthritis Rheum* **41**: 910–917.
35. Onodera S, Nishihira J, Iwabuchi K, *et al.* (2002) Macrophage migration inhibitory factor up-regulates matrix metalloproteinase-9 and -13 in rat osteoblasts. Relevance to intracellular signaling pathways. *J Biol Chem* **277**: 7865–7874.
36. Morand EF, Tapley P, Hall P, *et al.* (2005) Efficacy of a novel, orally active MIF antagonist in a model of rheumatoid arthritis. *Inflamm Res* **54**(Suppl 2): S90.

16

Adhesion-dependent Signaling and Growth Control

Robert Mitchell

INTRODUCTION

Integrins are heterodimeric cell adhesion receptors composed of individual α and β subunits. Integrin ligands include a variety of extracellular matrix (ECM) proteins as well as counter receptors on neighboring cells. Signal transduction initiated by cell membrane-associated integrin binding to ECM contributes to several key biological processes, including differentiation, neoangiogenesis, immune regulation, cell motility and proliferative control. Functional cooperativity between integrins and growth factor receptor-dependent signal transduction is well documented and many of the scaffolding proteins, adaptor molecules and signaling enzymes involved have been delineated.

Central to the contributions by integrins to cellular homeostasis are the Rho family of small GTPases. RhoA, for instance, is strongly activated by integrin-dependent adhesion and is responsible for cooperative integrin + growth factor-induced G1 phase cyclin D1 expression. The activity

J. G. Brown Cancer Center, University of Louisville, Delia Baxter Research Bldg., 580 S. Preston St. Louisville, KY 40202. Tel.: 502-852-7698; Fax: 502-852-5679; E-mail: robert.mitchell@louisville.edu of another Rho family member, Rac1, is also tightly controlled by integrin-dependent adhesion and ensuing membrane microdomain stabilization.

Of the many stimuli reported to induce macrophage migration inhibitory factor (MIF) secretion, the integrin ligation-mediated release of cellular MIF is one of the most tightly regulated. This chapter will discuss the prior as well as current studies that have sought to investigate the contribution of MIF to integrin-dependent signal transduction pathways and the biochemical and phenotypic consequences of integrin-induced MIF secretion.

BACKGROUND

The major ECM proteins include collagen, vitronectin, laminin and fibronectin. Heterodimeric integrin receptors binding to these matrix proteins include at least 24 distinct $\alpha\beta$ subunit combinations made up of at least 18 α subunits and 8 β subunits.[1] When cells attach to substrates containing ECM proteins, integrin-rich membrane domains are formed and constitute what is known as focal adhesions or focal contacts.[2] The clustering of integrins in focal contacts leads to functional interactions with the actin cytoskeleton through a variety of actin-associated proteins as well as scaffolding molecules. It is through these actin-associated adhesive foci that integrins initiate signaling either individually or in combination with other extracellular mediators such as growth factors.

The description of integrins coupling to the ERK pathway of MAP kinases was first reported by Chen and colleagues.[3] The authors reported that quiescent fibroblasts plated onto extracellular matrix proteins such as fibronectin or laminin had a strong stimulation of ERK enzymatic activity peaking at 10 min but remaining elevated for as long as one hour. In contrast, fibroblasts plated onto poly-D-lysine, a substrate that does not cluster integrins, were entirely deficient in ERK stimulation. Despite extensive studies, the mechanism of integrin-mediated ERK activation is still somewhat controversial. While several reports have shown that the activation of the protooncogene Ras is required for adhesion-dependent ERK stimulation,[4,5] others have suggested that Ras is not an essential component of this activation.[6,7] A subsequent report demonstrated how these inconsistencies might be resolved.[8] Howe and Juliano reported that there are two phases of ERK activation initiated by integrin ligation, the first being

Ras-dependent and the second being Ras-independent. The first is the initial, acute phase which requires a Ras-mediated Raf-1 membrane localization, while a later "sustained" phase of ERK activation is independent of Ras but requires the protein kinase C activation of Raf-1.[8] These findings revealed that Raf-1 is an essential component of adhesion-dependent signaling to ERK MAP kinase, but proximal signaling to Raf is complex and not fully understood.

MIF-DEPENDENT MODULATION OF SUSTAINED ERK MAP KINASE ACTIVATION

Not long after the publication of the Howe paper, it was first reported that both endogenous and exogenously-added MIF promotes the proliferation of quiescent murine fibroblasts.[9] The addition of serum to quiescent fibroblasts was found to result in the rapid release of preformed cellular MIF, and the serum-induced cell cycle progression of NIH 3T3 fibroblasts required secreted, autocrine-acting MIF for maximal proliferation. In order to determine if MIF was *sufficient* for mesenchymal cell cycle progression, recombinant MIF was tested for its respective ability to stimulate the proliferative potential of NIH 3T3 fibroblasts. Indeed, rMIF was found to moderately promote DNA synthesis in quiescent cells that was found to correspond closely to the *sustained* activation of ERK MAP kinase.[9] Further studies revealed that rMIF-mediated DNA synthesis required the activation of ERK MAP kinase as MEK1 inhibitors blocked the effect. The surprising aspect of this study was that MIF, a soluble protein previously thought to be primarily involved in innate and adaptive immune responses, was able to activate the highly mitogenic ERK MAP kinase pathway in a *sustained* fashion. In contrast to other growth factors that generally induce only a short, relatively acute, stimulation of the ERK pathway, MIF was found to promote ERK activation for up 24 hours post addition.[9]

INTEGRIN-MEDIATED PKC ACTIVATION INDUCES MIF-DEPENDENT SIGNALING TO ERK

As the dynamics of ERK activation between MIF and traditional growth factors were reminiscent of the prior report describing the differences between acute vs. sustained ERK activation by integrin ligation,[8] studies were undertaken to investigate if MIF participated in integrin-dependent signal transduction. Because the underlying hypothesis was that secreted MIF acts as an autocrine growth factor to promote integrin-initiated ERK activation, a model system similar to that used by Howe and Juliano[8] was utilized. Liao and colleagues reported that quiescent NIH 3T3 fibroblasts plated onto fibronectin result in the rapid secretion of preformed cellular MIF, while cells held in suspension do not.[10] Further studies revealed that immunodepletion of secreted MIF with anti-MIF monoclonal antibodies inhibits the sustained phase, but not the acute phase, of integrin-induced MAP kinase activation.

The activation of PKC isoforms in response to integrin-mediated signaling processes is well established.[11–15] Moreover, Howe and Juliano had reported that PKC is required for the "sustained" phase of Raf, MEK and ERK pathway activation. As integrin-mediated ERK activation is thought to be an important component of adhesion-dependent cell proliferation, it would follow that the role of PKC in this MIF-dependent ERK activation deserves closer attention. Interestingly, earlier studies had reported that PKC activity was necessary for MIF expression/secretion in primary human eosinophils and primary neurons.[16,17] Because Howe had reported that the sustained phase, but not the acute phase, of integrin-mediated ERK activation requires protein kinase C (PKC) activation, a functional role for PKC in integrin-induced MIF secretion was evaluated. To test the hypothesis that integrin-dependent PKC activation results in MIF secretion and the subsequent autocrine activation of MAP kinase, a broad spectrum PKC inhibitor was employed. The pretreatment of NIH-3T3 fibroblasts with PKC inhibitors was found to inhibit adhesion-induced MIF secretion with maximal effect, displaying almost 100% inhibition.[10] Consistent with prior studies, a dose-dependent inhibition of MAP kinase phosphorylation by PKC inhibition was also observed during cell adhesion to fibronectin.[8] In theory, if the PKC-dependent release of MIF was responsible for modulating the sustained activation of MAP kinase, then recombinant MIF addition to PKC-inhibited cells would restore adhesion-stimulated MAP kinase activation. Intriguingly, the supplementation of rMIF to PKC-inhibited cells adhering to fibronectin fully restored the capacity of integrins to activate signaling to MAP kinase.[10] These studies supported the very novel conclusion that PKC-dependent MIF secretion is at least partially responsible for integrin-stimulated sustained MAP kinase activation.

While it is generally accepted that PKC activates Raf-1 kinase activity, there is still speculation as to the exact mechanism of this stimulation.[18–21] From the studies discussed above, it is likely that the activation of integrins results in the PKC-dependent release of MIF from cells and that, once released, MIF acts to contribute to PKC-mediated ERK stimulation. The importance of this pathway extends beyond the normal physiological integrin-mediated processes mentioned here. For example, malignant cell transformation requires signals that allow a cell to mimic or bypass adhesion-generated signals. The oncogene Ras^{V12}, for instance, is thought to achieve anchorage-independence at least partially by activating the downstream effectors Raf-1, MEK and ERK. Interestingly, it has been suggested that Ras requires PKC for its transformation capability.[22,23] Moreover, some constitutively active PKC isoforms are sufficient to transform cells via a Raf-dependent mechanism.[24–26] These studies would suggest an important role for MIF in integrin signaling and cell transformation if, indeed, it contributes to the activation of "PKC-mediated" Raf-1 activation.

Ueffing *et al.* demonstrated that the PKC association with Raf-1 stimulates the production of autocrine growth factors which in turn are required for PKC-mediated Raf-1 and subsequent MEK and ERK activation.[21] In fact, this report supports the finding that activated PKC-mediated MIF release participates in the adhesion-induced PKC regulation of Raf, MEK and ERK stimulation,[10] and one can speculate that MIF, at least partially, accounts for this PKC-dependent growth factor activity.

Phospholipase C and PI 3-kinase enzymes have also been shown to be regulated by integrin-mediated adhesion,[27,28] and both of these pathways stimulate PKC activation.[29–31] Hence, upstream regulators of PKC activation in response to cell adhesion may indirectly affect MIF release and therefore ERK activation via MIF action. Studies designed to examine the precise role of individual PKC isoforms as well as the respective roles of PLC and/or PI 3-kinase pathways involved in integrin-mediated MIF release would be important first steps in delineating this important integrin-initiated signaling pathway.

MIF IN INTEGRIN-DEPENDENT CELL CYCLE PROGRESSION

The studies described above demonstrate a unique interdependence between MIF and integrin-initiated signal transduction. However, many of these studies were performed in the decidedly non-physiologic settings of growth factor starvation and/or substrate withdrawal. Obviously, normal cells rarely, if ever, find themselves in such a restricted environment.

In order to investigate integrin signaling contributions in more physiologically relevant conditions, Roovers and colleagues studied the relative influence of integrin-dependent adhesion in growth factor-mediated cell cycle progression.[32] In an elegant set of studies, the authors demonstrated that fibroblast $\alpha5\beta1$-integrin-dependent adhesion to fibronectin is necessary for growth factor-induced G1 phase cyclin D1 expression, retinoblastoma (Rb) inactivation and cell cycle progression into the S phase.[32] Interestingly, the requirement for this cooperative signaling between growth factors and integrin ligation in cyclin D1 transcription is dependent upon the sustained activation of ERK MAP kinase. In contrast to the study described above, where only *integrin*-dependent ERK activation kinetics was examined,[8] this report reveals that growth factors require integrin-initiated signaling to sustain MEK1-dependent ERK activation long enough to commence cyclin D1 transcription and subsequent S phase progression. Finally, the authors reported that the forced overexpression of constitutively active MEK1 in anchorage-restricted fibroblasts rescues the cyclin D1 defect, thereby validating the argument that integrin-dependent ERK activation is responsible for growth factor-induced cyclin D1 expression.[32]

Prior studies had already established that *both* growth factors and integrin-mediated cell adhesion to ECM (independently) result in the rapid release of preformed MIF.[9,10] In an effort to determine whether MIF participates in integrins' contribution to growth factor-induced sustained ERK activation and cyclin D1 expression, MIF-deficient mouse embryonic fibroblasts (MEFs) were examined. Liao and colleagues reported that $MIF^{-/-}$ fibroblasts are refractory to growth factor plus adhesion-induced sustained, but not acute, MAP kinase activation and cyclin D1 expression.[10] Moreover, Ckd4 activity, which is fully dependent upon cyclin D1 expression, was found to be heavily reliant upon functional MIF. The authors finally showed that MIF-deficiency results in impaired Rb inactivation, E2F-dependent transactivation and, as a consequence, defective cell cycle progression.[10]

It is important to note that while several intracellular signaling effectors have been found to be required for integrin-initiated signaling (reviewed in Ref. 33), MIF is the first secreted molecule reported to play an important, if not necessary, participatory role in adhesion-dependent signal transduction. Another intriguing aspect to these findings is that even in the presence of growth factor-enriched fetal calf serum or a combination of insulin and platelet-derived growth factor (PDGF), MIF exerts an extremely important influence over mitogen-induced cell cycle progression.[10] These studies revealed a unique and previously unrecognized mechanistic pathway for an MIF-dependent contribution to cell cycle modulation in normal cells. The next challenge was to identify how extracellular MIF was linked to this distal signaling pathway shown to induce the sustained activation of ERK MAP kinase culminating in cyclin D1 transcription and cell cycle progression.

A REQUIREMENT FOR RHO GTPASE IN MIF-MEDIATED ERK ACTIVATION

Around the time that the Liao paper was published, two reports were published demonstrating an important modulatory role for Rho GTPase in growth factor/integrin-dependent cell cycle progression from the mid-G1 phase into the S phase.[34,35] Specifically, it was reported that the activation of Rho and subsequently Rho kinase by integrins and growth factors is essential for the sustained activation of MAP kinase and cyclin D1 transcription. Interestingly, Rho was found to be responsible for maintaining the correct timing of cyclin D1 expression, in part by negatively regulating Rac1 activity. Further studies revealed that Rho activation contributes to the adhesion/growth factor-dependent regulation of sustained MAP kinase activity by promoting the formation of stress fibers and focal adhesions, which then allows for the recruitment and clustering of integrins.[35] In this model, Rho, Rho kinase and enhanced myosin light chain phosphorylation enhance stress fiber formation/integrin clustering and, ultimately, the sustained activation of the canonical Raf-1, MEK and ERK Map kinase pathway.

With these studies in mind, our laboratory undertook an investigation to determine whether MIF, now known to be both necessary and sufficient for sustained ERK MAP kinase activation, cyclin D1 expression and cell cycle progression, regulates Rho GTPase activity in an integrin-dependent and/or independent fashion. These findings were summarized by Swant and colleagues where it was revealed that the addition of recombinant MIF to NIH 3T3 fibroblasts induces the activation of RhoA GTPase, but had no apparent effect on other Rho family members.[36] Concomitant with Rho activation by MIF was the subsequent and sustained phosphorylation of focal adhesion kinase (FAK), myosin light chain (MLC) and ERK MAP kinase. While rMIF had previously been shown to stimulate the sustained activation of ERK and the subsequent cell cycle progression of murine fibroblasts,[9] Swant and colleagues additionally showed that cyclin D1 transcription was coordinately increased by rMIF treatment. Perhaps more importantly, it was revealed that the inhibition of Rho, its immediate downstream effector Rho kinase and myosin light chain kinase (MLCK), individually resulted in a loss of sustained ERK activation and cyclin D1 expression induced by rMIF.[36] These results placed Rho and Rho kinase directly upstream of ERK activation and suggested for the first time that the MIF-induced stimulation of Rho was responsible for its observed influence on sustained ERK activation and subsequent cyclin D1 transcription.

Although these results clearly implicated Rho as an important effector in rMIF-induced cell cycle progression, a more important question remained: what is the contribution of endogenous MIF to integrin-dependent cooperativity with growth factor-induced Rho activation, sustained ERK activity, cyclin D1 transcription and subsequent G1 phase progression? Recall that a prior report demonstrated that integrin-induced Rho activation is responsible for Rho kinase and MLC kinase-dependent stress fiber formation leading to a *re-clustering* of integrins, FAK activation and the *sustained* stimulation of the canonical ERK MAP kinase pathway.[35] Because it had previously been shown that secreted MIF is necessary for integrin + growth factor-induced sustained ERK activation and cyclin D1 expression, it would stand to reason that endogenous MIF is responsible for integrin-induced, Rho and Rho kinase-mediated, signaling to cyclin D1. To test this,

MIF+/+ and $MIF^{-/-}$ fibroblasts were examined for their respective abilities to activate Rho by cell adhesion to fibronectin. Not surprisingly, Rho activation by cell adhesion, especially at later time points, was severely diminished in the MIF-deficient cells when compared with the wildtype MIF cells. Also, consistent with the loss of Rho activity in $MIF^{-/-}$ cells, stress fiber formation, as assessed by F-actin staining with FITC-labeled phalloidin, was significantly reduced in $MIF^{-/-}$ cells.[36] Accompanying the decreased Rho activation and stress fiber formation associated with loss of MIF, sustained ERK phosphorylation, cyclin D1 expression and Rb inactivation were all substantially reduced, as is consistent with prior studies.[10] Finally, to further confirm that Rho was responsible for the loss of cyclin D1 and subsequent Rb inactivation, transient transfections with a constitutively active allele of Rho, Rho^{V14}, fully rescued the defective transcription of cyclin D1 as well as Rb inactivation in $MIF^{-/-}$ cells.[36] To summarize, MIF is both sufficient and necessary to induce an elaborate signal transduction pathway culminating in sustained ERK activity, cyclin D1 expression, Rb inactivation and cell cycle progression (Fig. 1, top panel). This pathway includes the effectors, Rho, Rho kinase, integrins and focal adhesion kinase.[36]

Because of the inherent importance of an endogenous, soluble mediator in stimulating signaling pathways leading to cyclin D1 expression, it is not unreasonable to speculate that MIF may be an important participant in malignant disease processes. Although the subject is covered in much greater detail elsewhere in this book, a short summary of the work done in mesenchymal cell malignant transformation will aid further discussion on the role of MIF in adhesion-dependent processes.

MIF IN MALIGNANT TRANSFORMATION OF MURINE FIBROBLASTS

Two previously published studies clearly demonstrate that MIF acts as an important participant in cell growth processes and phenotypic malignant transformation.[37,38] The first of these investigations revealed that MIF-deficient primary mouse embryonic fibroblasts are completely resistant to cell transformation induced by a combination of a p53 gain of function

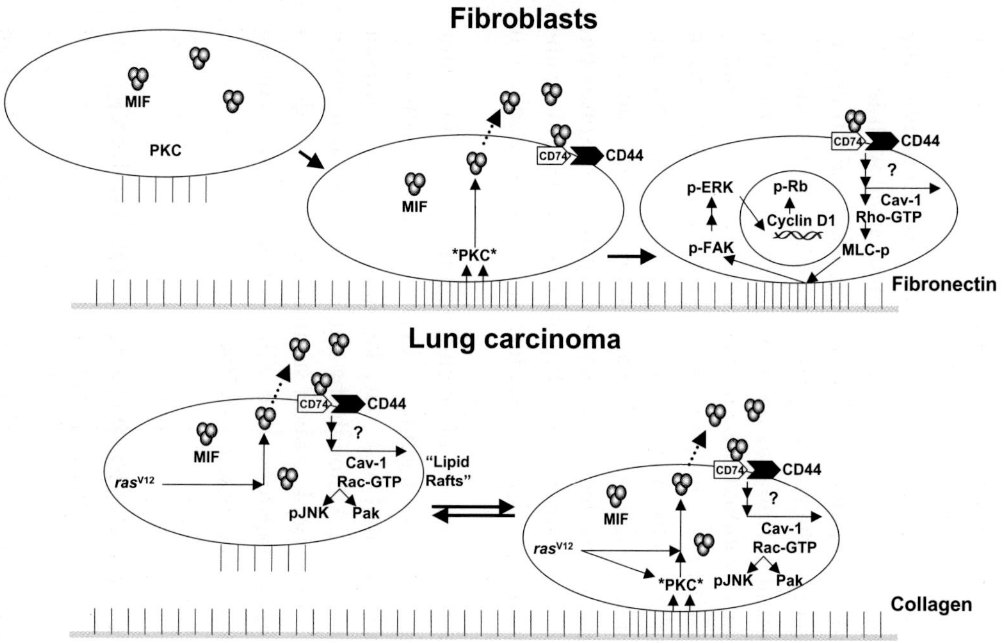

Fig. 1 Scheme of adhesion-dependent MIF signaling and its contribution to cell division, migration and anchorage-independent growth. When cell surface integrins (blue lines) bind to ligands such as fibronectin or collagen (red lines) downstream signaling is initiated. Integrin-ECM interaction in fibroblasts triggers the rapid secretion of MIF in a PKC-dependent manner. Once released, MIF contributes to adhesion-induced Rho GTPase activation leading to sustained ERK MAP kinase activity and subsequent cyclin D1 transcription. MIF-dependent G1 phase cyclin D1 expression is responsible for MIF-mediated cell cycle progression of murine fibroblasts (top panel). Malignant lung adenocarcinoma cells express and secrete high levels of MIF in both adherent and non-adherent conditions. Once released, MIF promotes caveolin-containing lipid raft stabilization that is permissive for Rac1 GTPase to bind to and activate downstream effectors such as PAK and JNK. MIF-dependent Rac1 activity in NSCLC cells contributes to lung adenocarcinoma malignant phenotypes, including migration, invasion, adhesion and anchorage-independence (bottom panel).

mutant ($p53^{H175}$ — as discussed above) and oncogenic H-ras^{V12}.[37] Moreover, it was found that the classic combinations of c-*myc* and H-ras^{V12} and adenoviral E1A plus H-ras^{V12} were also unable to efficiently transform MIF-deficient cells. More importantly, the re-introduction of MIF to oncogene-expressing $MIF^{-/-}$ fibroblasts fully restored the oncogenic transformation capacity of these cells.[37]

The second report also describes a central role for MIF in effecting malignant processes.[38] More importantly, in this study utilizing an entirely different MIF-deficient mouse strain and gene targeting strategy, similar results were found. Moreover, in the very first description of MIF promoting *de novo* carcinogenesis, it was found that $MIF^{-/-}$ mice are partially refractory to benzo[*a*]pyrene (B[*a*]P)-mediated mesenchymal tumor growth.[38] Taken together, these studies reveal a critical and previously uncharacterized role for MIF in contributing to malignant cell growth, oncogenic cell transformation and *de novo* tumorigenesis.

Certainly one of the more important turning points in MIF-associated biology was the finding that MIF regulates the activity of the p53 tumor suppressor negatively.[39] Hudson and colleagues demonstrated that some growth inhibitory and pro-apoptotic functions of p53 could be counteracted by the addition of exogenous, or retrovirally overexpressed, MIF. These studies were confirmed in macrophages both *in vitro* and *in vivo*.[40] Existing evidence also supports the conclusion that MIF-dependent p53 inactivation is responsible for its apparent malignant growth requirements observed in mesenchymal cells.[37,38] The findings by Fingerle-Rowson *et al.* demonstrate that p53 loss through gene deletion abrogates the requirement for MIF in cell replication and malignant transformation in some settings.[38]

However, it is important to note that the inactivation of endogenous, cellular p53 by the gain of function mutants ($p53^{H175}$) actually potentiates the requirement for MIF in cell cycle progression and oncogenic cell transformation.[10,36,37] The enhanced cell growth requirement for MIF observed in the presence of mutated p53 is thought to be due to a greater dependency of MIF-dependent cyclin D1 expression and subsequent Rb inactivation.[10] Although still unclear, it is possible that this discrepancy may lay in the fact that the p53 gain of function mutants possess oncogenic properties that are independent of endogenous p53 inactivation and may require the action of MIF.[41,42] Based on this, prior observations, and the fact that p53 mutations are found in over 50% of all human tumors, there is the distinct possibility that MIF contributes to cell cycle dysregulation of a large number of clinically relevant human cancers by promoting Rb inactivation through the modulation of cyclin D1 expression and Cdk4 activation.

MODULATION OF HUMAN LUNG CANCER CELL ANCHORAGE-INDEPENDENCE BY MIF

It is important to keep in mind that the vast majority of studies that seek to investigate integrin-dependent signal transduction have two things in common: (1) nearly all the published works related to adhesion-induced influences on cell behavior have been done in rodent fibroblasts; and (2) the impetus behind most of the published works on adhesion-mediated signal transduction is the attempt to identify the pathways that cancer cells have usurped in order to subsist in restrictive environments. For example, the studies that investigate integrin-dependent signaling requirements in normal cells provide us important clues about pathways that need to be disrupted or overridden in malignant cells in order to achieve anchorage-independence. More importantly, signaling through these adhesion-dependent pathways and the resultant phenotypes can vary substantially between different cell types. For example, when a non-transformed fibroblast is removed from its normal substrate (i.e. fibronectin), the cell will simply stop growing and undergo cell cycle arrest.[32,43] In stark contrast, when a non-transformed epithelial cell is removed from its normal substrate (i.e. collagen, laminin), it undergoes a well-characterized program of cell death or apoptosis.[44,45] The apoptotic program induced by the disruption of epithelial cell-matrix interactions has been coined "anoikis".[44] More importantly, the malignant transformation of epithelial cells by viral oncoproteins or human oncogenes rescues the detached epithelial cell from apoptosis, thereby allowing the cell not only to survive, but also to proliferate in an anchorage-independent environment. Therefore, it goes without saying that the identification of intra- and extracellular mediators that facilitate the escape of malignant cells from cytostatic or cytotoxic signals associated with anchorage-independence is crucial.

As all of the studies described above clearly point to MIF as being an important component of normal, non-transformed cell integrin-mediated signaling, our laboratory undertook an investigation to determine if MIF facilitates anchorage-independence in established human lung adenocarcinoma cells.[46] Small hairpin RNAs (shRNAs) designed against human MIF were used to silence MIF expression in human lung adenocarcinoma cell lines. The cell line used for this study, A549, readily forms colonies in soft agar, demonstrating that these cells have devised mechanisms that can override the apoptosis program that is normally initiated when non-transformed epithelial cells are removed from their normal substrate.[44] The relative anchorage-independent potential of cells that had been depleted of MIF were compared with cells that had been transfected with a nonsense shRNA (normal MIF expression). Intriguingly, cells transfected with control, nonsense shRNA had more than 75% more soft agar colonies than cells transfected with MIF-specific shRNA. These results suggest a potent contributing role of MIF in the anchorage-independent growth of human lung adenocarcinoma cells and likely indicate that MIF plays an important participatory role in lung cancerogenesis.

Our laboratory and others have shown that MIF expression in human lung tumors is significantly higher than in normal tissue.[46–48] As a result of this high MIF expression found in lung cancer cells, it is possible that MIF is constitutively expressed and secreted in these transformed cells *even in the absence of adhesion.* From earlier studies, we know that MIF is sufficient for activating signaling pathways previously attributed to cell-matrix interactions.[10,36] If MIF participates in stimulating signaling pathways initiated by integrin ligation in epithelial cells, it is likely that the transformed epithelial cells utilize the constitutive secretion of MIF to subsist in adhesion-restricted environments. In other words, the overexpression of MIF by lung tumor cells may provide the signals necessary to achieve anchorage-independence. As discussed earlier, the primary consequence of a non-transformed epithelial cell-matrix disruption is a cell death program previously coined anoikis.[44] In contrast, tumorigenic or transformed epithelial cells are largely resistant to anoikis. To investigate whether the loss of MIF results in increased cell death in the normally anoikis-resistant A549 cells,[49] MIF-containing and MIF-depleted cells were held in suspension (anchorage withdrawal) for 16 hours. Consistent with the inability of A549 cells lacking MIF to grow in soft agar, adhesion-restricted MIF-depleted cells were nearly 50% less viable than the cells containing MIF, suggesting an important functional contribution by MIF in maintaining anoikis resistance (unpublished observations). Consistent with defective integrin-associated signaling pathways, it was also discovered that lung cancer cells lacking MIF are almost completely defective in adhesive, migratory and invasive properties.[46] Together, these results strongly suggested that integrin-dependent signaling is compromised in MIF-deficient cells.

DEFECTIVE RAC, BUT NOT RHO, ACTIVITY IN LUNG TUMOR CELLS LACKING IN MIF

Because prior studies on murine fibroblasts had revealed a central role for MIF in modulating RhoA activation induced by integrin ligation, similar studies were undertaken in human lung adenocarcinoma cells. Much to our surprise, and in contrast to murine fibroblasts, human lung cancer cells lacking MIF have similar levels of GTP-loaded Rho as the control cells.[46] Intriguingly, while no differences were observed with respect to Rho activity, MIF-depleted cells displayed a remarkably defective activation of Rac and downstream effectors of Rac. Of the three Rho GTPase family members, Rac, Rho and Cdc42, Rac is arguably the most important in promoting and maintaining growth and invasive phenotypes in tumor cells.[50–52] More importantly, Rac is also a central participant in anoikis resistance in both normal and cancer cell lines,[53,54] and is reported to be necessary for anchorage-independence in a variety of cell types and human tumors.[55,56]

Because of the well documented and central importance of Rac in all of the phenotypes observed to be defective in MIF-depleted cells, we predicted that the restoration of Rac activation in MIF-depleted cells would rescue the observed defective phenotypes. Constitutively active Rac mutants which have been used experimentally for many years are extremely well characterized.[56] The mutation of glycine at position 12 to valine (Rac^{V12}) renders Rac insensitive to its intrinsic GTPase activity, hence it is "stuck" in the GTP bound form. Because only GTP bound small GTPases are capable of binding effectors and initiating signaling, Rac^{V12} mutants are considered "constitutively active."

If defective Rac activation was indeed responsible for the observed loss of migration and invasion in MIF-depleted cells, the introduction of a constitutively active allele of Rac (Rac^{V12}) should, in theory, rescue this effect. Very surprisingly, however, this was not the case. While the stable Rac^{V12}-expressing A549 cell lines were nearly twice as motile as the vector control cells, MIF depletion rendered these cells completely defective in their migratory and invasive potential.[46] Even more intriguing was the fact that pulldown experiments using GST-Pak-binding domain (GST-PBD) of the Rac effector Pak, revealed that both endogenous Rac and ectopically expressed Rac^{V12} were unable to bind to this effector fusion protein in cells lacking MIF. Generally, this PBD pulldown assay is considered to be an indicator of GTP loading of Rac because, almost invariably, GTP-loaded Rac binds to, and activates, its effectors. Therefore, Rac^{V12} should be constitutively loaded with GTP.

Prior studies by del Pozo and colleagues had demonstrated that non-transformed cells held in suspension (i.e. no integrin signaling) are unable to induce the activation of the Rac1 effector, PAK, even when Rac was properly loaded with GTP.[57,58] Similar to the findings observed with MIF depletion, this group also reported that Rac^{V12} does not rescue the deficient PAK activation phenotype observed in suspension cells.[57] The same group went on to demonstrate that the requirement for integrin ligation in Rac activation lies in the ability of Rac to localize to caveolin-containing lipid raft membrane micro-domains, the partitioning of which is governed by the phosphorylation-state of caveolin-1.[59,60] The observation that Rac^{V12} activity requires MIF, coupled with the fact that MIF participates in integrin-mediated signaling that in turn mediates Rac and Rac^{V12} activity, strongly suggested that Rac targeting to lipid rafts might represent a point of control by MIF.

REQUIREMENT FOR MIF IN CAVEOLIN TARGETING TO MEMBRANE MICRO-DOMAINS

To reiterate, at this juncture it was hypothesized that the MIF-dependent modulation of Rac was due to its previously described participation in integrin-dependent signaling processes. Specifically, the working hypothesis was that integrin-dependent lipid raft formation and the subsequent Rac membrane translocation and effector activation, require MIF. To test this, MIF-containing and MIF-deficient cells were analyzed for lipid raft components and proper assembly. Intriguingly, cells lacking MIF had substantially reduced caveolin-1 and flotillin-2 in sucrose-sedimented lipid raft fractions, and this decrease was associated with a corresponding increase of these lipid raft markers in cytosolic fractions.[46] Most importantly, Rac membrane localization to lipid raft fractions was found to be severely compromised in MIF-deficient vector *and* Rac^{V12}-expressing cells. This finding likely explains how MIF promotes Rac activity/effector activation and the subsequent phenotypic effects on lung cancer cell migration, invasion, adhesion and anchorage-independent growth (Fig. 1, bottom panel). In a broader sense, the MIF-dependent modulation of integrin-mediated lipid raft stabilization may help to explain the previously described contributions by MIF to adhesion-dependent signal transduction.

IMPLICATIONS AND DISCUSSION

The studies described in this chapter implicate MIF as being a central participant in integrin-dependent signaling processes and the myriad of phenotypes associated with them. To the best of our knowledge, no other secreted cytokine, lymphokine or growth factor has ever been shown to play such a key role in adhesion-dependent biological processes. Of even more relevance is the fact that tumor cell-derived MIF is now shown to participate in integrin-initiated signaling mechanisms usurped by malignant cells in order to acquire anchorage-independence and metastatic potential.[46] If these findings also hold true for other cancers, MIF may represent a unique and previously undiscovered target for future anti-cancer therapeutic strategies. Indeed, several small molecule antagonists have now been described that block several of the MIF-dependent inflammatory, immune and malignant disease processes.[46,61–66]

These findings underscore just how important detailed studies into MIF-dependent contributions to adhesion-mediated pathophysiologic consequences can be. Combined with both the past and future studies described in the other chapters of this book, the true importance and biologic significance of this unique molecule will soon be realized.

REFERENCES

1. Giancotti FG. (2000) Complexity and specificity of integrin signaling. *Nat Cell Biol* **2**: E13–E14.
2. Burridge K, Chrzanowska-Wodnicka M. (1996) Focal adhesions, contractility, and signaling. *Annu Rev Cell Dev Biol* **12**: 463–518.
3. Chen Q, Kinch MS, Lin TH, *et al.* (1994) Integrin-mediated cell adhesion activates mitogen-activated protein kinases. *J Biol Chem* **269**: 26602–26605.
4. Schlaepfer DD, Hunter T. (1997) Focal adhesion kinase over-expression enhances Ras-dependent integrin signaling to ERK2/mitogen-activated protein kinase through interactions with and activation of c-Src. *J Biol Chem* **272**: 13189–13195.
5. Clark EA, Hynes RO. (1996) Ras activation is necessary for integrin-mediated activation of extracellular signal-regulated kinase 2 and cytosolic phospholipase A2 but not for cytoskeletal organization. *J Biol Chem* **271**: 14814–14818.
6. Chen Q, Lin TH, Der CJ, Juliano RL. (1996) Integrin-mediated activation of MEK and mitogen-activated protein kinase is independent of Ras. *J Biol Chem* **271**: 18122–18127.
7. Lin TH, Chen Q, Howe A, Juliano RL. (1997) Cell anchorage permits efficient signal transduction between Ras and its downstream kinases. *J Biol Chem* **272**: 8849–8852.
8. Howe AK, Juliano RL. (1998) Distinct mechanisms mediate the initial and sustained phases of integrin-mediated activation of the Raf/MEK/mitogen-activated protein kinase cascade. *J Biol Chem* **273**: 27268–27274.
9. Mitchell RA, Metz CN, Peng T, Bucala R. (1999) Sustained mitogen-activated protein kinase (MAPK) and cytoplasmic phospholipase A2 activation by macrophage migration inhibitory factor (MIF). Regulatory role in cell proliferation and glucocorticoid action. *J Biol Chem* **274**: 18100–18106.
10. Liao H, Bucala R, Mitchell RA. (2003) Adhesion-dependent signaling by macrophage migration inhibitory factor (MIF). *J Biol Chem* **278**: 76–81.
11. Schwartz MA, Lechene C. (1992) Adhesion is required for protein kinase C-dependent activation of the Na^+/H^+ antiporter by platelet-derived growth factor. *Proc Natl Acad Sci USA* **89**: 6138–6141.
12. Miranti CK, Ohno S, Brugge JS. (1999) Protein kinase C regulates integrin-induced activation of the extracellular regulated kinase pathway upstream of Shc. *J Biol Chem* **274**: 10571–10581.

13. Chun JS, Ha MJ, Jacobson BS. (1996) Differential translocation of protein kinase C epsilon during HeLa cell adhesion to a gelatin substratum. *J Biol Chem* **271**: 13008–13012.
14. Cybulsky AV, Carbonetto S, Cyr MD, *et al.* (1993) Extracellular matrix-stimulated phospholipase activation is mediated by beta 1-integrin. *Am J Physiol* **264**: C323–C332.
15. Xu J, Clark RA. (1997) A three-dimensional collagen lattice induces protein kinase C-zeta activity: role in alpha2 integrin and collagenase mRNA expression. *J Cell Biol* **136**: 473–483.
16. Rossi AG, Haslett C, Hirani N, *et al.* (1998) Human circulating eosinophils secrete macrophage migration inhibitory factor (MIF) potential role in asthma. *J Clin Invest* **101**: 2869–2874.
17. Busche S, Gallinat S, Fleegal MA, *et al.* (2001) Novel role of macrophage migration inhibitory factor in angiotensin II regulation of neuromodulation in rat brain. *Endocrinology* **142**: 4623–4630.
18. Kolch W, Heidecker G, Kochs G, *et al.* (1993) Protein kinase C alpha activates Raf-1 by direct phosphorylation. *Nature* **364**: 249–252.
19. Marais R, Light Y, Mason C, *et al.* (1998) Requirement of Ras-GTP-Raf complexes for activation of Raf-1 by protein kinase C. *Science* **280**: 109–112.
20. Cai H, Smola U, Wixler V, *et al.* (1997) Role of diacylglycerol-regulated protein kinase C isotypes in growth factor activation of the Raf-1 protein kinase. *Mol Cell Biol* **17**: 732–741.
21. Ueffing M, Lovric J, Philipp A, *et al.* (1997) Protein kinase C-epsilon associates with the Raf-1 kinase and induces the production of growth factors that stimulate Raf-1 activity. *Oncogene* **15**: 2921–2927.
22. Bjorkoy G, Perander M, Overvatn A, Johansen T. (1997) Reversion of Ras- and phosphatidylcholine-hydrolyzing phospholipase C-mediated transformation of NIH 3T3 cells by a dominant interfering mutant of protein kinase C lambda is accompanied by the loss of constitutive nuclear mitogen-activated protein kinase/extracellular signal-regulated kinase activity. *J Biol Chem* **272**: 11557–11565.
23. Bjorkoy G, Overvatn A, Diaz-Meco MT, *et al.* (1995) Evidence for a bifurcation of the mitogenic signaling pathway activated by Ras and phosphatidylcholine-hydrolyzing phospholipase C. *J Biol Chem* **270**: 21299–21306.
24. Perletti GP, Folini M, Lin HC, *et al.* (1996) Over-expression of protein kinase C epsilon is oncogenic in rat colonic epithelial cells. *Oncogene* **12**: 847–854.
25. Wang QJ, Acs P, Goodnight J, *et al.* (1998) The catalytic domain of PKC-epsilon, in reciprocal PKC-delta and -epsilon chimeras, is responsible for conferring tumorgenicity to NIH3T3 cells, whereas both regulatory and catalytic

domains of PKC-epsilon contribute to *in vitro* transformation. *Oncogene* **16**: 53–60.

26. Cacace AM, Ueffing M, Philipp A, *et al.* (1996) PKC epsilon functions as an oncogene by enhancing activation of the Raf kinase. *Oncogene* **13**: 2517–2526.
27. Nakanishi H, Brewer KA, Exton JH. (1993) Activation of the zeta isozyme of protein kinase C by phosphatidylinositol 3,4,5-trisphosphate. *J Biol Chem* **268**: 13–16.
28. Toker A, Meyer M, Reddy KK, *et al.* (1994) Activation of protein kinase C family members by the novel polyphosphoinositides PtdIns-3,4-P2 and PtdIns-3,4,5-P3. *J Biol Chem* **269**: 32358–32367.
29. Liu M, Simon MI. (1996) Regulation by cAMP-dependent protein kinease of a G-protein-mediated phospholipase C. *Nature* **382**: 83–87.
30. Takeda H, Matozaki T, Takada T, *et al.* (1999) PI 3-kinase gamma and protein kinase C-zeta mediate RAS-independent activation of MAP kinase by a Gi protein-coupled receptor. *EMBO J* **18**: 386–395.
31. Le Good JA, Ziegler WH, Parekh DB, *et al.* (1998) Protein kinase C isotypes controlled by phosphoinositide 3-kinase through the protein kinase PDK1. *Science* **281**: 2042–2045.
32. Roovers K, Davey G, Zhu X, *et al.* (1999) Alpha5beta1 integrin controls cyclin D1 expression by sustaining mitogen-activated protein kinase activity in growth factor-treated cells. *Mol Biol Cell* **10**: 3197–3204.
33. Lee JW, Juliano R. (2004) Mitogenic signal transduction by integrin- and growth factor receptor-mediated pathways. *Mol Cells* **17**: 188–202.
34. Welsh CF, Roovers K, Villanueva J, *et al.* (2001) Timing of cyclin D1 expression within G1 phase is controlled by Rho. *Nat Cell Biol* **3**: 950–957.
35. Roovers K, Assoian RK. (2003) Effects of Rho kinase and actin stress fibers on sustained extracellular signal-regulated kinase activity and activation of G(1) phase cyclin-dependent kinases. *Mol Cell Biol* **23**: 4283–4294.
36. Swant JD, Rendon BE, Symons M, Mitchell RA. (2005) Rho GTPase-dependent signaling is required for macrophage migration inhibitory factor-mediated expression of cyclin D1. *J Biol Chem* **280**: 23066–23072.
37. Petrenko O, Fingerle-Rowson G, Peng T, *et al.* (2003) Macrophage migration inhibitory factor deficiency is associated with altered cell growth and reduced susceptibility to Ras-mediated transformation. *J Biol Chem* **278**: 11078–11085.
38. Fingerle-Rowson G, Petrenko O, Metz CN, *et al.* (2003) The p53-dependent effects of macrophage migration inhibitory factor revealed by gene targeting. *Proc Natl Acad Sci USA* **100**: 9354–9359.
39. Hudson JD, Shoaibi MA, Maestro R, *et al.* (1999). A proinflammatory cytokine inhibits p53 tumor suppressor activity. *J Exp Med* **190**: 1375–1382.

40. Mitchell RA, Liao H, Chesney J, *et al.* (2002) Macrophage migration inhibitory factor (MIF) sustains macrophage proinflammatory function by inhibiting p53: regulatory role in the innate immune response. *Proc Natl Acad Sci USA* **99**: 345–350.
41. Sigal A, Rotter V. (2000) Oncogenic mutations of the p53 tumor suppressor: the demons of the guardian of the genome. *Cancer Res* **60**: 6788–6793.
42. van Oijen MG, Slootweg PJ. (2000) Gain-of-function mutations in the tumor suppressor gene p53. *Clin Cancer Res* **6**: 2138–2145.
43. Roovers K, Assoian RK. (2000) Integrating the MAP kinase signal into the G1 phase cell cycle machinery. *Bioessays* **22**: 818–826.
44. Frisch SM, Francis H. (1994) Disruption of epithelial cell-matrix interactions induces apoptosis. *J Cell Biol* **124**: 619–626.
45. Frisch SM, Screaton RA. (2001) Anoikis mechanisms. *Curr Opin Cell Biol* **13**: 555–562.
46. Rendon BE, Teneng I, Zhao M, *et al.* (2006) Regulation of human lung adenocarcinoma cell migration and invasion by MIF: role of Rac1 GTPase and lipid raft assembly. *Submitted*
47. Tomiyasu M, Yoshino I, Suemitsu R, *et al.* (2002) Quantification of macrophage migration inhibitory factor mRNA expression in non-small cell lung cancer tissues and its clinical significance. *Clin Cancer Res* **8**: 3755–3760.
48. Kayser K, Nwoye JO, Kosjerina Z, *et al.* (2003) Atypical adenomatous hyperplasia of lung: its incidence and analysis of clinical, glycohistochemical and structural features including newly defined growth regulators and vascularization. *Lung Cancer* **42**: 171–182.
49. Wei L, Yang Y, Zhang X, Yu Q. (2002) Anchorage-independent phosphorylation of p130(Cas) protects lung adenocarcinoma cells from anoikis. *J Cell Biochem* **87**: 439–449.
50. Hauck CR, Sieg DJ, Hsia DA, *et al.* (2001) Inhibition of focal adhesion kinase expression or activity disrupts epidermal growth factor-stimulated signaling promoting the migration of invasive human carcinoma cells. *Cancer Res* **61**: 7079–7090.
51. Chan AY, Coniglio SJ, Chuang YY, *et al.* (2005) Roles of the Rac1 and Rac3 GTPases in human tumor cell invasion. *Oncogene* **24**: 7821–7829.
52. Titus B, Schwartz MA, Theodorescu D. (2005) Rho proteins in cell migration and metastasis. *Crit Rev Eukaryot Gene Expr* **15**: 103–114.
53. Coniglio SJ, Jou TS, Symons M. (2001) Rac1 protects epithelial cells against anoikis. *J Biol Chem* **276**: 28113–28120.
54. Zugasti O, Rul W, Roux P, *et al.* (2001) Raf-MEK-Erk cascade in anoikis is controlled by Rac1 and Cdc42 via Akt. *Mol Cell Biol* **21**: 6706–6717.

55. Regala RP, Weems C, Jamieson L, *et al.* (2005) Atypical protein kinase Ciota plays a critical role in human lung cancer cell growth and tumorigenicity. *J Biol Chem* **280**: 31109–31115.
56. Qiu RG, Chen J, Kirn D, *et al.* (1995) An essential role for Rac in Ras transformation. *Nature* **374**: 457–459.
57. del Pozo MA, Price LS, Alderson NB, *et al.* (2000) Adhesion to the extracellular matrix regulates the coupling of the small GTPase Rac to its effector PAK. *EMBO J* **19**: 2008–2014.
58. del Pozo MA, Kiosses WB, Alderson NB, *et al.* (2002) Integrins regulate GTP-Rac localized effector interactions through dissociation of Rho-GDI. *Nat Cell Biol* **4**: 232–239.
59. del Pozo MA, Alderson NB, Kiosses WB, *et al.* (2004) Integrins regulate Rac targeting by internalization of membrane domains. *Science* **303**: 839–842.
60. del Pozo MA, Balasubramanian N, Alderson NB, *et al.* (2005) Phospho-caveolin-1 mediates integrin-regulated membrane domain internalization. *Nat Cell Biol* **7**: 901–908.
61. Dios A, Mitchell RA, Aljabari B, *et al.* (2002) Inhibition of MIF bioactivity by rational design of pharmacological inhibitors of MIF tautomerase activity. *J Med Chem* **45**: 2410–2416.
62. Lubetsky JB, Dios A, Han J, *et al.* (2002) The tautomerase active site of macrophage migration inhibitory factor is a potential target for discovery of novel anti-inflammatory agents. *J Biol Chem* **277**: 24976–24982.
63. Swope M, Sun HW, Blake PR, Lolis E. (1998) Direct link between cytokine activity and a catalytic site for macrophage migration inhibitory factor. *EMBO J* **17**: 3534–3541.
64. Senter PD, Al Abed Y, Metz CN, *et al.* (2002) Inhibition of macrophage migration inhibitory factor (MIF) tautomerase and biological activities by acetaminophen metabolites. *Proc Natl Acad Sci USA* **99**: 144–149.
65. Al Abed Y, Dabideen D, Aljabari B, *et al.* (2005) ISO-1 binding to the tautomerase active site of MIF inhibits its pro-inflammatory activity and increases survival in severse sepsis. *J Biol Chem* **280**: 36541–36544.
66. Nicoletti F, Creange A, Orlikowski D, *et al.* (2005) Macrophage migration inhibitory factor (MIF) seems crucially involved in Guillain-Barre syndrome and experimental allergic neuritis. *J Neuroimmunol* **168**: 168–174.

Index